Adsorption

Adsorption: Theoretical Aspects and Environmental Applications offers an in-depth yet accessible exploration of adsorption processes, from fundamental theories to practical applications in environmental management and pollution control. Featuring updated content on equilibrium, kinetics, thermodynamics, and fixed-bed dynamics, this new edition integrates two decades of experimental research and adds coverage on multielement adsorption systems, common data analysis errors, and nonlinear isotherm modeling to provide crucial insights. Designed for undergraduate and postgraduate students, as well as industry practitioners, this book serves as a vital resource for mastering adsorption techniques, optimizing experimental methodologies, and advancing sustainable solutions for wastewater treatment and environmental challenges.

Key Features:

- Includes end-of-chapter problems and a solutions manual, helping students to develop skills to model, analyze, and design systems involving adsorption.
- Explores key topics, including adsorption equilibrium, kinetics, thermodynamics, and fixed-bed adsorption, with updated content on multielement systems and mass transfer dynamics.
- Includes numerous case studies that cover a variety of areas, such as water treatment, carbon capture, air emissions control, and environmental remediation, providing a comprehensive understanding.

Adsorption

Theoretical Aspects and Environmental Applications

Ronaldo Ferreira do Nascimento,
Ari Clecius Alves de Lima,
Carla Bastos Vidal,
Diego de Quadros Melo,
Giselle Santiago Cabral Raulino, and
Vicente de Oliveira Sousa Neto

CRC Press
Taylor & Francis Group
Boca Raton London New York

CRC Press is an imprint of the
Taylor & Francis Group, an **informa** business

Designed cover image: Shutterstock

First edition published 2026
by CRC Press
2385 NW Executive Center Drive, Suite 320, Boca Raton FL 33431

and by CRC Press
4 Park Square, Milton Park, Abingdon, Oxon, OX14 4RN

CRC Press is an imprint of Taylor & Francis Group, LLC

ISBN: 9781041102991 (hbk)
ISBN: 9781041102984 (pbk)
ISBN: 9781003654391 (ebk)

DOI: 10.1201/9781003654391

Typeset in Times
by codeMantra

Access the Instructor and Student Resources: routledge.com/9781041102991

Contents

Acknowledgments

This book would not have been possible without the collaboration of several partners over the years. There have been many partnerships and contributions from professors, students, and professionals in the field who have added knowledge to the group at the Trace Analysis Laboratory, the LAT Family.

Special thanks go to professors: Adonay Loiola (UFC), Alcineia Oliveira (UFC), Aldaléa Marques (UFMA), André Bezerra (UFC), André Oliveira (UFC), Antoninho Valentini (UFC), Antônio Gomes (UFC), Cícero Bezerra (UFMA), Diana Azevedo (UFC), Douglas Franco (in memoriam, IQSC-USP), Edilberto Silveira (UFC), Edmar Marques (UFMA), Elisane Longhnotti (UTFPR), Francisco Belmino (UFC), Francisco Dias (UFC), Galberto Martins (URCA), Helena Becker (UFC), José Capelo (UFC), Luiz Lopes (UFC), Luiz Vasconcellos (UFC), Marcos Sasaki (UFC), Maria Goretti (in memoriam, UFC), Moisés Basto (UFC), Murilo Luna (UFC), Odair Ferreira (UEL), Pierre Fachine (UFC), Raimundo Nonato (URCA), Rivelino Cavalcante (UFC), Selma Mazzetto (UFC), and Walter Costa (UFMA). We can't forget to mention the pioneering adsorption students in our research group: Francisco Wagner, André Gadelha, Jefferson Ribeiro, Marcelo (*in memoriam*), Maria Aparecida, and Francisco Cláudio.

We would also like to thank our group colleagues, who directly or indirectly contributed to the publication of this book: Allen Lopes, André Henrique, Cícero Moura, Clêrton Gomes, Fátima Itana, Fernando Pedro, Eliezer Fares, Larissa Lucena, Jefferson Ribeiro, Jhonyson Guedes, Juliene Oliveira, Igor Martins, Luiza Sena, Fábio Erlan, Pablo Gordiano, Francileide Oliveira, Raquel Bessa, Edipo Silva, Lucas Amorim, Renata Silva, Rouse Costa, Tatiana Fernandes, Sarah Moreira, Thiago Coutinho, Vitor Paulo, Edmilson Almeida, Ana Kátia Braz, Marcos Vinício, Leila Sousa, and all those who were once part of the LAT Family, thank you very much!

Special thanks go to our partner laboratories: LANÁGUA, LABPOA, LABOSAN, and LRX.

Finally, we would like to thank our family members for their patience, love, dedication, and care: Tecia Carvalho, Maria Tecia, and Antônio Lucas; Camila Lino, Maria Júlia, and Maria Alves; José and Sônia; Werther and Benício; and Daniel and Yan.

Preface

Adsorption: Theoretical Aspects and Environmental Applications aims to present in a didactic and comprehensible way important aspects in the study of separation processes by adsorption, from the fundamentals of the equilibrium and kinetics of the solute in the adsorbent particles, to the important parameters in the dynamics of adsorption in a fixed bed. Experimental data obtained from projects carried out by the research group at the Trace Analysis Laboratory (LAT), linked to the Department of Hydraulic Engineering and Environmental Sanitation at the Federal University of Ceará, are used to illustrate and exemplify the items covered where appropriate. These research projects have been carried out over the last 25 years (2000–2025), with the collaboration of numerous fellow researchers, teachers, postgraduate students, and undergraduate students.

In this new edition, the chapters have been updated with recent relevant topics and supported by examples from the literature and mainly from the authors themselves. After a brief introduction (Chapter 1), designed to familiarize the reader with some important aspects of adsorption, Chapters 2 and 3 address the fundamental points of the adsorption process, namely, the phenomena related to adsorption equilibrium and kinetics at the adsorbent particle level. Chapter 2, "Adsorption equilibrium", has been updated, adding the topic "equilibrium for multielement systems", with various examples, as well as a study of adsorption selectivity.

Chapter 4 deals with the thermodynamic aspects of the adsorption process. This chapter has been updated with the topic "Errors in estimating thermodynamic parameters in adsorption". This is important because many scientific studies in the literature do not take into account the dimensionality of the constants acquired through isotherm models, resulting in an erroneous determination of the value of the thermodynamic equilibrium constants. A new chapter, now Chapter 5, "Main errors in fitting theoretical models", has been added, which does a didactic job of portraying the main common errors in data analysis in batch adsorption processes, for example, the linearization of isotherm models. Chapter 6 deals with the resistance to mass transfer to adsorbent particles in a fixed bed, followed by a presentation, in Chapter 7, of the methods most commonly used in experimental investigations, relating them to fundamental parameters and statistical analyses applied to the calculations of the parameters investigated in the adsorption process (Chapter 8).

Chapters 6 and 7, "Fixed-bed adsorption" and "Experimental methodology" (Chapters 5 and 6 in the first edition), have been revised with the aim of providing a better understanding and adding a discussion of the behavior of breakthrough curves in multielement systems. Chapter 8, "Statistical tools for adsorption experiments", has been updated with the topic: "Fitting nonlinear isotherms to experimental adsorption data" using the Origin software tool, because, despite the frequent use of Microsoft Office Excel software to obtain data and graphs, this is a valuable alternative tool. Finally, Chapter 9 covers the main techniques for characterizing adsorbent materials. It is hoped that the reader will gain an introductory knowledge of the various aspects of adsorption as a separation phenomenon. This chapter has been updated with the inclusion of the topic "Analysis by Raman spectroscopy". As you know, Raman spectroscopy is a technique that covers various types of materials and is a fundamental technique for two-dimensional materials, stimulating the exploration of various types of materials, especially those that form layers.

Since research into theories and environmental applications of adsorptive processes is a broad and current area of study, we believe that the content presented here is of relevant interest to the areas of hydraulic engineering, environmental sanitation, and environmental chemistry, as well as to students, researchers, and other professionals in related areas.

About the Authors

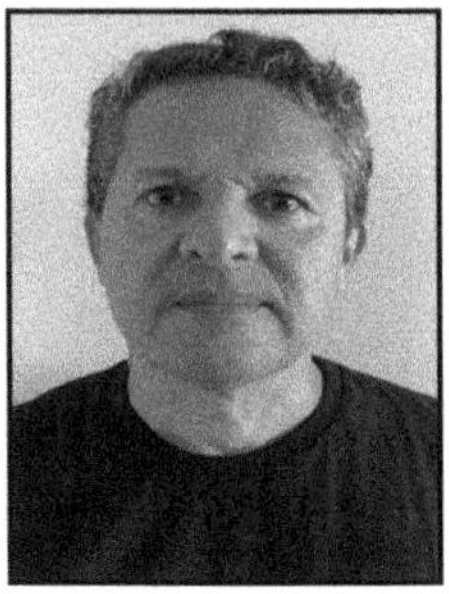

Ronaldo Ferreira do Nascimento holds a degree in Industrial Chemistry from the Federal University of Maranhão (1991) and a PhD in Analytical Chemistry from the Institute of Chemistry of São Carlos—USP (1997). He is a Full Professor at the Department of Analytical Chemistry and Physical Chemistry at the Federal University of Ceará. He has experience in the area of Chemistry, with an emphasis on separation (chromatographic methods), environmental chemistry (trace analysis), and adsorption (treatment of aqueous effluents).

Giselle Santiago Cabral Raulino holds a degree in Industrial Chemistry from the Federal University of Ceará (2006) and a master's degree and a doctorate in Civil Engineering – Environmental Sanitation (2011; 2016). She is currently an EBTT professor at the Federal Institute of Education, Science and Technology of Ceará. She has experience in the field of Chemistry, with an emphasis on environmental chemistry, trace element analysis, chemometrics, and adsorption using low-cost materials.

Carla Bastos Vidal holds a degree in Technology in Chemical Processes from the Federal Institute of Education, Science and Technology of Ceará – IFCE (2008). She holds a degree in Chemistry from the State University of Ceará (2011) and a master's (2011) and a PhD (2015) in Civil Engineering with the area of concentration in environmental sanitation at the Federal University of Ceará. She is a postdoctoral fellow in analytical chemistry (2016–2019). She has experience in the area of environmental sanitation, with an emphasis on environmental chemistry, working mainly on the following topics: advanced treatments of industrial effluents and WWTPs, equilibrium modeling and adsorption kinetics, and synthesis of low-cost adsorptive materials.

Ari Clecius Alves de Lima holds a degree in Chemical Engineering from the Federal University of Ceará (2003), a master's degree in Civil Engineering (water resources) from the Federal University of Ceará (2009), and a PhD in Civil Engineering (water resources) from the Federal University of Ceará (2015). He has experience in the area of sanitary engineering, with an emphasis on environmental sanitation, working mainly on the following topics: treatment of water and effluents by advanced oxidative processes and adsorption, solid waste with a focus on composting and biogas production and purification, chromatographic techniques, and statistical tools such as design of experiments, multivariate analysis, and machine learning. He is currently the coordinator of the Laboratory of Resíduos Sólidos e Efluentes da Fundação Núcleo de Tecnologia Industrial do Estado do Ceará.

Diego de Quadros Melo holds a bachelor's degree in Chemistry from the State University of Piauí (2009), a master's and a PhD in Analytical Chemistry from the Federal University of Ceará (2010–2015). He is currently an EBTT professor at the Federal Institute of Education, Science and Technology of the Sertão Pernambucano. He has experience in the field of chemistry, with an emphasis on chromatographic methods, environmental chemistry, adsorption, and chemistry teaching.

Vicente de Oliveira Sousa Neto holds a degree in Chemistry from the Federal University of Ceará (1998) and a PhD in Civil Engineering focused on environmental sanitation from the same university (2012). He has completed a postdoctoral fellowship in physics at the Federal University of Ceará (2019). He is currently a professor in the Department of Chemistry at the State University of Ceará, within the Faculty of Education, Sciences and Letters of the Central Sertão (FECLESC) in Quixadá. His expertise includes environmental chemistry, with a focus on bioadsorption (effluent treatment using modified and low-cost bioadsorbents), materials chemistry with an emphasis on developing thin films for photovoltaic cells and creating nanostructured oxides based on transition metals for pollutant removal in water, and computational chemistry, particularly molecular docking studies.

1 Adsorption
Basic Principles

Vicente de Oliveira Sousa Neto, Carla Bastos Vidal, Antonia Mayza de Morais França, Ari Clecius Alves de Lima, Giselle Cabral Raulino, Diego de Quadros Melo, and Ronaldo Ferreira do Nascimento

INTRODUCTION

The chemical contamination of water by a wide range of organic and inorganic pollutants—including toxic metals, BTEX, PAHs, anions, among others—has necessitated the development of technologies for removing these pollutants from liquid and gaseous waste. When present in trace quantities, these substances often resist biological degradation methods or are not effectively removed by physicochemical treatment methods. Consequently, adsorption has emerged as one of the most prevalent methods for this purpose, gaining recognition as an important separation and purification process in recent decades. This method has attracted scientific interest since the beginning of the century due to its technological and biological significance, along with its practical applications in industry and environmental protection, making it a valuable tool across multiple sectors (Cooney, 1999; Al-Ahmed, 2024). Furthermore, adsorption stands as one of the most efficient processes for water and wastewater treatment, employed by industries to reduce levels of toxic compounds in their effluents (De Oliveira et al., 2020; Manimegalai et al., 2024).

Environmental Impact of Adsorption Technology

Adsorption is a mass transfer process that examines the capacity of certain solids to concentrate specific substances from liquid or gaseous fluids onto their surfaces, enabling the separation of fluid components. Since the adsorbed components accumulate on the external surface, the greater this external surface area per unit of solid mass, the more favorable the adsorption process becomes. Consequently, adsorbents are typically porous solid particles (Ruthven, 1984; Shaji & Zachariah, 2017). The species that accumulates at the material interface is called the adsorbate, while the solid surface where the adsorbate accumulates is termed the adsorbent (Ruthven, 1984).

DOI: 10.1201/9781003654391-1

Adsorption separation processes operate through three distinct mechanisms: steric, equilibrium, and kinetic mechanisms. In the steric mechanism, the adsorbent material's pores have specific dimensions that permit certain molecules to enter while excluding others. Equilibrium mechanisms depend on the varying capacities of different solids to accommodate different adsorbate species, with some compounds being preferentially adsorbed over others. The kinetic mechanism is based on the differing diffusivities of various species within the adsorbent pores (Do, 1998), which will be explored in greater detail in Chapter 3.

Chemisorption and Physisorption

Depending on the nature of the forces involved, adsorption can be classified by its intensity into two types: physical adsorption and chemical adsorption (Ruthven, 1984; Adamson & Gast, 1997; Liu et al., 2019). In the case of physical adsorption, the attachment of the adsorbate to the surface of the adsorbent involves relatively weak interactions, primarily due to van der Waals forces, which are similar to molecular cohesion forces (Ruthven, 1984; Liu et al., 2019).

In contrast, chemisorption involves chemical bonds that are much stronger than those in physisorption, as they result essentially from the exchange or sharing of electrons between the adsorbate molecules and the surface of the adsorbent (Liu et al., 2019).

The concepts of chemisorption and physisorption are distinct; however, the two adsorption mechanisms are not completely independent. The distinction between whether a species is physically or chemically adsorbed is often unclear (Ruthven, 1984; Adamson & Gast, 1997; Liu et al., 2019), since both processes can frequently be described using the principles of physical adsorption. In general, the differences between physical adsorption and chemical adsorption can be summarized as follows:

- Chemical adsorption is highly specific, and not all solid surfaces have active sites capable of chemically adsorbing the adsorbate. It should be noted that not all molecules present in the fluid can be chemically adsorbed—only those capable of binding to the active sites. Physical adsorption, unlike chemical adsorption, is nonspecific (Liu et al., 2019).
- From a thermodynamic perspective, the heat associated with physisorption is generally below 1 kcal·mol^{-1} (4.184 kJ·mol^{-1}), comparable to the energy involved in condensation or vaporization. In contrast, the heat of adsorption in chemical adsorption is comparable to the heat of reaction and typically ranges from 20 to 50 kcal·mol^{-1} (Smith, 1981; Kim et al., 2006). According to Yu et al. (2001), the absolute magnitude of the change in free energy (ΔG_o) for physical adsorption is generally lower than that for chemisorption. The former ranges from −20 to 0 kJ mol^{-1}, whereas the latter falls between −80 and −400 kJ mol^{-1} (Jaycock & Parfitt, 1981; Yu et al., 2001).
- Similarly, the standard enthalpy change (ΔH_o) for physisorption is generally within the range of −2.1 to −20.9 kJ mol^{-1}, whereas chemisorption exhibits enthalpy changes from −80 to −200 kJ mol^{-1} (Húmpola et al., 2013; Xiao et al., 2019).

- Another characteristic of physical adsorption is that it occurs across the entire surface of the adsorbent, and is therefore considered non-localized. In contrast, chemical adsorption occurs only at active sites and is thus considered localized (Saleh, 2021).
- Another way to interpret physical adsorption is to consider that it occurs when the intermolecular attraction forces between the molecules in the fluid phase and the solid surface are stronger than the attractive forces among the fluid molecules themselves. This type of adsorption is rapid and reversible, resulting from weak intermolecular attraction forces between the adsorbent and the adsorbed molecules. It is also important to note that, since there is no bond formation or breaking, the chemical nature of the adsorbate remains unchanged. A further distinction of physisorption is the possibility of forming multiple layers of adsorbed molecules (Vidal et al., 2012).
- Adsorption rates, on the other hand, are not reliable criteria for distinguishing between chemical and physical adsorption. Chemical adsorption can be fast if the activation energy is low or negligible and slow if the activation energy is high. Physical adsorption is generally fast, but it may be slow when it involves diffusion into a porous medium.

A positive standard entropy change (ΔS°) indicates increased randomness in the arrangement of the adsorbate at the solid/solution interface, whereas a negative ΔS° reflects a more ordered configuration. It is hypothesized that variations in ΔS° are linked to the displacement of water molecules adsorbed on the surface by the incoming adsorbate species (Húmpola et al., 2013). Adsorption processes are always exothermic, resulting in a negative enthalpy change, which correlates with a decrease in entropy. This is attributed to the fact that molecules in their gaseous or dissolved states possess greater freedom of movement compared to those that are adsorbed onto the surface.

Factors Influencing the Adsorption Process

Adsorption phenomena result from a combination of the types of forces involved in physical and chemical adsorption. Accordingly, several factors influence the adsorption process, such as the surface area, the properties of the adsorbent and the adsorbate, the system temperature, the nature of the solvent, and the pH of the medium (Rápó & Tonk, 2021). The characteristics of the adsorbent include surface area, pore size, density, functional groups present on the surface, and material hydrophobicity. Conversely, the nature of the adsorbate is determined by its polarity, molecular size, solubility, and acidity or basicity. Operational conditions primarily involve temperature, pH, and solvent nature (Cooney, 1999; Ruthven, 1984). Only some of these factors will be described here.

Surface Area

The intensity of adsorption is proportional to the specific surface area, since adsorption is a surface phenomenon. In the case of larger particles, diffusion resistance is lower; however, a considerable portion of the internal surface of the particle may not be accessible for adsorption. (Thomas & Crittenden, 1998; Xiao et al., 2019).

Adsorbent Properties

The physicochemical nature of the adsorbent is a determining factor, as both the adsorption capacity and rate depend on the specific surface area, porosity, specific pore volume, pore size distribution, the functional groups present on the adsorbent's surface, and the nature of the precursor material (Aljamali et al., 2021).

Adsorbate Properties

Particle size is a factor that influences the adsorption process, as the adsorption rate is always dependent on intraparticle transport. The polarity of the adsorbate also plays a role, since a polar adsorbate may exhibit greater affinity either for the solvent or the adsorbent, depending on the polarity involved. As an example, polar functional groups—such as hydroxyls, carboxylic acids, and amines—present in various adsorbent materials, like lignocellulosic substances, show strong affinity for metal ions, promoting better interaction between these ions and the surface of the adsorbent (Ngah & Hanafiah, 2008).

Temperature

In adsorption processes, the temperature of the system affects mainly the adsorption rate constant. An increase in temperature can lead to an increase in kinetic energy and mobility of the adsorbate species, as well as increase the rate of intraparticle diffusion of the adsorbate (Wilke & Chang, 1955). According to Khattri and Singh (2009), this factor can affect the solubility and chemical potential of the adsorbate. Thus, a change in process temperature leads to a change in adsorption capacity (Khattri & Singh, 2009). In general, temperature has two important effects on the adsorption process. The first effect is related to the diffusion rate of adsorbate molecules throughout the outer boundary layer.

The second effect is that increasing the temperature decreases the viscosity of the solution. The other effect is related to the adsorption equilibrium state for a given adsorbate (Doğan et al., 2006). Additionally, an increase in temperature can produce pore clearance within the adsorbent's structure, allowing larger molecules to penetrate (Doğan et al., 2006).

The speed of various physical–chemical processes increases considerably with temperature. This dependence can be seen in the adsorption velocity. By monitoring adsorption over time at different temperatures, the adsorption rate constants are obtained. pH and potential zero charge (PZC). pH affects adsorption as it determines the degree of distribution of the chemical species. The intensity of this effect can be greater or less depending on the adsorbent, since the charges on the surface of the adsorbent depend on its composition and the characteristics of the surface (De Oliveira et al., 2020; Gonçalves et al., 2022). A convenient index of the tendency of a surface to become positively or negatively charged as a function of pH is the pH value required for the net charge of the adsorbent to be zero, the so-called zero-charge point, pH(pzc). For pH values lower than pH(pzc), the surface charge is positive, and the adsorption of anions is favored; and, for pH values higher than pH(pzc), the surface charge is negative, and the adsorption of cations is favored (Appel et al., 2003). The theory behind the technique of determining the charge point zero assumes that the protons H^+ and the hydroxyl groups OH constitute potential-determining ions. The adsorbent in solution can adsorb OH– or H+ ions. The surface charge of each particle will depend on the pH of the solution. Thus, the surface groups of each active site of the adsorbent can dissociate or associate protons from the solution, depending on the properties of the adsorbent and the pH of the solution. Consequently, the surface of the active sites becomes positively charged when it associates with a proton solution under acidic conditions. Under alkaline conditions, the surface becomes negatively charged and protons are lost to the solution under alkaline conditions (Čerović et al., 2007; Bakatula et al., 2018).

Effect of pH on the Surface Charges of Adsorbents

As discussed in the previous section, pH is a fundamental parameter in the adsorption process because it determines the charge on the surface of the adsorbent and governs electrostatic interactions between the adsorbent and the adsorbate (Bautista-Toledo et al., 2005).

The pH at the zero-charge point (pH_(pzc)) is the pH below which the surface of the adsorbent is positive, while above this value, it is negative (Ayranci et al., 2005). The importance of this variable in adsorption is that the charges of the adsorbate and the adsorbent must be opposite for there to be greater electrostatic interaction between. If the charges are equal, the adsorption process will be impaired due to electrostatic repulsion (Bautista-Toledo et al., 2005).

Now, let us analyze how a chemical modification can affect the PZC. Figure 1.1 shows the effect on the displacement of pH (pzc) in three natural adsorbents, two of which are chemically modified: coconut shell (BC); coconut shell modified with tartaric acid (BCATAlc), and coconut shell modified with polymerized formaldehyde (BCFP). As can be seen, the chemical modifications to the adsorbents brought about significant changes in pH (pzc) values. In the sample, Unmodified BC has a pH (pzc) of 4.5. This value decreased to 2.7 for the BCATAlc and BCFP samples, respectively.

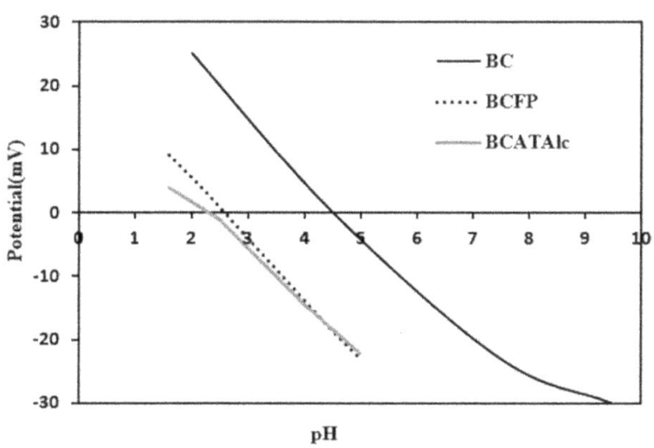

FIGURE 1.1 Surface charge potential of adsorbents as a function of pH. (This file by Vicente Neto is licensed under the Creative Commons Attribution 4.0 International license. https://upload.wikimedia.org/wikipedia/commons/2/2d/Surface_charge_potential.jpg.)

Species Equilibrium

The bonds between ligands and the central atom in complex ions are predominantly coordinate rather than covalent, given that metal cations are inherently electron-deficient, exhibiting an electropositive character. Consequently, most metal cations tend to attract polar molecules and negatively charged ions containing lone electron pairs. This attraction is so strong that, except for elements from Groups 1 and 2 of the periodic table, all other metal cations occur in solution as complex cations rather than as free ions. This implies that, instead of simple cations, complex ions are formed by the metal cation and solvent molecules. Examples include $[Al(H_2O)_6]^{3+}$, $[Fe(H_2O)_6]^{2+}$, $[Cu(H_2O)_4]^{2+}$, when water is the solvent. In the presence of other solvents, such as liquid ammonia, glacial acetic acid, ether, acetone, etc., a cation undergoes solvation in a manner analogous to hydration in water.

In adsorption studies, determining the types of metal species present in the aqueous medium is essential. It is crucial to note that species distribution depends on pH, and variations in pH favor the formation of one species over another. This factor must be analyzed in conjunction with the previously discussed zero-charge point (PZC). Materials with a positive surface charge exhibit a higher affinity for negative species, while the same principle applies to negatively charged surfaces interacting preferentially with positively charged species.

Stability Constant

The stability of complexes is determined by the metal-ligand (M-L) bond energy. The bonding that occurs is quantitatively characterized by equilibrium constants, which describe the stability of the complexes. For a general example:

$$M^{y+} + xL \leftrightarrow [ML_x]^{y+} \tag{1.1}$$

where M^{y+} represents the metal cation, L denotes the neutral donor group, and $[ML_x]^{y+}$ is the resulting complex ion.

The absolute stability constant is obtained by applying the law of mass action to the chemical equilibrium represented by the following equation:

$$K = \frac{[ML_x]^{y+}}{[M^{y+}][L]^x} \tag{1.2}$$

This constant defines the relationship between the concentration of the metal complex ion and the concentrations of free metal ions and ligands in solution. Consequently, the higher the value of K, the greater the stability of the metal complex.

When considering the stepwise formation of complexes involving successive reaction stages, as illustrated in the following example:

$$M + L \leftrightarrow ML \; k_1 = \frac{[M]}{[M][L]} \tag{1.3}$$

$$ML + L \leftrightarrow ML_2 \; k_2 = \frac{[ML_2]}{[ML][L]} \tag{1.4}$$

$$ML_2 + L \leftrightarrow ML_3 \; k_3 = \frac{[ML_3]}{[ML_2][L]} \tag{1.5}$$

------ ---- ↔ ------ ------=----------

$$ML_{n-1} + L \leftrightarrow ML_n k_n = \frac{[ML_n]}{[ML_{n-1}][L]} \tag{1.6}$$

The constants k_1, k_2, k_3, and k_n are referred to as partial formation constants, and their product represents the absolute or total formation constant. By multiplying all the equations above, the stability constant or formation constant of the complex (β) is obtained:

$$\beta_1 = k_1; \beta_2 = k_1 \cdot k_2; \beta_3 = k_1 \cdot k_2 k_3; \beta_n = k_1 \cdot k_2 k_3 \ldots k_n$$

In aqueous solutions, the equilibrium distribution of Zn(II) species in a cyanide medium can be represented by reactions 1.7–1.10, where β_1–β_4 are the stability constants of the respective complexes (Yngard et al., 2007).

$$Zn^{+2} + CN^- \leftrightarrow \left[Zn(CN)^+\right] \; \beta_1 = \frac{[CuCN]}{\left[Cu^+\right]\left[CN^-\right]} = 10^{5.34} \tag{1.7}$$

$$Zn^{+2} + 2CN^- \leftrightarrow \left[Zn(CN)_2\right] \; \beta_2 = \frac{\left[Zn(CN)_2\right]}{\left[Zn^+\right]\left[CN^-\right]^2} = 10^{11.97} \tag{1.8}$$

$$Zn^{+2} + 3CN^- \leftrightarrow \left[Zn(CN)_3^-\right] \; \beta_3 = \frac{\left[ZnCN_3^-\right]}{\left[Zn^+\right]\left[CN^-\right]^3} = 10^{16.05} \tag{1.9}$$

$$Zn^{+2} + 4CN^- \leftrightarrow \left[Zn(CN)_4^{2-}\right] \; \beta_4 = \frac{\left[Zn(CN)_4^{2-}\right]}{\left[Zn^+\right]\left[CN^-\right]^4} = 10^{19.62} \tag{1.10}$$

In aqueous solutions, the equilibrium of Cu(II) cyano-complexes can be represented by reactions 1.11–1.16, where k_1–k_4 are the equilibrium constants (Lu et al., 2002; Alonso-González et al., 2010).

$$Cu^+ + CN^- \leftrightarrow CuCN \; K_1 = \frac{[CuCN]}{\left[Cu^+\right]\left[CN^-\right]} = 3.16 \times 10^{19} \tag{1.11}$$

$$CuCN + CN^- \leftrightarrow CuCN_2^- \; K_2 = \frac{\left[CuCN_2^-\right]}{[CuCN]\left[CN^-\right]} = 3.39 \times 10^4 \tag{1.12}$$

$$CuCN_2^- + CN^- \leftrightarrow CuCN_3^{2-} \; K_3 = \frac{\left[CuCN_3^{2-}\right]}{\left[CuCN_2^-\right]\left[CN^-\right]} = 4.17 \times 10^4 \tag{1.13}$$

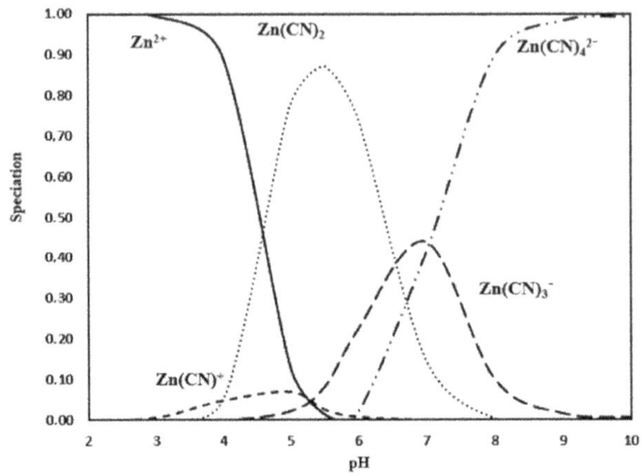

FIGURE 1.2 Distribution of Zn(II) in the Zn(II)/CN−/OH− system as a function of pH. (Equilibrium constant data obtained from Yngard et al. (2007). This file by Vicente Neto is licensed under the Creative Commons Attribution 4.0 International license. https://upload.wikimedia.org/wikipedia/commons/4/41/Illustrates_the_speciation_of_Zn2%2B.jpg.)

$$CuCN_3^{2-} + CN^- \leftrightarrow CuCN_4^{3-} \quad K_4 = \frac{\left[CuCN_4^{3-}\right]}{\left[CuCN_3^{2-}\right]\left[CN^-\right]} = 50.1 \tag{1.14}$$

$$CuCN \leftrightarrow Cu^+ + CN^- \quad K_{sp(5)} = \left[Cu^+\right]\left[CN^-\right] = 10^{-20} \tag{1.15}$$

$$Cu_2O + H_2O \rightarrow 2Cu^+ + 2OH^- \quad K_{sp(6)} = \left[Cu^+\right]^2\left[OH^-\right]^2 = 10^{-29.5} \tag{1.16}$$

Figures 1.2 and 1.3 present two distinct approaches for representing the species distribution of Zn(II) and Cu(II) cyano-complexes, respectively. A close relationship is observed between species distribution and the pH of the medium. For Zn(II), species distribution was expressed in fractional terms, whereas for copper ions, the logarithm of concentration was employed. When selecting the pH for the application of an adsorbent, special care must be taken to ensure that the chosen conditions do not lead to metal precipitation, as such a process would be undesirable and would render the use of the adsorbent unnecessary.

Effect of pH on Species Distribution and Surface of the Adsorbent

Figure 1.4 shows the distribution of chemical species for the metals Cu^{2+}, Zn^{2+}, and Cd^{2+} as the pH of the medium varies. It is observed that for the Cu^{2+} ion at pH 5.5, there is a predominance of 57% in the form of $[Cu(H_2O)_6]^{2+}$; 38.85% as $[CuR(H_2O)_5]^+$; and 4.15% as $[CuR_2(H_2O)_4]$. It is also evident that +2 charge complexes are predominant for both Cd^{2+} and Zn^{2+} ions. The distribution of species for Cd^{2+} is 77.10% as $[Cd(H_2O)_6]^{2+}$ and 21.70% as $[CdR(H_2O)_5]^+$; while for Zn^{2+}, it is 92.63% as $[Zn(H_2O)_6]^{2+}$ and 7.16% as $[ZnR(H_2O)_5]^+$. This indicates that, on average,

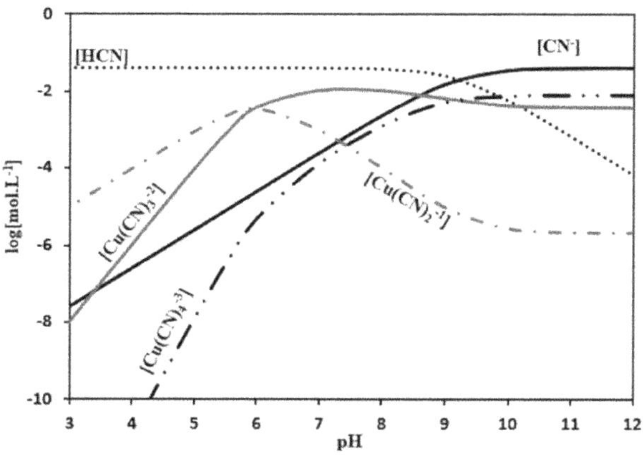

FIGURE 1.3 Distribution of Cu(II) in the Cu(II)/CN−/OH− system as a function of pH. (Equilibrium constant data obtained from Alonso-González et al. (2010). This file by Vicente Neto is licensed under the Creative Commons Attribution 4.0 International license. https://upload.wikimedia.org/wikipedia/commons/3/35/Illustrates_the_speciation_of_Cu2%2B.jpg)

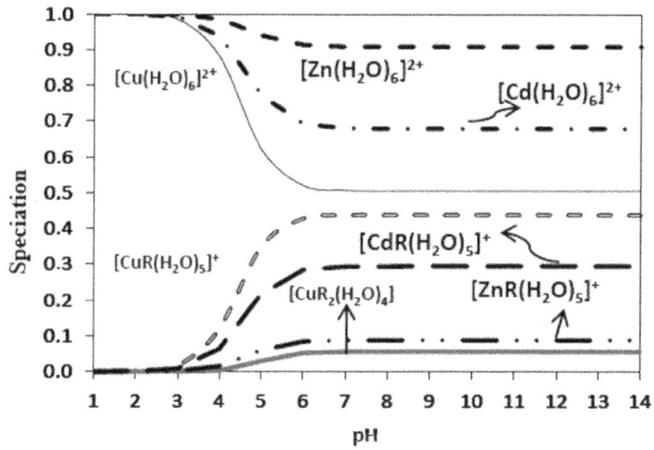

FIGURE 1.4 Species distribution in solution based on pH, where R = acetate group (CH_3COO^-). Acetic acid concentration: 5×10^{-3} M. (Complex constants were taken from Harris (2003). This file by Vicente Neto is licensed under the Creative Commons Attribution 4.0 International license. https://commons.wikimedia.org/wiki/File:Metal_species_distribution_in_solution.jpg)

95.85% of copper, 98.80% of Cd^{2+}, and 99.79% of Zn^{2+} are positively charged, favoring the access of metal complexes to the negative surface (pH of the medium > pH_{pzc}) of the adsorbent.

The distribution of species for Cd^{2+} and Zn^{2+} was 77.10% $\left[Cd(H_2O)_6^{2+}\right]$ and 21.70% $\left[CdR(H_2O)_5^{+}\right]$, while Zn^{2+} had 92.63% $\left[Zn(H_2O)_6^{2+}\right]$ and 7.16% $\left[Zn(H_2O)_5^{+}\right]$.

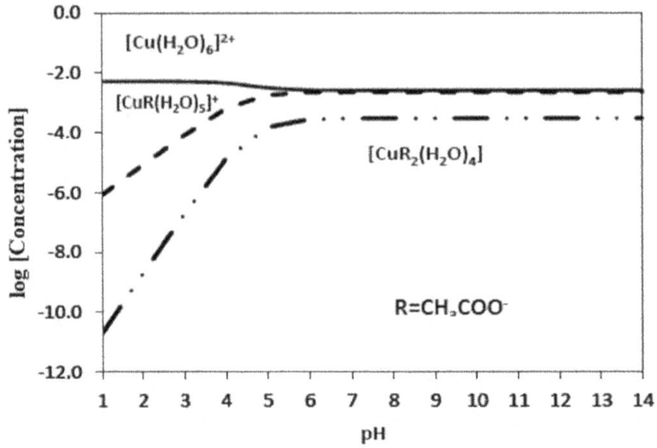

FIGURE 1.5 Illustrates the speciation of Cu^{2+}, highlighting the concentration of chemical species for Cu^{2+}. (Complex constants were taken from Harris (2003). This file by Vicente Neto is licensed under the Creative Commons Attribution 4.0 International license. https://commons.wikimedia.org/wiki/File:Illustration_of_Cu%C2%B2%E2%81%BA_speciation.jpg)

This means that, on average, 95.85% of copper, 98.80% of Cd^{2+}, and 99.79% of Zn^{2+} are positively charged, which helps these metal complexes interact with negatively charged surfaces (when the medium's pH is higher than the adsorbent's PZC). Figure 1.5 shows a graph of Cu^{2+} species concentration, revealing that at pH 5.5, the concentration of positively charged species is up to 100 times higher than that of neutral species.

SUMMARY

The nature of adsorption can be classified as either physical or chemical, depending on the forces involved. Physisorption is governed by weak van der Waals forces, leading to reversible, multilayer adsorption that is nonspecific and non-localized. In contrast, chemisorption involves stronger chemical bonds, occurs only at specific active sites, and is generally irreversible and monolayer in nature. Thermodynamic parameters such as enthalpy, free energy, and entropy differ significantly between the two processes, with chemisorption typically exhibiting higher energy changes and greater specificity. While distinctions between the two mechanisms exist, overlaps often occur, complicating classification.

Several factors influence adsorption efficiency. The surface area of the adsorbent directly affects adsorption capacity. Adsorbent properties, including porosity and functional groups, and adsorbate characteristics, such as polarity and molecular size, also play significant roles. Temperature impacts the mobility, diffusion, and equilibrium of the adsorbate, thereby altering adsorption rates and capacities. Lastly, pH influences the surface charge of the adsorbent, governed by the point of zero charge (pHpzc), which determines the affinity for anionic or cationic species. Understanding these variables is crucial for optimizing adsorption processes across different applications.

In aqueous systems, metal cations typically form complex ions due to their electropositive and electron-deficient nature, favoring interactions with polar molecules and anions bearing lone electron pairs. These interactions result in solvation complexes, with water forming hydration shells such as $[Al(H_2O)_6]^{3+}$ and $[Cu(H_2O)_4]^{2+}$. The equilibrium distribution of metal species in solution, especially under varying pH, plays a critical role in adsorption studies. The pH not only influences species formation but also interacts with the zero-charge point (PZC), dictating the electrostatic affinity between charged adsorbent surfaces and ionic species.

The stability of metal complexes is governed by their equilibrium constants, reflecting the strength of metal-ligand bonding. Stability constants can be derived from single-step or multi-step complexation reactions, with the overall formation constant representing the product of individual equilibrium steps. Case studies involving Zn(II) and Cu(II) cyano-complexes illustrate the species distribution across different pH values and ligand concentrations. Accurate knowledge of species equilibrium and complex stability is vital for predicting adsorption behavior and avoiding undesirable metal precipitation, ensuring effective adsorbent selection and process design.

EASY LEVEL—SOLVED QUESTIONS

1. At pH 2.0, calculate the percentage of Zn^{2+} that exists as the free ion, assuming minimal complexation under such acidic conditions.

   Solution: At this low pH, acetic acid remains mostly undissociated, yielding negligible CH_3COO^-. Therefore, virtually no complexation occurs.

   Answer: Approximately 100% of Zn^{2+} remains uncomplexed at pH 2.0.
2. Determine the concentration of acetate ions (CH_3COO^-) at pH 5.0, given acetic acid's $Ka = 1.8 \times 10^{-5}$ and $[CH_3COOH] = 5.0 \times 10^{-3}$ M.

   Solution: From $k_a = ([H^+][CH_3COO^-]/[CH_3COOH])$, $[H^+] = 10^{-5}$ M at pH 5.

   Substituting:

   $1.8 \times 10^{-5} = (10^{-5}\, x)/(5 \times 10^{-3} - x) \approx (10^{-5} \times x)/(5 \times 10^{-3})$

   Solving for x: $x \approx 9.0 \times 10^{-4}$ M

   Answer: $[CH_3COO^-] \approx 9.0 \times 10^{-4}$ M
3. Explain why metal-acetate complexation increases with pH.

   Solution: Higher pH values promote deprotonation of acetic acid, increasing CH_3COO^- availability for complex formation.

   Answer: Because more acetate ions are available for binding as the pH increases.
4. At what approximate pH does $Cd(CH_3COO)_2$ surpass free Cd^{2+} as the dominant species?

   Solution: From the speciation diagram, $Cd(CH_3COO)_2$ becomes predominant around pH 6.

   Answer: $pH \approx 6.0$
5. If the total Cu^{2+} concentration is 1.0×10^{-3} M at pH 8.0 and Cu exists exclusively as $Cu(CH_3COO)_2$, calculate the complex concentration.

   Solution: All Cu^{2+} is converted to $Cu(CH_3COO)_2$.

   Answer: $[Cu(CH_3COO)_2] = 1.0 \times 10^{-3}$ M

MEDIUM LEVEL—SOLVED QUESTIONS

6. Given that 75% of Cu^{2+} is complexed as $Cu(CH_3COO)^+$ at pH 6.0 and $[Cu]_{total} = 2.0 \times 10^{-3}$ M, determine $[Cu^{2+}]$.

   Solution: Free $Cu^{2+} = 25\% \times 2.0 \times 10^{-3} = 5.0 \times 10^{-4}$ M

   Answer: 5.0×10^{-4} M
7. Using $K_1 = 10^3$ for $Zn^{2+} + CH_3COO^- \rightleftharpoons Zn(CH_3COO)^+$, calculate $[Zn(CH_3COO)^+]$ if $[Zn^{2+}] = 5.0 \times 10^{-4}$ M and $[CH_3COO^-] = 1.0 \times 10^{-3}$ M.

   Solution: $[Zn(CH_3COO)^+] = K_1 \times [Zn^{2+}] \times [CH_3COO^-] = 10^3 \times 5.0 \times 10^{-4} \times 1.0 \times 10^{-3} = 5.0 \times 10^{-4}$ M

   Answer: 5.0×10^{-4} M
8. Provide a chemical explanation for why $Zn(CH_3COO)_2$ forms at lower pH than $Cu(CH_3COO)_2$.

   Solution: Zn^{2+} has a higher affinity for acetate than Cu^{2+}, resulting in earlier complex formation.

   Answer: Stronger interaction between Zn^{2+} and acetate allows complexation at lower pH.
9. Estimate the amount of acetate needed to fully complex 1.5×10^{-3} M Cu^{2+} into $Cu(CH_3COO)_2$ at pH 10.

   Solution: 1 mole of $Cu(CH_3COO)_2$ consumes 2 moles of acetate: Required acetate $= 2 \times 1.5 \times 10^{-3} = 3.0 \times 10^{-3}$ M

   Answer: 3.0×10^{-3} M
10. Describe the trend in metal-acetate complex formation from pH 3 to 9.

    Solution: With increasing pH, acetate concentration increases, favoring the formation of mono- and di-acetate complexes.

    Answer: Complexation increases in both quantity and coordination number as pH rises.

DIFFICULT LEVEL—SOLVED QUESTIONS

11. With $K_1 = 10^{2.8}$ and $K_2 = 10^{2.5}$ for Cd^{2+} complexation, calculate $[Cd(CH_3COO)_2]$ from 1×10^{-3} M Cd^{2+} and 1×10^{-3} M CH_3COO^-.

    Solution: First: $[Cd(CH_3COO)^+] = K_1 \times [Cd^{2+}] \times [CH_3COO^-] = 10^{2.8} \times 10^{-3} \times 10^{-3} = 6.3 \times 10^{-4}$ M

    Then: $[Cd(CH_3COO)_2] = K_2 \times [Cd(CH_3COO)^+] \times [CH_3COO^-]$
    $= 10^{2.5} \times 6.3 \times 10^{-4} \times 10^{-3} \approx 2.0 \times 10^{-4}$ M

    Answer: 2.0×10^{-4} M
12. At pH 7.5, if 20% of Cd^{2+} remains uncomplexed from a 1×10^{-3} M solution, identify the dominant complexed species.

    Solution: Complexed fraction $= 80\% = 8 \times 10^{-4}$ M, given the pH and graph, $Cd(CH_3COO)_2$ predominates.

    Answer: Major species: $Cd(CH_3COO)_2$
13. Derive the total metal concentration expression (M_t) in terms of stepwise formation constants β_1 and β_2 and $[CH_3COO^-]$.

Solution: $M_t = [M^{2+}] + \beta_1[L][M^{2+}] + \beta_1\beta_2[L]^2[M^{2+}]$

Answer: $M_t = [M^{2+}](1 + \beta_1[L] + \beta_1\beta_2[L]^2)$

14. Explain mathematically how $[M^{2+}]$ varies with increasing pH, assuming constant acetate concentration.

Solution: As pH rises, additional CH_3COO^- forms, increasing complexation. Thus, $[M^{2+}]$ decreases due to equilibrium shift.

Answer: $[M^{2+}] \propto 1/(1 + \beta_1[L] + \beta_1\beta_2[L]^2)$; it decreases with pH.

15. If total $Cu^{2+} = 10^{-3}$ M and $\beta_1 = 10^3$, $\beta_2 = 10^2$, calculate the concentrations of Cu^{2+}, $Cu(CH_3COO)^+$, and $Cu(CH_3COO)_2$ at $[CH_3COO^-] = 1.0 \times 10^{-3}$ M.

Solution: Let $x = [Cu^{2+}]$ Then:

$Cu(CH_3COO)^+ = \beta_1 \times x \times [L]$

$Cu(CH_3COO)_2 = \beta_1\beta_2 \times x \times [L]^2$

Total $Cu = x + \beta_1 x[L] + \beta_1\beta_2 x[L]^2$

$= x(1 + \beta_1[L] + \beta_1\beta_2[L]^2)$

Solving for x: $x = 10^{-3}/(1 + 10^3 \times 10^{-3} + 10^3 \times 10^2 \times 10^{-6})$

$= 10^{-3}/(1 + 1 + 0.1) = 4.76 \times 10^{-4}$

LEARNING EXERCISES

Easy Level

1. At 25°C, determine the hydrogen ion concentration in a 0.01 mol L^{-1} acetic acid solution, given $Ka = 1.8 \times 10^{-5}$.

Solution: $CH_3COOH \rightleftharpoons CH_3COO^- + H^+$

$Ka = [H^+]^2/[CH_3COOH] \approx x^2/0.01$

$x^2 = 1.8 \times 10^{-5} \times 0.01 = 1.8 \times 10^{-7}$

$x = \sqrt{(1.8 \times 10^{-7})} \approx 1.34 \times 10^{-4}$ mol L^{-1}

Answer: $[H^+] = 1.34 \times 10^{-4}$ mol L^{-1}

2. If the total concentration of Zn^{2+} is 2.0×10^{-3} mol L^{-1} and no complexation occurs at pH 3, determine the concentration of free Zn^{2+}.

Solution: Since no complexation takes place, the total concentration is equal to the free ion concentration.

Answer: $[Zn^{2+}] = 2.0 \times 10^{-3}$ mol L^{-1}

Intermediate Level

3. A solution contains 1.0×10^{-3} mol/L of Cu^{2+} at pH 7, and 80% is complexed as $Cu(CH_3COO)^+$. Calculate the concentration of free Cu^{2+}.

Solution: 20% remains free → $0.20 \times 1.0 \times 10^{-3} = 2.0 \times 10^{-4}$ mol L^{-1}

Answer: $[Cu^{2+}] = 2.0 \times 10^{-4}$ mol L^{-1}

4. Given $Kf = 10^4$ for $Zn^{2+} + CH_3COO^- \rightleftharpoons Zn(CH_3COO)^+$, calculate $[Zn(CH_3COO)^+]$ if $[Zn^{2+}] = 1.0 \times 10^{-3}$ mol L^{-1} and $[CH_3COO^-] = 2.0 \times 10^{-3}$ mol L^{-1}.

Solution: $[Zn(CH_3COO)^+] = Kf \times [Zn^{2+}] \times [CH_3COO^-] = 10^4 \times 1.0 \times 10^{-3} \times 2.0 \times 10^{-3} = 2.0 \times 10^{-2}$ mol L^{-1}

Answer: $[Zn(CH_3COO)^+] = 2.0 \times 10^{-2}$ mol L^{-1}

Advanced Level

5. Consider the reactions: $Cd^{2+} + CH_3COO^- \rightleftharpoons Cd(CH_3COO)^+$ $K_1 = 10^{2.6}$
   $Cd(CH_3COO)^+ + CH_3COO^- \rightleftharpoons Cd(CH_3COO)_2$ $K_2 = 10^{2.2}$
   If $[Cd^{2+}] = 5.0 \times 10^{-4}$ mol L^{-1} and $[CH_3COO^-] = 1.0 \times 10^{-3}$ mol L^{-1}, calculate $[Cd(CH_3COO)_2]$.
   Solution:
   $[Cd(CH_3COO)^+] = K_1 \times [Cd^{2+}] \times [CH_3COO^-]$
   $= 10^{2.6} \times 5.0 \times 10^{-4} \times 1.0 \times 10^{-3} \approx 2.0 \times 10^{-4}$ mol L^{-1}
   $[Cd(CH_3COO)_2] = K_2 \times [Cd(CH_3COO)^+] \times [CH_3COO^-]$
   $= 10^{2.2} \times 2.0 \times 10^{-4} \times 1.0 \times 10^{-3} \approx 3.2 \times 10^{-5}$ mol L^{-1}
   Answer: $[Cd(CH_3COO)_2] \approx 3.2 \times 10^{-5}$ mol L^{-1}
6. Derive the expression for the total concentration of metal M, considering the following species: M^{2+}, $M(CH_3COO)^+$, $M(CH_3COO)_2$, and the cumulative formation constants β_1 and β_2.
   Solution: $M_{total} = [M^{2+}] + \beta_1[L][M^{2+}] + \beta_2[L]^2[M^{2+}] = [M^{2+}](1 + \beta_1[L] + \beta_2[L]^2)$
   Answer: $M_{total} = [M^{2+}](1 + \beta_1[L] + \beta_2[L]^2)$

Applied Questions

The point of zero charge (PZC) is a fundamental parameter in surface chemistry, representing the pH at which the net surface charge of an adsorbent is zero. Under these conditions, the number of positive and negative charges on the material's surface is balanced, resulting in no electrostatic attraction or repulsion toward charged species in solution.

In environmental remediation, especially for the removal of ionic contaminants, the PZC serves as a guiding reference for optimizing pH conditions. When the solution pH is lower than the PZC, the adsorbent surface becomes positively charged, favoring the removal of anions. Conversely, at pH values above the PZC, the surface acquires a negative charge, enhancing the adsorption of cationic species. This relationship allows researchers and engineers to tailor operational conditions for maximum adsorption efficiency based on the nature of the target contaminant.

Furthermore, the understanding of PZC is essential in selecting appropriate materials for specific environmental applications, such as wastewater treatment and soil decontamination. It enables the prediction and control of surface–adsorbate interactions, contributing to more efficient and sustainable remediation strategies. Thus, PZC is not merely a theoretical concept, but a critical tool in the design and optimization of environmentally relevant adsorption processes.

Application-Based Questions: Adsorption and PZC

Based on the PZC graph provided, here are five formal, application-based questions related to the removal of cationic adsorbates using the point of zero charge (PZC):

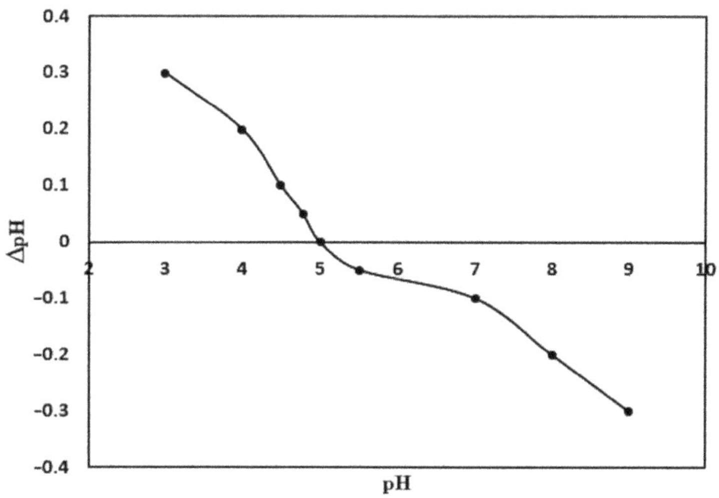

According to the graph, what is the approximate PZC of the adsorbent, and how does this value influence its interaction with cationic species at varying pH levels?

1. When the solution pH is below the PZC, the adsorbent surface tends to be positively charged. Discuss how this surface condition may affect the adsorption efficiency for cationic contaminants.
2. In an adsorption experiment targeting cationic species, how would adjusting the solution pH above the PZC alter the electrostatic interactions, and what implications would this have for adsorption performance?
3. The ΔpH data suggest shifts in the acid–base behavior of the adsorbent surface. How can such changes be interpreted in the context of optimizing pH conditions for cationic and anionic species removal?
4. Considering the provided PZC profile, propose an ideal pH range for maximizing the uptake of a divalent cationic adsorbate, and justify your selection using evidence from the graph.
5. Using the PZC profile, propose a suitable pH range for maximizing the adsorption of a divalent cationic contaminant and justify your recommendation based on surface charge considerations.

EXERCISES

Multiple-Choice Questions

1. **What is a characteristic of chemical adsorption?**
   a. It is nonspecific and occurs on all solid surfaces.
   b. Only molecules capable of binding to the active site can be adsorbed.
   c. All molecules present in the fluid are chemically adsorbed.
   d. It does not require active sites to occur.

2. **Why can't all solid surfaces carry out chemical adsorption?**
   a. Because chemical adsorption requires specific active sites.
   b. Because chemical adsorption is nonspecific.
   c. Because all solid surfaces have active sites.
   d. Because there are no suitable solid surfaces for chemical adsorption.
3. **What is the main difference between physical adsorption and chemical adsorption?**
   a. Both are highly specific.
   b. Chemical adsorption is nonspecific, while physical adsorption is specific.
   c. Physical adsorption is nonspecific, while chemical adsorption is highly specific.
   d. Both occur without the need for active sites.
4. **What is required for a molecule to be chemically adsorbed?**
   a. Any molecule can be adsorbed.
   b. The molecule must be capable of binding to the solid surface's active site.
   c. The molecule must be in a different physical state.
   d. There is no need to bind to the active site.
5. **Why doesn't chemical adsorption occur with all molecules present in the fluid?**
   a. Because not all molecules can bind to the active sites.
   b. Because chemical adsorption occurs only in the liquid state.
   c. Because all molecules are capable of physical adsorption.
   d. Because larger molecules are always excluded.

Questions on Adsorption Thermodynamics

6. **What is the typical range for standard Gibbs free energy change (ΔG°) in physisorption?**
   a. 0 to −20 kJ mol^{-1}
   b. −80 to −400 kJ mol^{-1}
   c. −20 to −80 kJ mol^{-1}
   d. −400 to −800 kJ mol^{-1}
7. **What is the main thermodynamic property that sets chemisorption apart from physisorption?**
   a. Temperature range
   b. Standard enthalpy change (ΔH°)
   c. Specific heat capacity
   d. Molecular weight of adsorbates
8. **Why is physisorption considered nonspecific?**
   a. Because it happens with big enthalpy changes.
   b. Because it doesn't need active sites.
   c. Because it involves highly specific chemical reactions.
   d. Because it requires energy input to occur.

9. **What is the typical range for standard enthalpy change (ΔHo) in chemisorption?**
   a. −2.1 to −20.9 kJ mol^{-1}
   b. −80 to −200 kJ mol^{-1}
   c. 0 to −20 kJ mol^{-1}
   d. −20 to −80 kJ mol^{-1}
10. **Which process usually involves larger energy changes?**
   a. Physisorption
   b. Chemisorption
   c. Both involve the same energy changes
   d. None of the above

SUBJECTIVE QUESTIONS ON THERMODYNAMICS OF ADSORPTION

1. What is the primary thermodynamic distinction between physisorption and chemisorption?
2. Why does chemisorption exhibit a significantly broader range of Gibbs free energy change (ΔGo) compared to physisorption?
3. Why are enthalpy changes (ΔHo) significant in distinguishing adsorption mechanisms?
4. How does adsorption specificity impact its applications in materials science?
5. How do the energy requirements of physisorption and chemisorption compare?
6. How do molecular interactions dictate whether physisorption or chemisorption occurs?
7. Can you provide an example where physisorption is preferable to chemisorption and explain why?
8. What makes chemisorption valuable in catalytic processes?
9. What factors influence the enthalpy values associated with physisorption and chemisorption?
10. Why are adsorption mechanisms crucial for environmental applications?

SUBJECTIVE QUESTIONS ON ADSORBENT AND ADSORBATE PROPERTIES

1. How do the physical and chemical properties of an adsorbent influence adsorption capacity and rate?
2. How does particle size impact molecular diffusion and transport mechanisms during adsorption?
3. How does the polarity of an adsorbate affect its affinity for solvents and adsorbents? Can you provide examples?
4. Why do functional groups such as hydroxyl and carboxyl groups enhance metal adsorption?

5. How do lignocellulosic materials interact with metals during adsorption, and what factors contribute to strengthening this process?

OBJECTIVE QUESTIONS ON PH AND POINT OF ZERO CHARGE

1. **How does pH change adsorption?**
   a. By messing with the temperature.
   b. By deciding which chemical species are around.
   c. By taking away active sites on the adsorbent.
   d. By altering the adsorbent's physical structure.
2. **Why does pH_{PZC} matter for surface charge?**
   a. It defines how porous the adsorbent is.
   b. It's the pH where the adsorbent has no net charge.
   c. It controls the adsorbate's chemical polarity.
   d. It shows the adsorbent's maximum adsorption ability.
3. **Why does the surface charge become positive below the pH_{PZC}?**
   a. Because the adsorbent grabs H^+ ions.
   b. Because the adsorbent loses H^+ ions.
   c. Because the adsorbent picks up OH^- ions.
   d. Because the adsorbent breaks up active sites.
4. **What happens to the surface charge when pH is above pH_{PZC}?**
   a. It turns positive, making it easier for anions to stick.
   b. It stays neutral, so adsorption doesn't really happen.
   c. It becomes negative, which helps cations get adsorbed.
   d. It doesn't change, no matter what kind of adsorbate is around.
5. **How does pH regulate electrostatic interactions in adsorption?**
   a. By adjusting how porous the adsorbent is.
   b. By deciding the charge of both the adsorbent and the adsorbate.
   c. By changing the physical state of the adsorbent.
   d. By stripping away functional groups from the adsorbent.

OBJECTIVE QUESTIONS: METAL SPECIATION

Refer to Figure 1.2 to answer the questions.

1. At which pH level does Zn^{2+} exhibit the highest speciation?
   a. pH 2
   b. pH 5
   c. pH 7
   d. pH 10
2. Which zinc cyanide complex has the highest speciation at pH 7?
   a. Zn^{2+}
   b. $Zn(CN)_2$
   c. $Zn(CN)_3^-$
   d. $Zn(CN)_4^{2-}$

3. How does the speciation $Zn(CN)_4^{2-}$ change as the pH increases from 2 to 10?
   a. It remains constant throughout the pH range
   b. It gradually increases, becoming dominant at higher pH
   c. It decreases steadily as pH rises
   d. It peaks at pH 5 and then declines
4. At approximately what pH does $Zn(CN)_2$ reach its maximum speciation?
   a. pH 3
   b. pH 5.5
   c. pH 7
   d. pH 9
5. How does the speciation of $Zn(CN)_3^-$ compare to $Zn(CN)_4^{2-}$ at pH 8?
   a. $Zn(CN)_3^-$ is more dominant than $Zn(CN)_4^{2-}$.
   b. $Zn(CN)_4^{2-}$ has a higher speciation than $Zn(CN)_3^-$.
   c. Both species exhibit equal speciation.
   d. Neither species is present at pH 8.

Refer to Figure 1.2 to answer the questions.

1. When does the $[Cu(H_2O)_6]^{2+}$ complex start seriously dropping?
2. Which metal complex is the predominant species at pH 7?
3. At pH 12, which species has the lowest concentration?
4. Between pH 4 and 6, which complex becomes more prevalent?
5. Approximately how much of $[Zn(H_2O)_6]^{2+}$ remains at pH 3?

REFERENCES

Adamson, A. W., & Gast, A. P. (1997). *Physical Chemistry of Surfaces*. 6th ed. Wiley, New York.

Al-Ahmed, Z. A. (2024). Surface methodology for optimized adsorption of hazardous organic pollutant from aqueous solutions via novel magnetic metal organic framework: Kinetics, isotherm study, and DFT calculations. *Journal of Molecular Liquids*, 409, 125507. https://doi.org/10.1016/j.molliq.2024.125507

Aljamali, N. M., Aldujaili, R. A. B., & Alfatlawi, I. O. (2021). Physical and chemical adsorption and its applications. *International Journal of Thermodynamics and Chemical Kinetics*, 7(2), 1–8.

Alonso-González, O., Nava-Alonso, F., Uribe-Salas, A., & Dreisinger, D. (2010). Use of quaternary ammonium salts to remove copper–cyanide complexes by solvent extraction. *Minerals Engineering*, 23(8), 765–770. https://doi.org/10.1016/j.mineng.2010.05.017

Appel, C., Ma, L. Q., Rhue, R. D., & Kennelley, E. (2003). Point of zero charge determination in soils and minerals via traditional methods and detection of electroacoustic mobility. *Geoderma*, 113(1—2), 77–93. https://doi.org/10.1016/S0016-7061(02)00316-6

Ayranci, E., Hoda, N., & Bayram, E. (2005). Adsorption of benzoic acid onto high specific area activated carbon cloth. *Journal of Colloid and Interface Science*, 284(1), 83–88. https://doi.org/10.1016/j.jcis.2004.10.033

Bakatula, E. N., Richard, D., Neculita, C. M., & Zagury, G. J. (2018). Determination of point of zero charge of natural organic materials. *Environmental Science and Pollution Research International*, 25(8), 7823–7833. https://doi.org/10.1007/s11356-017-1115-7

Bautista-Toledo, I., Ferro-García, M. A., Rivera-Utrilla, J., Moreno-Castilla, C., & Vegas Fernández, F. J. (2005). Bisphenol A removal from water by activated carbon. Effects of carbon characteristics and solution chemistry. *Environmental Science & Technology*, 39(16), 6246–6250. https://doi.org/10.1021/es0481169

Čerović, L. S., Milonjić, S. K., Todorović, M. B., Trtanj, M. I., Pogozhev, Y. S., Blagoveschenskii, Y., & Levashov, E. A. (2007). Point of zero charge of different carbides. *Colloids and Surfaces A: Physicochemical and Engineering Aspects*, 297(1–3), 1–6. https://doi.org/10.1016/j.colsurfa.2006.10.012

Cooney, D. O. (1999). *Adsorption Design for Wastewater Treatment*. CRC Press, Florida.

De Oliveira, F. F., Moura, K. O., Costa, L. S., Vidal, C. B., Loiola, A. R., & Do Nascimento, R. F. (2020). Reactive adsorption of parabens on synthesized micro- and mesoporous silica from coal fly ash: pH effect on the modification process. *ACS Omega*, 5(7), 3346–3357. https://doi.org/10.1021/acsomega.9b03537

Do, D. D. (1998). *Adsorption Analysis: Equilibria and Kinetics*. Imperial College Press, London.

Doğan, M., Alkan, M., Demirbaş, Ö., Özdemir, Y., & Özmetin, C. (2006). Adsorption kinetics of maxilon blue GRL onto sepiolite from aqueous solutions. *Chemical Engineering Journal*, 124(1–3), 89–101. https://doi.org/10.1016/j.cej.2006.08.016

Gonçalves, R. G. L., Lopes, P. A., Pochapski, D. J. et al. (2022). Effect of pH, ionic strength, and temperature on the adsorption behavior of Acid Blue 113 onto mesoporous carbon. *Environmental Science and Pollution Research*, 29, 77188–77198. https://doi.org/10.1007/s11356-022-21193-y

Harris, D. C. (2003). *Quantitative Chemical Analysis*. 6th ed. W.H. Freeman and Company, New York.

Húmpola, P. D., Odetti, H. S., Fertitta, A. E., & Vicente, J. L. (2013). Thermodynamic analysis of adsorption models of phenol in liquid phase on different activated carbons. *Journal of the Chilean Chemical Society*, 58, 1541–1544. https://dx.doi.org/10.4067/S0717-97072013000100009

Jaycock, M. J., & Parfitt, G. D. (1981). *Chemistry of Interfaces* (pp. 12–13). Ellis Horwood, Chichester.

Khattri, S. D., & Singh, M. K. (2009). Removal of malachite green from dye wastewater using neem sawdust by adsorption. *Journal of Hazardous Materials*, 167(1–3), 1089–1094. https://doi.org/10.1016/j.jhazmat.2009.01.101

Kim, S. J., Chung, J. H., Kim, T. Y., & Cho, S. Y. (2006). Biosorption of heavy metals and cyanide complexes on biomass. In H.-K. Rhee, I.-S. Nam, & J. M. Park (Eds.), *Studies in Surface Science and Catalysis* (Vol. 159, pp. 141–144). Elsevier. https://doi.org/10.1016/S0167-2991(06)81553-3

Liu, L., Luo, X.-B., Ding, L., & Luo, S.-L. (2019). Application of nanotechnology in the removal of heavy metal from water. In X. Luo, & F. Deng (Eds.), *Nanomaterials for the Removal of Pollutants and Resource Reutilization* (pp. 83–147). https://doi.org/10.1016/B978-0-12-814837-2.00004-4

Lu, J., Dreisinger, D. B., & Cooper, W. C. (2002). Thermodynamics of the aqueous copper-cyanide system. *Hydrometallurgy*, 66(1–3), 23–36. https://doi.org/10.1016/S0304-386X(02)00081-6

Manimegalai, G., Selvaraj, K., Yu, B., Kumar, M., Muthaiah, C., Panneerselvam, S., Arulraj, A., Mangalaraja, R. V., Habila, M., Allohedan, H., Yuli Yanto, D. H., & Arumugam, D. B. (2024). Acorus calamus-mediated $CuFe_2O_4$/reduced graphene oxide (AcL-CF-G) nanocomposite and its versatile biomedical and environmental remediation applications. *Journal of Industrial and Engineering Chemistry*. https://doi.org/10.1016/j.jiec.2024.12.019

Ngah, W. S. W., & Hanafiah, M. A. K. M. (2008). Biosorption of copper ions from dilute aqueous solutions on base treated rubber (*Hevea brasiliensis*) leaves powder: Kinetics, isotherm, and biosorption mechanisms. *Journal of Environmental Sciences*, 20(10),1168–1176. https://doi.org/10.1016/s1001-0742(08)62205-6

Rápó, E., & Tonk, S. (2021). Factors affecting synthetic dye adsorption; desorption studies: A review of results from the last five years (2017–2021). *Molecules*, 26(17). https://doi.org/10.3390/molecules26175419

Ruthven, D.M. (1984) *Principle of Adsorption and Adsorption Processes*, John Wiley & Sons, New York.

Saleh, T. A. (2021). Surface and morphological characterization of hybrid materials. In T. A. Saleh (Ed.), *Polymer Hybrid Materials and Nanocomposites* (pp. 241–283). William Andrew Publishing, Norwich.

Shaji, A., & Zachariah, A. K. (2017). Surface area analysis of nanomaterials. In S. Thomas, R. Thomas, A. K. Zachariah, & R. K. Mishra (Eds.), *Thermal and Rheological Measurement Techniques for Nanomaterials Characterization* (pp. 197–231). https://doi.org/10.1016/B978-0-323-46139-9.00009-8

Smith, J. M. (1981). *Chemical Engineering Kinetics*. 3rd ed. McGraw Hill, New York.

Thomas, W. J., & Crittenden, B. (1998). Adsorbents. In W. J. Thomas, & B. Crittenden (Eds.), *Adsorption Technology & Design* (pp. 8–30). Butterworth-Heinemann, Oxford.

Vidal, C. B., Raulino, G. S. C., Barros, A. L., Lima, A. C. A., Ribeiro, J. P., Pires, M. J. R., & Nascimento, R. F. (2012). BTEX removal from aqueous solutions by hdtma-modified y zeolite. *Journal of Environmental Management*, 112, 178–185. https://doi.org/10.1016/j.jenvman.2012.07.026

Wilke, C. R., & Chang, P. (1955). Correlation of diffusion coefficients in dilute solutions. *AIChE Journal*,1, 264–270.

Xiao, D., Ding, W., Zhang, J., Ge, Y., Wu, Z., & Li, Z. (2019). Fabrication of a versatile lignin-based nano-trap for heavy metal ion capture and bacterial inhibition. *Chemical Engineering Journal*, 358, 310–320. https://doi.org/10.1016/j.cej.2018.10.037

Yngard, R., Damrongsiri, S., Osathaphan, K., & Sharma, V. K. (2007). Ferrate (VI) oxidation of zinc-cyanide complex. *Chemosphere*, 69(5), 729–735. https://doi.org/10.1016/j.chemosphere.2007.05.017

Yu, Y., Zhuang, Y.-Y., & Wang, Z.-H. (2001). Adsorption of water-soluble dye onto functionalized resin. *Journal of Colloid and Interface Science*, 242(2), 288–293. https://doi.org/10.1006/jcis.2001.7780

2 Adsorption Equilibrium

*Vicente de Oliveira Sousa Neto,
Diego de Quadros Melo,
Antonia Mayza de Morais França,
Allen Lopes de Barros,
Ari Clecius Alves de Lima, Carla Bastos Vidal,
Giselle Santiago Cabral Raulino,
and Ronaldo Ferreira do Nascimento*

INTRODUCTION

ADSORPTION EQUILIBRIUM

Studying adsorption equilibrium is important for figuring out how adsorption-based separation processes work. Basically, when a certain amount of solid (the adsorbent) meets a liquid that has an adsorbable solute (the adsorbate), adsorption keeps going until equilibrium happens. In simple terms, adsorbate molecules or ions move from the liquid to the adsorbent's surface until things balance out, meaning, the solute concentration in the liquid phase (Ce) stops changing.

At this point, equilibrium is reached, and the adsorbent's adsorption capacity (qe) can be measured. To do that, scientists use a set mass of adsorbent and different starting concentrations of adsorbate. They plot adsorption capacity (q) against C_e based on experimental data. Using isotherm equation models, they can express the q vs. C_e relationship mathematically and figure out the adsorbent's max adsorption capacity in the lab (Cooney, 1999). To determine the adsorption isotherm, a known amount of adsorbent is added to different solutions with varying starting concentrations. Once equilibrium is reached, researchers measure the final solute concentration in the liquid and figure out how much the adsorbent has adsorbed (Figure 2.1).

Nonequilibrium conditions can seriously mess with experimental results, leading to incomplete or misleading data (Figure 2.2). If equilibrium isn't achieved, the adsorbate may not have enough time to fully interact with the adsorbent's surface, which means lower observed adsorption capacities.

Additionally, nonequilibrium conditions can cause inconsistencies and variability in the measurements since the adsorption process isn't stable yet. So, a graph of adsorption capacity versus equilibrium concentration can be created.

It is important to know how to determine these values. To obtain C_e values, once equilibrium has been reached, the adsorbent is separated from the solution using a membrane filter, filter paper, or centrifugation. The supernatant solution is then analyzed to find the residual concentration of the adsorbate (C_e). This can be measured

DOI: 10.1201/9781003654391-2

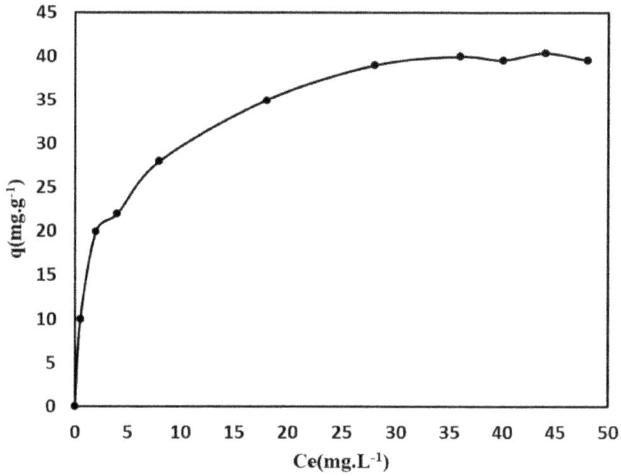

FIGURE 2.1 Illustration of the adsorption study based on the concentration effect: Experimental conditions: Acetate buffer, pH = 5.5, T = 28°C, adsorbent dosage = 2 $g{\cdot}L^{-1}$, contact time = 2 h. (This file by Vicente Neto is licensed under the Creative Commons Attribution 4.0 International License. https://commons.wikimedia.org/wiki/File:Illustration_of_the_adsorption_study_based_on_the_concentration_effect.jpg.)

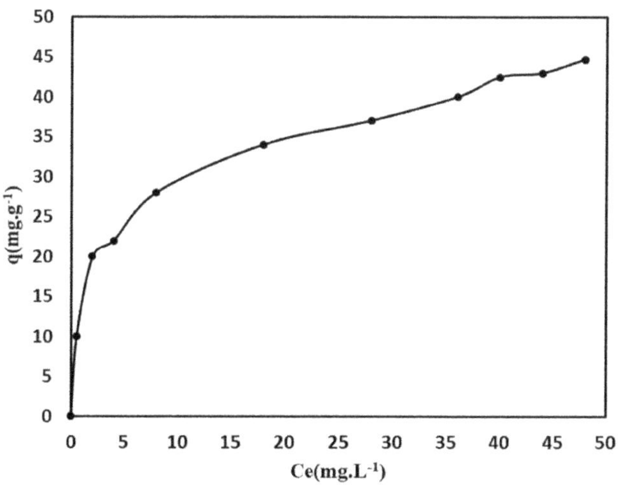

FIGURE 2.2 Illustration of the adsorption study based on the concentration effect: adsorption isotherm under nonequilibrium condition. (This file by Vicente Neto is licensed under the Creative Commons Attribution 4.0 International License. https://commons.wikimedia.org/wiki/File:Adsorption_isotherm_under_nonequilibrium_condition.jpg.)

using different analytical techniques, depending on the adsorbate, like gas chromatography, liquid chromatography, UV-visible spectrometry, atomic absorption spectrometry, emission spectrometry, or other suitable methods. To get the (q) values, a mass balance is needed—basically, the amount of adsorbate adsorbed on the adsorbent should match the amount removed from the solution, as shown in Equation 2.1. Nonequilibrium conditions can significantly affect experimental results by providing incomplete or misleading data (Figure 2.2).

If equilibrium is not achieved, the adsorbate may not have sufficient time to fully interact with the adsorbent surface, resulting in lower observed adsorption capacities. Additionally, nonequilibrium conditions may lead to inconsistencies and variability in the measurements, as the adsorption process is still ongoing and not stabilized. Thus, a graph of adsorption capacity versus equilibrium concentration can be generated. It is necessary to understand how to determine the values of these variables. To obtain the values of Ce, once equilibrium is achieved, the adsorbent is separated from the solution using a membrane filter, filter paper, or centrifugation. The supernatant solution is then analyzed to determine the residual concentration of the adsorbate Ce. This can be determined using analytical techniques, depending on the adsorbate used, such as gas chromatography, liquid chromatography, ultraviolet or visible spectrometry, atomic absorption spectrometry, emission spectrometry, or other suitable methods. To obtain the values of (q), a mass balance must be performed, where the amount of adsorbate adsorbed on the adsorbent should be equal to the amount of adsorbate removed from the solution, as expressed in Equation 2.1.

$$q = \frac{(C_o - C_e)V}{m} \tag{2.1}$$

where,

q: Adsorption capacity (mg g^{-1});
C_0: Initial concentration of adsorbate (mg L^{-1});
C_e: Concentration of adsorbate at steady (mg L^{-1});
V: Volume of solution (L);
m: Mass of adsorbent (g)

After determining (q) and C_e, a graph of (q) versus C_e (with (q) on the vertical axis and C_e on the horizontal axis) can be made, resulting in an adsorption isotherm graph. The term isotherm means the experiments are done at a constant temperature (i.e., under isothermal conditions). Of course, the experiments can be repeated at different constant temperatures, generating another set of (q) versus C_e data for each one (Kinniburgh, 1986). That way, the study looks at how temperature affects adsorption.

In short, isotherms are diagrams showing how the equilibrium concentration on the solid adsorbent changes with the partial pressure or concentration of the liquid phase at a set temperature. These graphs can take different shapes, giving useful insights into the adsorption mechanism (Figure 2.3).

Looking at different types of isotherms can reveal important details about the adsorption process (Do, 1998). A linear isotherm shows that the amount of adsorbate held per unit of adsorbent is directly proportional to the adsorbate's equilibrium concentration in the liquid phase.

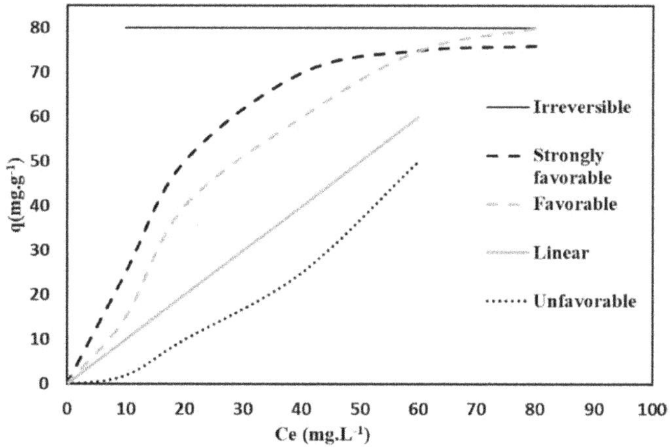

FIGURE 2.3 Illustration of the classification of equilibrium isotherms. (This figure by Vicente Neto is licensed under the Creative Commons Attribution 4.0 International License. https://commons.wikimedia.org/wiki/File:Classification_of_equilibrium_isotherms.jpg.)

Meanwhile, the isotherm area under equilibrium presents a favorable status where the amount of adsorbate attached to the adsorbent at the equilibrium state is still fairly large.

There are many models of adsorption isotherms, with each having a conceptual framework of its own. It is important to choose the model wisely for a particular study, because looking at the right output of the analyses, we can find results that will make sense and will be relevant in practical terms. Furthermore, comparing the theoretical results from the model with some experimental data is crucial for establishing the model itself. Analysis of this kind offers an opportunity to compare and to the revision of model parameters and improve understanding of the behavior of adsorption under particular conditions.

This approach is key to getting meaningful insights and pushing forward the development and optimization of adsorbents.

In the next section, we will go over some of the main adsorption models, along with a detailed breakdown of their pros and cons, key parameters, practical significance, and how they can be used to develop new adsorbents.

Langmuir Model

The Langmuir model was first used to assess and compare how well gases adsorb onto a solid surface. This adsorption process is based on a reversible chemical reaction where adsorption and desorption rates balance out and reach equilibrium (Langmuir, 1918).

The model's main ideas are:

- All molecules/species adsorb onto specific sites on the adsorbent.
- Each site can only hold one chemical species at a time.
- The adsorption energy is the same for every site.
- There's no interaction between species on neighboring sites.

However, the Langmuir model doesn't take into account things like variations in molecular size, screening or blocking effects, multilayer adsorption, or partial molecule mobility—all of which can impact the adsorption process (Marczewski, 2011; Hu et al., 2023). This model was specifically tweaked to describe adsorption equilibrium in solid/solution systems, summed up in the following expression (Equation 2.2):

$$q = \frac{q_{max} k_L C_e}{1 + k_L C_e} \tag{2.2}$$

where:

- q_e (mg g^{-1}) denotes the quantity of solute adsorbed per unit mass of adsorbent at equilibrium.
- C_e (mg L^{-1}) represents the equilibrium concentration of the solute.
- q_{max} (mg g^{-1}) signifies the maximum adsorption capacity of the adsorbent.
- k_L (L mg^{-1}) is the equilibrium constant for adsorption.

k_L (L mg^{-1}) is the equilibrium constant, and it is tied to the free energy of adsorption, showing how strongly the adsorbent surface and the solute interact. q_{max} is the constant that represents the adsorbent's max adsorption capacity.

When $k_L C_e \ggg 1$, the isotherm is super favorable. But if $k_L C_e < 1$, it ends up almost linear. Based on experimental data, k_L and q_{max} can be worked out using linear and nonlinear regression methods. The linear form of Equation 2.2 can take five different formats, as shown in Table 2.1.

Even with its limitations, the Langmuir equation has done a pretty good job of describing adsorption phenomena in most systems. Figure 2.4 shows how theoretical predictions from isotherm models can be tested. In this case, the tested model is the linear Type II version of the Langmuir model.

TABLE 2.1
Linear Forms of Langmuir Adsorption Isotherms

Type	Linear Form	Graph	Parameters	Equation
I	$\frac{1}{q_e} = \frac{1}{K_L q_{max} C_e} + \frac{1}{q_{max}}$	$1/q_e$ vs. $1/C_e$	$q_{max} = 1/(\text{linear coefficient})$ $k_L = \text{linear coefficient/slope}$	Equation 2.3
II	$\frac{C_e}{q_e} = \frac{1}{q_{max}} C_e + \frac{1}{q_{max} K_L}$	C_e/q_e vs. C_e	$q_{max} = 1/\text{slope}$ $k_L = \text{slope/linear coefficient}$	Equation 2.4
III	$q_e = -\left(\frac{1}{k_L}\right)\frac{q_e}{C_e} + q_{max}$	$1/q_e$ vs. q_e/C_e	$q_{max} = \text{linear coefficient}$ $k_L = -1/\text{slope}$	Equation 2.5
IV	$\frac{q_e}{C_e} = -K_L q_e + K_L q_{max}$	q_e/C_e vs. q_e	$q_{max} = \text{Linear coefficient/slope}$ $k_L = -\text{slope}$	Equation 2.6
V	$\frac{1}{C_e} = K_L q_{max} \frac{1}{q_e} - K_L$	$1/C_e$ vs. $1/q_e$	$q_{max} = \text{slope/–Linear Coefficient}$ $k_L = -\text{intercept}$	Equation 2.7

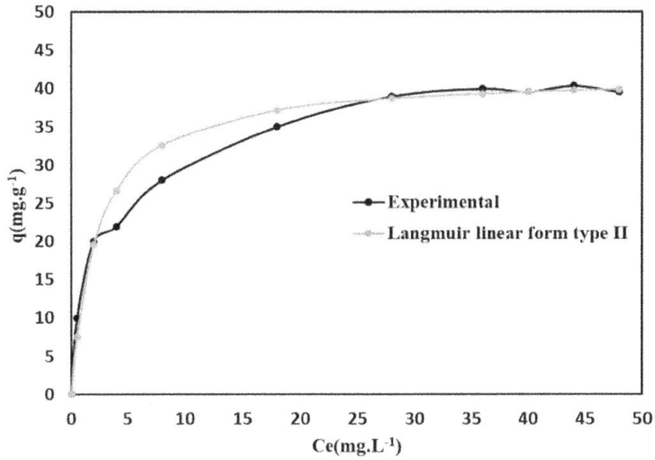

FIGURE 2.4 Illustration of the comparative data: theoretical Langmuir model and experimental isotherm under equilibrium conditions.

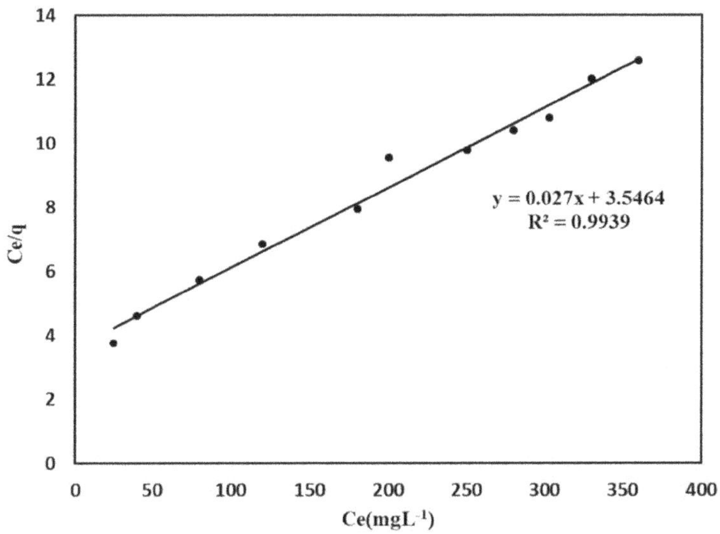

FIGURE 2.5 Graph of C_e/q versus C_e for determining Langmuir parameters. (Reproduced with permission from Melo, 2015. PhD thesis, Federal University of Ceará.)

Usually, the first two approaches are the most commonly used. So, based on Langmuir's Linear Type I Form (Table 2.1), plotting $1/q$ versus $1/C_e$ gives a straight line—normally obtained through a linear fit using the least-squares method—with a slope of $1/(k_L q_{max})$ and an intercept of $1/q_{max}$. By determining the slope and intercept values, it is easy to calculate the two parameters, k_L and q_{max}. Figure 2.5 shows the data from Melo et al. (2013) in a study on Cu^{2+} ion adsorption onto silica-EDTA spheres. The graph was obtained using the Langmuir Linear Type II Form (Table 2.1).

The experimental data in Figure 2.5 were collected like this: 25 mg of silica spheres functionalized with EDTA were added to a single-element solution (25 mL) inside 50 mL Erlenmeyer flasks. The metal ion concentrations ranged from 10 to 350 mg·L^{-1}.

The system was subjected to mechanical stirring at 250 rpm for a duration of 24 h, buffered at a pH of 5.5, and maintained at a room temperature of 28°C. Thus, using the angular (slope) and linear (intercept) coefficients, the calculated values of q_{max} and k_L are 37.03 and 7.60×10^{-3}, respectively.

It is important to emphasize that some authors favor the use of the nonlinear method over the linear one (Subramanyam & Das, 2014; Popoola, 2019; Prihatdini et al., 2023). According to Ho (2006), the application of the four linearized Langmuir equations results in inconsistencies in the parameters obtained when using the same experimental data sets. Therefore, Ho (2006) advocates for the use of the nonlinear method. Additional details regarding errors in linear transformations will be discussed in Chapter 5.

Melo et al. (2013) employed both linear and nonlinear methods to compare and estimate the validity of experimental equilibrium adsorption values with predicted equilibrium values. For the optimization procedure, an error function needs to be defined to evaluate the model's performance. In this study, the sum of squared errors (ERRSQ), a widely used error function in the literature, was adopted and is expressed by Equation 2.8 (Kumar & Sivanesan, 2006; Foo & Hameed, 2010; Popoola, 2019).

$$\text{ERRSQ} = \sum_{i=1}^{p} \left(q_{exp} - q_{cal}\right)^2 \tag{2.8}$$

where

q_{exp}: experimental adsorption capacity (mg·g^{-1});
q_{cal}: capacity calculated by the model (mg·g^{-1})

Since the nonlinear method is based on estimating parameter values, followed by the program calculating the best values to minimize error, it is important to use a strategic approach to avoid unnecessary repetitions. A significant strategy involves estimating parameters through the linear method and then submitting these values to the program Popoola (2019). This ensures that the program adjusts the obtained values to achieve the smallest possible error. Table 2.2 presents the parameters calculated for the Langmuir isotherm using a nonlinear method.

In the following topics, the adsorption results using the Langmuir model will be discussed based on the analysis of the Langmuir parameter k_L. Some questions need to be answered: What should we expect from the value of k_L? What is its relationship with removal efficiency? How can we graphically represent the efficiency of adsorption? All these questions will be discussed in the text.

The Langmuir Parameters

The parameters k_L (L mg^{-1}) and q_{max} (mg g^{-1}) depend on the adsorbent material and can be utilized to compare the performance of different adsorbents. The parameter

TABLE 2.2
Parameters of Nonlinear Langmuir Model for Different Metal Ions

		Parameters		
Metallic Ion	**Model**	q_{max}	k_L	**SSE**
Cu^{2+}		37.64	7.2×10^{-2}	1.57
Cd^{2+}	Nonlinear	23.67	2.9×10^{-2}	5.03
Zn^{2+}		36.38	1.2×10^{-2}	0.28

Source: Data obtained from Melo et al. (2013).

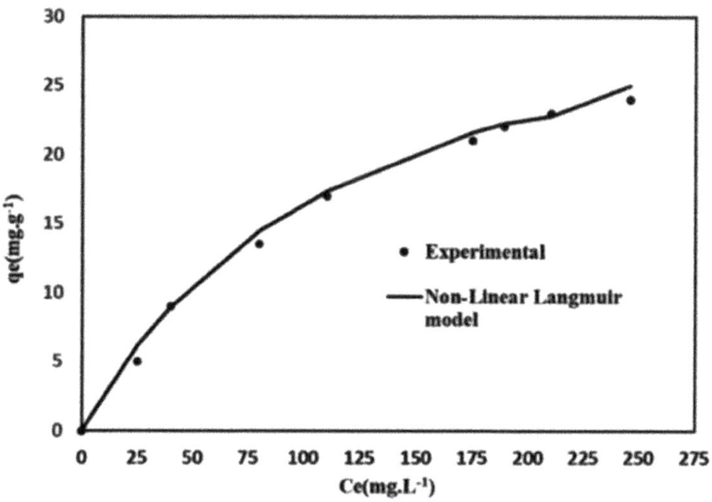

FIGURE 2.6 Isotherm for the adsorption of Cu^{2+} by Si-APTS-EDTA at pH 5.5, and maintained at a room temperature of 28°C. temperature. (Reproduced with permission from Melo et al., 2013.)

qmax (saturation limit) corresponds to the maximum adsorption capacity, while k_L (Langmuir constant) is related to the interaction forces between the adsorbate and the adsorbent. The higher the value of k_L, the better the performance of the adsorbent material. Dimensional analysis reveals the behavior of k_L within the system. Starting from Equation 2.2 (Figure 2.6):

$$q_e = \frac{q_{max} K_L C_e}{1 + K_L C_e} \quad (2.2)$$

Isolating k_L, the following steps are performed

a. $q_e + q k_L C_e = q_{máx} k_L C_e$
b. $q_e + q k_L C_e = q_{máx} k_L C_e$

c. $q_{máx}k_LC_e - q_ek_LC_e = q_e$

d. $k_LC_e\left(q_{máx} - q_e\right) = q_e$

e. $k_L = \dfrac{q_e}{C_e\left(q_{máx} - q_e\right)}$

f. $k_L \alpha \dfrac{1}{C_e}$

Considering C_e as the concentration of the chemical species at equilibrium, and therefore the species that was not adsorbed, it is observed that the lower its quantity, the higher the efficiency of the process and the greater the value of k_L.

Now, let us analyze the Langmuir k_L parameters from Table 2.2. In this table, we observe that the metals copper and zinc exhibit improved adsorption capacity, with values of 37.59 and 36.38 mg·g^{-1}, respectively. The k_L values are 7.2×10^{-3} L·mg^{-1} for Cu^{2+} and 1.2×10^{-2} L·mg^{-1} for Zn^{2+}.These data are consistent with our previous discussion, indicating that the copper removal capacity by the adsorbent is slightly greater than that of zinc. Likewise, the k_L value for copper is 1.67 times lower than that for zinc. The reader may observe that, although the k_L value for zinc removal is higher than that for copper removal, the q_{max} (36.38 mg·g^{-1}) for zinc was only slightly lower than that for copper (37.59 mg·g^{-1}). This suggests that zinc demonstrated good removal efficiency despite unfavorable adsorption conditions, with a higher k_L value than that of copper.

It is important to note that cadmium did not exhibit the same efficiency, as its k_L value was the highest and its adsorption capacity was the lowest. There is a subtlety in interpreting the results for zinc in comparison to copper.

The zinc adsorption results suggest that the adsorbent has the potential to remove more zinc, provided that more suitable conditions, yet to be investigated, are applied.

In the following sections, we will examine adsorption efficiency graphically, based on the characteristics of its curves. The reader should become familiar with the so-called characteristics of the curve. A linear adsorption process is graphically represented by a straight line, in which the adsorption capacity increases linearly with concentration. This type of adsorption is undesirable because the maximum adsorption capacity cannot be estimated.

However, this linear reference is widely used as a benchmark in the graphical analysis of adsorption curve characteristics in experiments. The more the curve deviates from the straight line toward the axes, the better the adsorption performance.

The Separation Factor RL

The characteristics of the Langmuir isotherm can be expressed in terms of the equilibrium parameter, or separation factor R_L, which is a dimensionless constant defined according to Equation 2.8 Magdy and Daifullah (1998).

$$R_L = \frac{1}{1 + K_L.C_o} \tag{2.9}$$

where C_o represents the initial concentration of the analyte in solution and k_L denotes the constant related to the adsorption energy (Langmuir constant). The R_L values indicate the nature of adsorption:

- If $R_L > 1$, adsorption is unfavorable;
- If $R_L = 1$, adsorption is linear;
- If $0 < R_L < 1$, adsorption is favorable;
- If $R_L = 0$, adsorption is irreversible.

Examining the characteristics of adsorption curves is crucial, as they significantly influence the breakthrough dynamics within a fixed-bed column (Chu, 2004). The Langmuir adsorption isotherm exhibits a distinctive plateau at elevated concentrations, reflecting a saturation state in which further adsorption ceases (Allen et al., 2004). Conversely, at lower concentrations, the isotherm simplifies into a linear form, adhering to Henry's law.

Figure 2.7 illustrates the separation factor obtained for the graphs in Figures 2.1 and 2.2, which refer to a system that has reached equilibrium and another that has not. The illustration shows that both experiments suggest that adsorption was favorable, considering that the R_L values fall within the range of 0–1.

By analyzing the two curves, it is evident that the lower curve (black line) exhibits better adsorption performance, as its R_L range is significantly narrower (0.1–0.019) compared to the R_L range of the upper curve (gray line), which varies from 0.17 to 0.035. Therefore, the smaller range is further from $R_L = 1$, a straight line that would be positioned above both curves (not shown). Consequently, in developing new

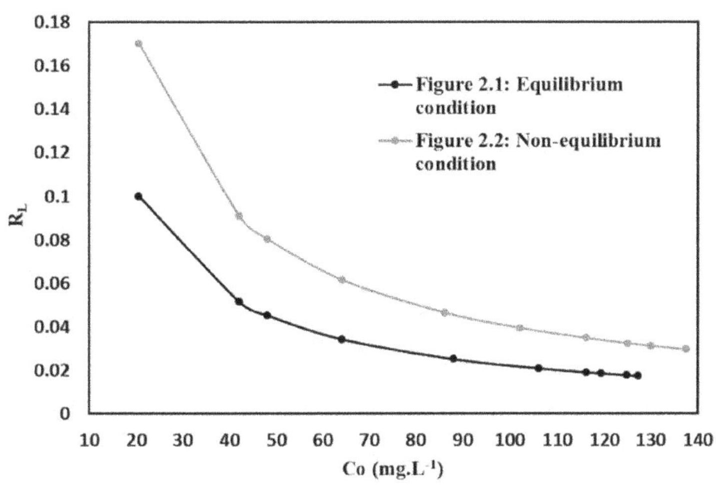

FIGURE 2.7 Illustration of the characteristic separation factor curve based on the Langmuir model. (This figure by Vicente Neto is licensed under the Creative Commons Attribution 4.0 International License. https://commons.wikimedia.org/wiki/File:Characteristic_separation_factor_curve.jpg.)

adsorbents or comparing them, we are primarily interested in adsorbents with narrow R_L ranges. When modifying an adsorbent, the expectation is that the separation factor curve shifts toward the axes.

Dimensionless Form of Langmuir Equation

Based on a similar concept, another dimensionless separation, also called the dimensional form of the Langmuir equation, can be applied to the performance study of adsorption curves (Cooney, 1999; Wu et al., 2010).

The Langmuir parameters, K_L and q_{max}, can be determined graphically using the linearized form of Equation 2.4 presented in Table 2.1. It is renumbered as Equation 2.9:

$$\frac{C_e}{q_e} = \frac{1}{q_{max}} C_e + \frac{1}{q_{max} k_L} \tag{2.4}$$

Let C_{ref} be a reference concentration assigned as the highest C_e during an adsorption process. In this case, the corresponding q_e is set to q_{ref}. From Equation 2.4, we have:

$$\frac{C_{ref}}{q_{ref}} = \frac{1}{q_{max}} C_{ref} + \frac{1}{q_{max} k_L} \tag{2.10}$$

Dividing Equation 2.10 by Equation 2.4, we obtain:

$$\frac{q_e}{q_{ref}} = \frac{C_e}{C_{ref}} \left[\frac{1 + k_L C_{ref}}{1 + k_L C_{ref} \left(\frac{C_e}{C_{ref}} \right)} \right] \tag{2.11}$$

Equation 2.11 is called the dimensionless form of the Langmuir equation, in which $k_L C_{ref}$ is the unique parameter when (q_e/q_{ref}) is plotted against (C_e/C_{ref}). It can be simplified to facilitate its use. To achieve this, we must initially replace Co from Equation 2.9 with C_{ref}, resulting in Equation 2.12.

$$R_L = \frac{1}{1 + K_L \cdot C_{ref}} \tag{2.12}$$

The combination of Equations 2.11 and 2.12 results in Equation 2.13:

$$\frac{q_e}{q_{ref}} = \frac{C_e}{C_{ref}} \left[\frac{1 + k_L C_{ref}}{R_L + (1 - R_L) \left(\frac{C_e}{C_{ref}} \right)} \right] \tag{2.13}$$

Figure 2.8 illustrates an experimental verification of the characteristic adsorption curve for an equilibrium condition, with a separation factor ranging from 0.100 to

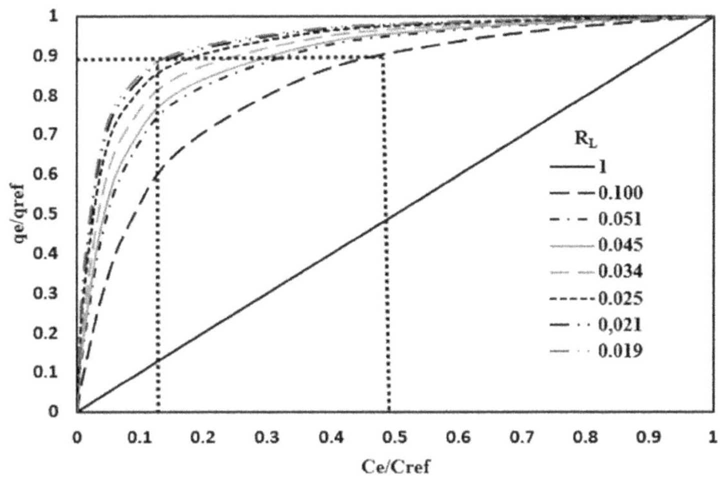

FIGURE 2.8 Illustration of the characteristic equilibrium curves based on the Langmuir equation. (This figure by Vicente Neto is licensed under the Creative Commons Attribution 4.0 International License. https://commons.wikimedia.org/wiki/File:Characteristic_equilibrium_curves.jpg.)

0.019, using the central line with an R_L value of 1 as a reference, indicating linear adsorption. This curve demonstrates good adsorption performance because it maintains a significant distance from the central line due to the low R_L values obtained experimentally. Comparatively, it shows slightly better performance than the adsorption curve in Figure 2.9. This advantage is subtle because the curves in this figure deviate more significantly from the central line, indicating a more efficient interaction between the adsorbate and the adsorbent as R_L decreases. Furthermore, the range of R_L separation factor obtained in Figure 2.8 is narrower than that obtained experimentally in Figure 2.9, allowing for a more accurate analysis of the adsorption conditions in equilibrium systems compared to nonequilibrium systems. These characteristics make the data in Figure 2.8 particularly reliable for use in developing new adsorbents.

Freundlich Model

The Freundlich isotherm is characterized by an empirical equation applicable to systems with surface heterogeneity. According to Suzuki (1990), there is no limit to the adsorption capacity, as the amount of adsorbate tends toward infinity, meaning it does not predict saturation. The nonlinear equation is expressed as:

$$q = k_F \cdot (C_e)^{1/n} \tag{2.14}$$

Its linearized form is expressed as Equation 2.15:

$$\ln(q_e) = \ln(k_F) + \frac{1}{n}\ln(C_e) \tag{2.15}$$

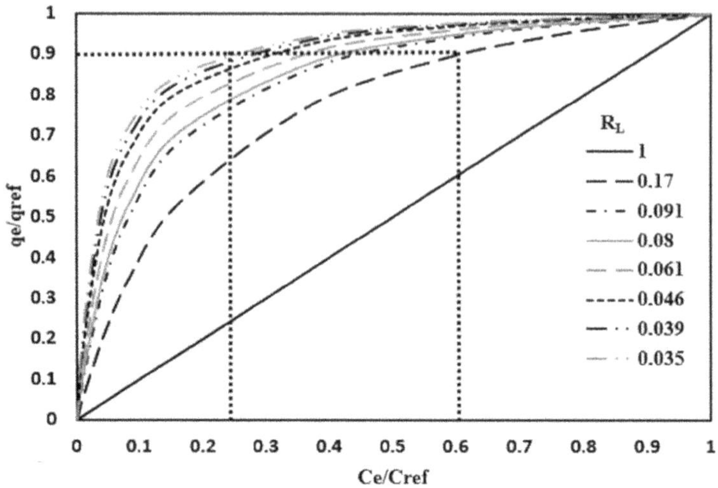

FIGURE 2.9 Illustration of the characteristic nonequilibrium curve based on the Langmuir equation. (This figure by Vicente Neto is licensed under the Creative Commons Attribution 4.0 International License. https://commons.wikimedia.org/wiki/File:Characteristic_non-equilibrium_curves.jpg.)

The parameters k_f and n represent the Freundlich constants and the affinity coefficient, respectively. In the linear form, they can be determined through the slope and intercept of the fitted linear curve. At saturation, qe becomes constant and independent of the increase in C_e. Therefore, the Freundlich equation no longer applies (Ruthven, 1984; Coles & Yong, 2006).

The Freundlich Parameters

The constant k_f ((mg g^{-1})(L mg^{-1})$^{1/n}$) relates to the adsorption capacity of the solid, and n indicates whether the process is favorable. Values of n within the range of 1–10 suggest favorable conditions for adsorption (Satyanarayana et al., 1999; Febrianto et al., 2009). When n is negative, it indicates that the solvent has a greater affinity for the adsorbent than for the adsorbate. Values of n between 0 and 1 suggest that adsorption is unfavorable (Febrianto et al., 2009). In general, it can be stated that the higher the value of n (the smaller the value of 1/n), the stronger the interaction between the adsorbate and the adsorbent. Conversely, when the value of 1/n equals 1, this indicates that adsorption is linear; that is, the energies are identical for all adsorption sites. When the value of 1/n exceeds one, the adsorbent demonstrates a greater affinity for the solvent, suggesting strong intermolecular attraction between them (Delle-Site, 2001).

It is important to note that the Freundlich equation cannot predict equilibrium adsorption data at extremely high concentration ranges. Furthermore, this equation does not simplify to the linear adsorption expression at very low concentrations. However, studies generally fall within a moderate concentration range, allowing the equation to be used for data fitting purposes (Cooney, 1999).

TABLE 2.3
Parameters for Nonlinear Freundlich Isotherm

		Parameters			
Metallic Ion	**Model**	**n**	**1/n**	**k_F**	**ERRSQ**
Cu^{2+}		1.78	0.56	1.10	2.64
Cd^{2+}	Não-linear	4.16	0.24	5.46	11.10
Zn^{2+}		1.85	0.54	1.51	12.03

Source: Data obtained from Melo et al. (2013).

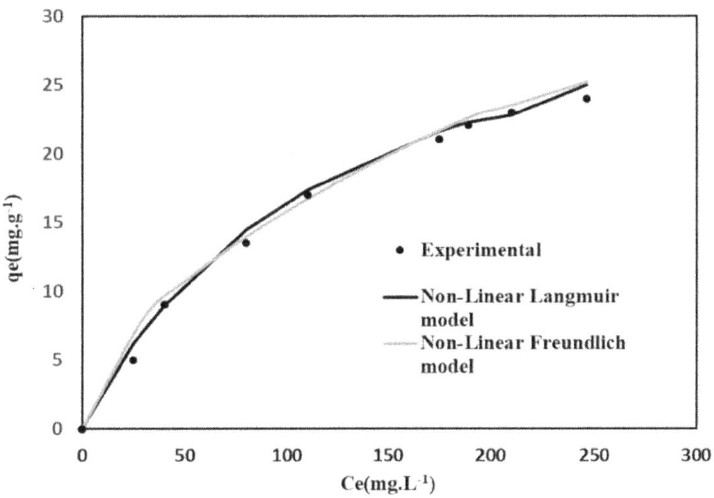

FIGURE 2.10 Comparison of nonlinear adsorption isotherms with the experimental data for the ion Cu^{2+}. Experimental conditions: adsorbent mass: 25 mg; aliquot: 25 mL; pH: 5.5; ambient temperature: 28°C. (Adapted from Melo et al., 2013.)

The Freundlich parameters, calculated through linear and nonlinear regression analysis along with the error function, are listed in Table 2.2. The linear and nonlinear models were compared based on both the R^2 value and the error function value. For all the ions studied, the R^2 values for both linear and nonlinear models are very similar. However, the error function value is lower for the nonlinear models, making them satisfactory in representing the experimental data.

The value of n shown in Table 2.3 for all ions was greater than 1, indicating that adsorption is favorable. These data coincide with the R_L values, confirming favorable adsorption. Figure 2.4 presents the graphical comparison of the Langmuir and the Freundlich models based on nonlinear regression analysis for the ions Cu^{2+}, Zn^{2+}, and Cd^{2+}.

According to Figure 2.10, the adsorption behavior is similar for the metallic ions studied at low concentrations, likely due to the high availability of adsorption sites. As the concentration of ions in the solution increases, the adsorption sites become less available.

In this study, the experimental adsorption capacity follows the order $Zn^{2+} > Cu^{2+} > Cd^{2+}$ Melo et al. (2013).

Temkin Isotherm

This equation accounts for adsorbent–adsorbate interactions and the uniform distribution of bonding energies (Foo & Hameed, 2010; Chu, 2021). Excluding extremely low or high concentrations, the model assumes that the adsorption heat of molecules in the layer tends to decrease linearly—rather than logarithmically—as the adsorbent coverage increases (Aharoni & Ungarish, 1977; Chu, 2021).

The Temkin equation (Temkin, 1940) was initially proposed to describe the adsorption of hydrogen onto platinum electrodes in an acidic medium, and it has the following form:

$$\theta = \frac{RT}{b_T} \ln(k_T C_e) \tag{2.16}$$

In Equation 2.16, θ is a dimensionless quantity defined by Equation 2.17.

$$\theta = \frac{q_e}{q_m} \tag{2.17}$$

By substituting Equation 2.17 in Equation 2.16, we obtain the correct form of the Temkin isotherm equation, as recently discussed by Zhou et al. (2022).

$$q_e = q_m \left(\frac{RT}{b_T} \right) \ln(k_T C_e) \tag{2.18}$$

where:

θ: coverage fraction of adsorbate on the adsorbent surface dimensionless;
q, q_m: are the equilibrium and saturated adsorption amount of solute on the solid surface, respectively, $mg{\cdot}g^{-1}$;
R: universal gas constant ($J{\cdot}mol^{-1}{\cdot}K^{-1}$);
T: temperature (K);
b: Temkin constant related to sorption heat ($J{\cdot}mol^{-1}$);
k_T: Temkin isotherm constant ($L{\cdot}mg^{-1}$);
C_e: equilibrium adsorbate concentration ($mg{\cdot}L^{-1}$).

The application of the Temkin isotherm has significantly expanded to encompass adsorption processes in liquid phases, focusing on environmental areas such as the removal of heavy metal ions and organic pollutants.

Table 2.4 is a list of the Temkin equation parameters for the adsorption of trans,trans-2,4-heptadienal (HDE) and trans,trans-2,4-decadienal (DDE) onto powdered activated carbon obtained by Li et al. (2015). Djelloul et al. (2017) employed Milk thistle seeds as a biosorbent for the removal of methylene blue (MB) from synthetic wastewater. Millar et al. (2015) successfully employed acid cation resin in the studies of Fe+2 ions exchange. Guechi and Hamdaoui (2016) employed potato

TABLE 2.4
Temkin Parameters for Pollutant Adsorption: Data Obtained from Equation 2.19

Sorbent	Species Pollutants	b ($J\cdot mol^{-1}$)	k_T ($L\cdot mg^{-1}$)	Erro Function APA	References
PAC[a]	HDE	14.2	0.155	7.35	Li et al. (2015)
	DDE	20.2	2.90	2.11	
	4-CP	28.69	8.30	2.73	
GAC[b]	Ph	23.15	5.871	5.54	Hamdaoui and Naffrechoux (2007)
	2-CP	23.37	0.911	5.54	
Milk thistle seeds	MB	116.10	0.1412	4.76	Djelloul et al. (2017)
Lanxess S108H	Fe(II)	19.21	8.30	32.92*	Millar et al. (2015)
Potato peel	Cu(II)	0.015	0.78	0.99**	Guechi and Hamdaoui (2016)
Lanxess S108H	U(VI)	14.64	5.37	44.34***	Liu et al. (2021)

Source: Table prepared by Vicente Neto.
Error functions-ARE*; r**; RMSE***.
[a] Powdered activated carbon.
[b] Granula activated carbon.

peel (PP) under various operating conditions in a batch process to remove Cu(II) ions from synthetic aqueous solutions. The data are summarized in Table 2.4, which presents the Temkin parameters.

Figure 2.11 presents a comparison between the experimental data and the theoretical data obtained through the application of the Temkin model with adapted data from Sousa Neto (2011)

Considerations on the Temkin Equation

The Temkin equation has not been used properly. Equation 2.20 is the most widely used form in adsorption studies.

$$q_e = \frac{RT}{b_T}\ln\left(k_T C_e\right) \tag{2.19}$$

However, this equation does not respect dimensional analysis. A more detailed discussion on this topic was provided by Zhou et al. (2022). We suggest that readers refer to that article. Here, we will briefly discuss the units in a didactic way without any mathematical formalism. To verify this, substituting the units of each parameter into Equation 2.19 yields:

$$q_e\left(mg\cdot g^{-1}\right) = \frac{R\left(J\cdot molK^{-1}\right)T(K)}{b_T\left(J\cdot mol^{-1}\right)}\ln\left[k_T\left(L\cdot g^{-1}\right)C_e\left(g\cdot L^{-1}\right)\right] \tag{2.20}$$

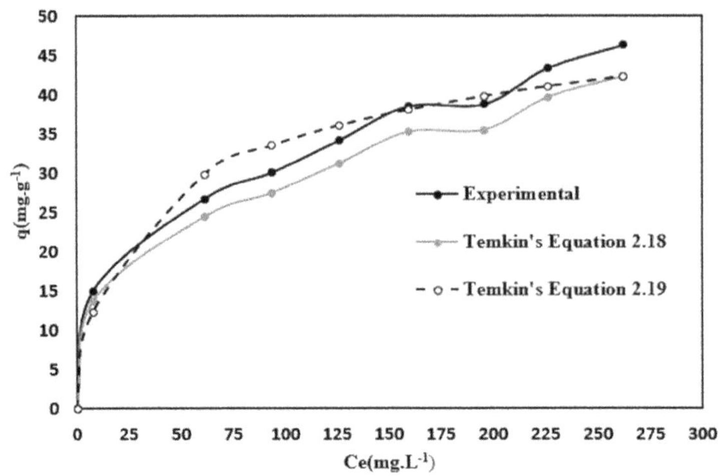

FIGURE 2.11 Comparison of the adsorption of the Temkin model with the experimental data for the ion Cu^{2+}.Experimental conditions: adsorbent mass: 25 mg; aliquot: 25 mL; pH: 5.5; ambient temperature: 28°C.). (Figure adapted with permission from Sousa Neto (2011), published in BioResources, an international open access journal.)

Equation 2.20 illustrates that the right-hand side is dimensionless due to the cancellation of units. However, the left-hand side retains dimensional properties because it represents distinct physical quantities, specifically milligrams of solute adsorbed per gram of adsorbent $mg \cdot g^{-1}$. According to fundamental principles of dimensional analysis in physics, a physically valid equation must exhibit dimensional consistency in all terms (Yan, 2012).

Since Equation 2.20 does not satisfy this criterion, it is an incorrect formulation and does not conform to Temkin's original model. According to Zhou et al. (2022), Equation 2.20 can be corrected by changing the units of k_t and C_e from $L \cdot g^{-1}$ and $g \cdot L^{-1}$ to $L \cdot mg^{-1}$ and $mg \cdot L^{-1}$, respectively, and by introducing θ into Equation 2.20. Zhou et al. (2022) explains how to correct Equation 2.20 to comply with dimensional analysis, aligning it with the original model proposed by Temkin, resulting in Equations 2.16 and 2.18. Figure 2.11 presents the comparison of the updated Temkin equation.

Redlich–Peterson Isotherms

The Redlich–Peterson equation (Redlich & Peterson, 1959) can be applied to adsorption processes over a wide range of concentrations. Regarding this variable, the model presents an exponential function in the denominator and a linear dependence in the numerator, making it applicable to both homogeneous and heterogeneous systems.

The isotherm equation has three parameters:

$$q_e = \frac{k_{RP} C_e}{1 + a_{RP} C_e^{\beta}} \tag{2.21}$$

where,

- q: Adsorption capacity at equilibrium ($mg{\cdot}g^{-1}$);
- C_e: Adsorbate concentration at equilibrium ($mg{\cdot}L^{-1}$);
- k_{RP} ($L{\cdot}g^{-1}$), $\mathbf{a_{RP}}$ $(L{\cdot}mg^{-1})^{\beta}$ are parameters of the Redlich–Peterson isotherm and β, an exponent varying from 0 to 1.

The empirical Redlich–Peterson equation incorporates characteristics from both the Langmuir and the Freundlich models, resembling the former at low concentrations when β approaches 1, and assuming the form of the latter in systems with high concentrations when β approaches zero (Foo & Hameed, 2010; Febrianto et al., 2009).

The Redlich–Peterson model is commonly used to predict equilibrium conditions involved in the biosorption of heavy metals. In most cases, the value of β is close to 1, indicating that the Langmuir isotherm model fits the experimental data well (Wang & Guo, 2020; Mozaffari Majd et al., 2022).

In some studies on heavy metal biosorption, the experimental data are well described by various models. For example, the biosorption of Cr^{+6} (Preetha & Viruthagiri, 2007), Cd^{+2} (Ho, 2006), Pb^{+2} (Ho, 2006), and Ni^{+2} (Padmavathy, 2008) has been documented using different types of adsorbents.

In a study on the adsorption of polycyclic aromatic hydrocarbons onto mesoporous organosilica, the data obtained for the acenaphthene compound were best described by the Redlich–Peterson model compared to the Freundlich, the Langmuir, and the Temkin isotherm models (Vidal et al., 2011).

The application of the Redlich-Peterson model is illustrated in the following example. Table 2.5 presents the results of an experimental study of adsorption isotherms in monoelement solutions of Cu^{2+}, Zn^{2+}, and Cd^{2+} ions using EDTA-functionalized silica spheres as adsorbents. The table shows the results for the nonlinear R-P model. The experimental conditions were an acetate buffer at pH 5.5 and a temperature of 28°C (Melo et al., 2013). According to Table 2.5, Redlich–Peterson isotherms showed good correlation with experimental data (Figure 2.12), as indicated by the high determination coefficients and low error function values.

TABLE 2.5
Redlich–Peterson Parameters for Isotherm Nonlinear Models for Monoelement Systems

Model	Parameter	Cu^{2+}	Zn^{2+}	Cd^{2+}	Reference
Redlich–Peterson	k_{RP}	0.27	0.41	0.92	Melo et al.,
	a_{RP}	7.7×10^{-3}	1.2×10^{-2}	6.1×10^{-2}	(2013)
	β	0.99	0.97	0.92	
	R^2	0.99	0.99	0.98	
	ERRSQ	1.53	0.85	6.73	

Source: Adapted from Melo et al., (2013).

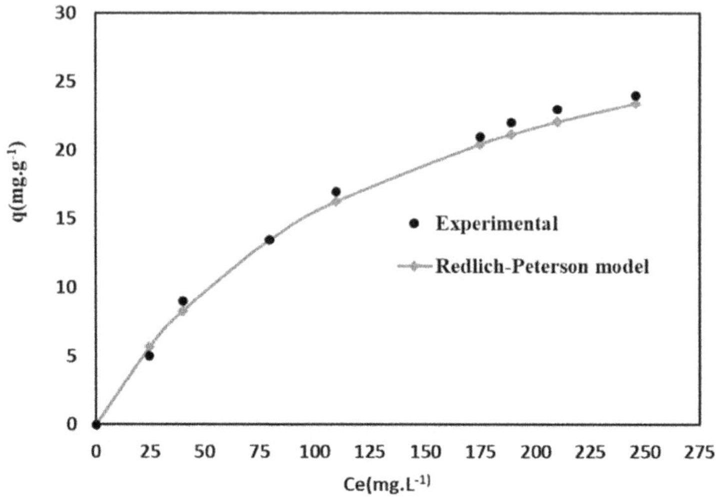

FIGURE 2.12 Comparison of the adsorption of the Redlich–Peterson model with the experimental data for the ion Cu^{2+}.Experimental conditions: adsorbent dosage: 1 $g{\cdot}L^{-1}$; pH: 5.5 (acetate buffer); orbital shaker device operating at 250 rpm; ambient temperature: 28°C. (Figure adapted with permission from Melo et al., 2013.)

Parameter Units

The determination of units for the Redlich–Peterson (R–P) parameters, k_{RP} and a_{RP}, has been a subject of significant ambiguity in the literature. Nevertheless, this task can be approached systematically.

The first critical step entails defining the units for the measured variables, C_e (liquid-phase concentration) and q (solid-phase concentration). Conventionally, C_e is reported in milligrams or millimoles of solute per liter of solution (mg L^{-1} or mmol L^{-1}), whereas q is quantified in milligrams or millimoles of solute per gram of adsorbent (mg g^{-1} or mmol g^{-1}).

Further clarity emerges upon analyzing Equation 2.22, where the denominator is inherently dimensionless. By algebraically rearranging this equation to Equation 2.23, the units of parameter k_{RP} become evident. Moreover, the term $a_{RP}C_e^{\beta}$ must also be dimensionless, which imposes a direct constraint on the units of a_{RP}. Specifically, a_{RP} adopts units reciprocal to those of C_e^{β}, yielding $(L{\cdot}mg^{-1})^{\beta}$ or $(mg\ L^{-1})^{-\beta}$, thereby ensuring dimensional homogeneity in the model.

$$\text{units of } k_{RP} = \frac{\text{units of q}}{\text{units of } C_e}\left(\text{unitless term} : 1 + a_{RP}C_e^{\beta}\right) \tag{2.22}$$

$$= \frac{mg \cdot g^{-1}}{mg \cdot L^{-1}} \tag{2.23}$$

$$= L \cdot g^{-1} \tag{2.24}$$

A notable concern arises from the recurrent misassignment of units for parameter k_{RP} as L·mg^{-1} in numerous studies, despite q being consistently measured in mg·g^{-1}.

It is important to emphasize that the correct unit designation for k_{RP} is L·g^{-1}, explicitly referencing the adsorbent mass in grams. Compounding this issue, parameter a_{RP} has frequently been erroneously reported with units of L·mg^{-1} rather than the dimensionally appropriate (L·mg^{-1})$^{\beta}$ formulation. Remarkably, this unit discrepancy persists even in highly cited review articles (Foo & Hameed, 2010; Al-Ghouti & Da'ana, 2020). Such persistent inaccuracies in fundamental dimensional analysis raise significant concerns about methodological rigor within the water treatment research community.

SUMMARY

Understanding adsorption equilibrium is essential for evaluating separation processes that rely on adsorption. Equilibrium is reached when the concentration of solute in the liquid phase (C_e) stabilizes, indicating balanced mass transfer between liquid and solid phases. Accurate data depend on achieving this equilibrium; otherwise, results may be misleading due to incomplete interaction between adsorbent and adsorbate. Analytical techniques such as chromatography and spectrometry are used to measure residual concentrations, while the adsorption capacity (q) is calculated via mass balance. Plotting q against C_e yields an adsorption isotherm, which reflects the adsorbent's behavior under varying concentrations and constant temperature. Different isotherm models provide insights into the nature of the adsorption mechanism, emphasizing the importance of selecting a suitable model for reliable interpretation. Comparing model predictions with experimental data allows refinement of theoretical assumptions and enhances comprehension of adsorption behavior under specific conditions.

The Temkin adsorption model, although widely used, has often been misapplied due to dimensional inconsistencies. Equation 2.20, in particular, does not satisfy the criteria required for a correct dimensional analysis, since the left-hand side has physical units while the right-hand side is dimensionless. To correct this, Zhou et al. (2022) recommend revising the units of the constants k_T and C_e to L mg^{-1} and mg L^{-1}, respectively, and introducing the factor θ. The Redlich–Peterson model presents challenges in defining units for the parameters k_{RP} and a_{RP}. A proper dimensional analysis reveals that k_{RP} should have units of L g^{-1}, while a_{RP} should adopt units inverse to C_e^{β}, such as (L mg^{-1})$^{\beta}$.

EXERCISES

Applied Questions

The Importance of Controlling Heavy Metal Discharge in Aquatic Ecosystems Post-Treatment

Effective regulation of heavy metal concentrations released into natural water sources following wastewater treatment is crucial for safeguarding environmental integrity and public health. Although many treatment facilities are equipped to reduce pollutant loads, residual levels of heavy metals such as lead, cadmium, mercury, and chromium often persist and are subsequently discharged into aquatic ecosystems. These metals are not biodegradable and tend to accumulate in sludge and biomass within biological treatment systems, thereby disrupting microbial communities essential to the efficient degradation of organic matter. Over time, toxicity can compromise biological reactor performance, reduce treatment efficiency, and lead to system failure.

Furthermore, once discharged into receiving waters, heavy metals bioaccumulate in aquatic organisms, triggering toxic effects such as reproductive failure, impaired growth, neurological damage, and mortality. Trophic transfer through food chains may pose broader ecological threats and increase human exposure through the consumption of contaminated fish and water. Thus, stringent monitoring and regulatory enforcement are indispensable to mitigate these persistent risks, preserve water quality, and maintain the functionality of biological treatment processes and aquatic biodiversity.

1. A study on pollutant removal at a metal plating factory was conducted under two different pH conditions. The data in the table refer to the adsorption of nickel at pH 3 and pH 5 at 28°C. Based on this information, answer the following questions.

pH = 5.3		pH = 3.3	
C_e ($mg{\cdot}L^{-1}$)	q_e ($mg{\cdot}g^{-1}$)	C_e ($mg{\cdot}L^{-1}$)	q_e ($mg{\cdot}g^{-1}$)
0.0	0	0	0
3.0	11	2.8	18.2
20.3	30	22	25.3
32.1	36	30	30.1
62.5	42	60	33.0
90.1	44	92	36.5
100.0	44.5	98	36.0
120.2	44.3	121	35.4

a. Obtain the Langmuir isotherms and the linear equations that describe the phenomenon at both pH levels. Which pH condition is more suitable, and why? Determine the Langmuir parameters.
b. Plot the graphs comparing the experimental and theoretical data.

Health Risks of Nitrate Contamination for Infants

1. Nitrate is a widespread contaminant that often enters drinking water supplies due to agricultural practices, including the excessive use of fertilizers and the improper disposal of animal waste. Once in the water system, nitrates can pose a serious health threat—especially to infants. When ingested by infants under 6 months of age, nitrates can be converted by

intestinal bacteria into nitrites, which interfere with the normal oxygen-carrying capacity of hemoglobin by forming methemoglobin. This condition, known as methemoglobinemia or "blue baby syndrome," reduces oxygen delivery to tissues and may lead to cyanosis, fatigue, difficulty breathing, and, in extreme cases, death. Because of these risks, regulatory agencies have set strict limits on nitrate concentrations in drinking water. Infants are particularly sensitive due to their underdeveloped enzyme systems, making nitrate exposure during this stage of life exceptionally dangerous. Safe water practices and regular water quality monitoring are essential to prevent such exposure and protect infant health.

2. The table below presents experimental data collected during a nitrate removal study involving a groundwater well at a rural household where an incident of infant methemoglobinemia ("blue baby syndrome") was reported. The study used an adsorbent material with a particle size between 80 and 100 mesh and a point of zero charge (PZC) of 6.8.

C_e (mg L^{-1})	q (mg g^{-1})
0.0	0.00
4.7	46.43
19.4	93.37
41.2	115.31
64.1	123.47
89.4	127.55
107.1	128.57

Answer the Following Questions Based on the Data and Background Provided

1. Explain the significance of the PZC value in selecting the pH for the adsorption experiment.
2. Estimate the nitrate removal capacity of the adsorbent based on the data.
3. Identify and justify the most suitable pH range for efficient nitrate adsorption in this system.
4. Discuss how the adsorbent's particle size might influence the adsorption efficiency.
5. Reflect on the relevance of this study for preventing nitrate-induced health risks in rural communities.

Objective Questions on Adsorption Equilibrium

1. What is the correct definition of adsorption equilibrium?
   a. The state where the adsorbate concentration in the liquid phase continuously changes.
   b. The state where the adsorbate molecules stop interacting with the adsorbent.

c. The state where the adsorbate concentration in the liquid phase (Ce) remains constant.
d. The state where the adsorbent dissolves in the liquid phase.

2. What technique is commonly used to separate the adsorbent from the solution after equilibrium is reached?
   a. Thermal decomposition
   b. Membrane filtration, filter paper, or centrifugation
   c. Electrolysis
   d. Crystallization
3. Which factor can lead to incorrect adsorption capacity measurements?
   a. Achieving adsorption equilibrium
   b. Using a known mass of adsorbent
   c. Nonequilibrium conditions
   d. Performing a mass balance
4. What type of graph is commonly used to analyze an adsorption equilibrium experimentally?
   a. Adsorption capacity (q) vs equilibrium concentration (C_e)
   b. Adsorption capacity (q) vs solution pH
   c. Adsorption capacity (q) vs reaction time
   d. Equilibrium concentration (C_e) vs adsorbent mass
5. Which analytical technique can be used to determine the residual adsorbate concentration (C_e)?
   a. X-ray diffraction
   b. Mass spectrometry
   c. Gas chromatography or ultraviolet-visible spectrometry
   d. Polymerization analysis

Objective Questions on the Separation Factor R_L

Question 1: What does the separation factor (R_L) indicate in adsorption studies?
   a. The amount of adsorbent required for equilibrium.
   b. The nature of adsorption (favorable, unfavorable, linear, or irreversible).
   c. The adsorption rate in a fixed-bed column.
   d. The molecular interaction between the adsorbate and adsorbent.

Question 2: If $R_L = 0$, what does it imply?
   a. Adsorption is irreversible.
   b. Adsorption follows Henry's law.
   c. Adsorption is unfavorable.
   d. Adsorption does not reach equilibrium.

Question 3: What happens to the Langmuir adsorption isotherm at high concentrations?
   a. It forms a plateau due to saturation.
   b. Adsorption ceases completely.
   c. The separation factor increases beyond 1.
   d. The adsorbent dissolves into the solution.

Question 4: What parameter determines the adsorption energy in Langmuir's equation?
a. C_0 (Initial concentration)
b. k_L (Langmuir constant)
c. q_e(Adsorption capacity)
d. Henry's constant

Question 5: Why is a narrow R_L range desirable in adsorption studies?
a. It indicates a higher adsorption performance.
b. It signifies that adsorption is unfavorable.
c. It means that equilibrium is impossible to achieve.
d. It suggests that adsorption will not follow Langmuir's model.

Subjective Questions on the Separation Factor R_L

Question 1: Explain the significance of the separation factor (R_L) in adsorption studies.

Question 2: Discuss how nonequilibrium conditions affect adsorption measurements.

Question 3: How does Langmuir's adsorption isotherm behave at different concentration levels?

Question 4: What role does the Langmuir constant (k_L) play in adsorption modeling?

Question 5: Why is the analysis of adsorption curves important for fixed-bed column dynamics?

Objective Questions on the Dimensionless Langmuir Equation

Question 1: What is the primary advantage of using the dimensionless form of the Langmuir equation?
a. It eliminates the need for equilibrium concentration calculations.
b. It facilitates graphical analysis of adsorption behavior.
c. It increases the adsorption capacity of the adsorbent.
d. It disregards the Langmuir constant in adsorption calculations.

Question 2: What role does C_{ref} play in the dimensionless Langmuir equation?
a. It determines the irreversible adsorption threshold.
b. It is the Langmuir constant related to adsorption energy.
c. It represents the reference concentration corresponding to the highest equilibrium concentration.
d. It describes the separation factor of the adsorption process.

Question 3: How is the adsorption equilibrium curve experimentally verified?
a. By measuring the separation factor (R_L) and plotting the adsorption curve.
b. By increasing the adsorbate concentration beyond equilibrium levels.
c. By modifying the adsorbent structure during adsorption.
d. By disregarding nonequilibrium adsorption states.

Question 4: Why is a narrower separation factor range desirable in adsorption studies?

a. It ensures higher adsorption efficiency.
b. It limits the adsorption of solutes in solution.
c. It prevents the adsorption equilibrium from being reached.
d. It simplifies the Langmuir equation calculations.

Question 5: What does the equation represent?

$$\left(\frac{q_e}{q_{ref}}\right) = \left(\frac{C_e}{C_{ref}}\right)\left[\frac{1}{R_L + (1 - R_L)\left(\frac{C_e}{C_{ref}}\right)}\right]$$

a. The simplified Langmuir equation accounting for R_L.
b. The mass balance equation for adsorption equilibrium.
c. The equation for irreversible adsorption conditions.
d. The correlation between adsorbate concentration and solution pH.

Subjective Questions on the Dimensionless Langmuir Equation

Question 1: Explain the significance of the dimensionless form of the Langmuir equation in adsorption studies.

Question 2: Describe the impact of the separation factor (R_L) on adsorption behavior.

Question 3: How does C_{ref} influence the dimensionless Langmuir equation?

Question 4: Why is experimental verification of adsorption equilibrium curves essential?

Question 5: How does the modification of adsorbents affect the separation factor curve?

Objective and Subjective Questions on the Freundlich Adsorption Model

Question 1: According to the Freundlich adsorption model, what assumption is made about the nature of the adsorbent surface?

a. It is homogeneous with identical adsorption sites.
b. It is heterogeneous with varying adsorption energies.
c. It strictly follows the Langmuir model.
d. It only applies to monolayer adsorption.

Answer: (b) It is heterogeneous with varying adsorption energies.

Question 2: What is the primary limitation of the Freundlich adsorption model when describing adsorption at high concentrations of adsorbate?

a. It assumes adsorption reaches saturation.
b. It predicts continuous adsorption increase, which is physically unrealistic.
c. It does not account for surface heterogeneity.

d. It requires specific temperature conditions to remain valid.

Answer: (b) It predicts continuous adsorption increase, which is physically unrealistic.

Question 3: In the linearized form of the Freundlich isotherm equation, what does the slope of the graph represent?

a. The adsorption capacity constant (k_F).
b. The equilibrium concentration of the adsorbate (C_e).
c. The inverse of the surface heterogeneity constant (1/n).
d. The maximum adsorption capacity of the adsorbent (q_{max}).

Subjective Questions with Sample Responses

Question 4: The Freundlich adsorption model is often used for nonideal adsorption systems. How does its empirical nature make it suitable for describing adsorption on heterogeneous surfaces?

Question 5: Unlike the Langmuir model, the Freundlich isotherm does not impose a monolayer adsorption limit. Discuss the implications of this in practical adsorption studies, particularly for systems with high concentrations of adsorbate.

Question 6: Considering that the Freundlich model does not account for saturation effects, how can researchers determine the validity range of the model when analyzing experimental adsorption data?

Question 7: The Freundlich equation incorporates an adsorption capacity constant (k_F) and a heterogeneity constant (1/n). What do these parameters reveal about the adsorption efficiency and surface properties of the adsorbent material?

Question 8: The model suggests an exponential distribution of adsorption sites with different adsorption energies. In your opinion, how does this assumption affect the accuracy and predictive power of adsorption models used in environmental applications?

Objective and Subjective Questions on the Temkin Adsorption Model

Question 1: The Temkin adsorption model assumes that the adsorption heat of molecules in the layer decreases in which manner as the adsorbent coverage increases?

a. Logarithmically.
b. Linearly.
c. Exponentially.
d. Quadratically.

Question 2: What is the primary purpose of introducing the dimensionless quantity θ into the Temkin equation?

a. To account for variations in equilibrium adsorption.
b. To align the equation with dimensional analysis principles.
c. To improve the accuracy of adsorption kinetics predictions.
d. To eliminate the dependence on temperature in adsorption calculations.

Question 3: According to Zhou et al. (2022), what correction is necessary for the widely used Temkin equation to conform to dimensional analysis?

a. Changing the units of k_T and C_e from $L{\cdot}g^{-1}$ and $g{\cdot}L^{-1}$ to $L{\cdot}mg^{-1}$ and $mg{\cdot}L^{-1}$.
b. Removing the universal gas constant (R) from the equation.
c. Introducing an additional adsorption equilibrium term.
d. Modifying the adsorption heat constant (b) to a temperature-dependent variable.

Question 4: The Temkin isotherm equation was originally developed to describe the adsorption of which element onto platinum electrodes?

a. Hydrogen.
b. Oxygen.
c. Carbon.
d. Iron.

Question 5: What is the role of the Temkin constant (b) in adsorption processes?

a. It represents the equilibrium concentration of the adsorbate.
b. It describes the sorption heat involved in adsorption.
c. It defines the maximum adsorption capacity.
d. It indicates the adsorbent's surface heterogeneity.

Subjective Questions with Sample Responses

Question 6: Discuss the significance of the Temkin isotherm equation in adsorption studies, particularly in liquid-phase environmental applications. How does it compare to other adsorption models, such as the Langmuir and the Freundlich isotherms?

Question 7: Dimensional analysis is fundamental in physics to ensure equation consistency. Based on Zhou et al. (2022), what are the consequences of using an equation that fails to comply with dimensional analysis, and how does the correction in Equation 2.19 resolve this issue?

$$q = q_m \frac{RT}{b} \text{In}\left(k_T C_e\right)$$

Question 8: Considering the experimental applications discussed in the text—such as the adsorption of heavy metals and organic pollutants onto biosorbents—how does the incorporation of θ improve the predictive accuracy of the Temkin model for real-world adsorption processes?

Objective and Subjective Questions on the Redlich–Peterson Adsorption Model

Objective Questions

Question 1: What is the correct unit designation for the Redlich–Peterson parameter k_{RP}?

a. $L{\cdot}mg^{-1}$.
b. $L{\cdot}g^{-1}$.

c. $mg{\cdot}g^{-1}$.
d. $mmol{\cdot}L^{-1}$.

Question 2: What constraint does dimensional homogeneity impose on the units of the parameter a_{RP} in the Redlich–Peterson model?
a. a_{RP} should have units of $(L{\cdot}mg^{-1})^{-\beta}$.
b. a_{RP} should be dimensionless.
c. a_{RP} should have units of $mg{\cdot}L^{-1}$.
d. a_{RP} should have the same units as C_e.

Question 3: Which of the following issues regarding R–P parameters has been frequently misreported in the literature?
a. The incorrect assignment of units for k_{RP} as $L{\cdot}mg^{-1}$ instead of $L{\cdot}g^{-1}$.
b. The incorrect assumption that a_{RP} is dimensionless.
c. The failure to distinguish between solid-phase and liquid-phase concentration units.
d. The omission of equilibrium conditions in R–P calculations.

Subjective Questions with Sample Responses

Question 4: The determination of units for R–P parameters has been a subject of ambiguity in adsorption studies. Discuss why dimensional homogeneity is essential in adsorption modeling and how incorrect unit assignments can lead to methodological errors.

Question 5: The Redlich–Peterson model combines characteristics of both the Langmuir and Freundlich isotherms. How does its semi-empirical nature contribute to adsorption modeling, and what are its advantages and limitations in real-world applications?

REFERENCES

Aharoni, C., & Ungarish, M. (1977). Kinetics of activated chemisorption. Part 2. Theoretical models. *Journal of the Chemical Society*, 73, 456–464. https://doi.org/10.1039/F19777300456

Al-Ghouti, M. A., & Da'ana, D. A. (2020). Guidelines for the use and interpretation of adsorption isotherm models: A review. *Journal of Hazardous Materials*, 393, 122383. https://doi.org/10.1016/j.jhazmat.2020.122383

Allen, S. J., McKay, G., & Porter, J. F. (2004). Adsorption isotherm models for basic dye adsorption by peat in single and binary component systems. *Journal of Colloid and Interface Science*, 280(2), 322–333. https://doi.org/10.1016/j.jcis.2004.08.078

Chu, K. (2004). Improved fixed bed models for metal biosorption. *Chemical Engineering Journal*, 97(2), 233–239. https://doi.org/10.1016/S1385-8947(03)00214-6

Chu, K. H. (2021). Revisiting the temkin isotherm: Dimensional inconsistency and approximate forms. *Industrial & Engineering Chemistry Research*, 60(35), 13140–13147. https://doi.org/10.1021/acs.iecr.1c01788

Coles, C. A., & Yong, R. N. (2006). Use of equilibrium and initial metal concentrations in determining Freundlich isotherms for soils and sediments. *Engineering Geology*, 85(1), 19–25. https://dx.doi.org/10.1016/j.enggeo.2005.09.023

Cooney, D. O. (1999). *Adsorption Design for Wastewater Treatment*. CRC Press, Florida.

Delle-Site, A. (2001). Factors affecting sorption of organic compounds in natural sorbent/water systems and sorption coefficients for selected pollutants. A review. *Journal of Physical and Chemical Reference Data*, 30(1), 187–439. https://doi.org/10.1063/1.1347984

Djelloul, C., Hasseine, A., & Hamdaoui, O. (2017). Adsorption of cationic dye from aqueous solution by milk thistle seeds: Isotherm, kinetic and thermodynamic studies. *Desalination and Water Treatment*, 78, 313–320. https://dx.doi.org/10.5004/dwt.2017.20920

Do, D. D. (1998). *Adsorption Analysis: Equilibria and Kinetics*. Imperial College Press, London.

Febrianto, J., Kosasih, A. N., Sunarso, J., Ju, Y. H., Indraswati, N., & Ismadji, S. (2009). Equilibrium and kinetic studies in adsorption of heavy metals using biosorbent: A summary of recent studies. *Journal of Hazardous Materials*, 162(2–3), 616–645. https://doi.org/10.1016/j.jhazmat.2008.06.042

Foo, K., & Hameed, B. (2010). Insights into the modeling of adsorption isotherm systems. *Chemical Engineering Journal*, 156(1), 2–10. https://dx.doi.org/10.1016/j.cej.2009.09.013

Guechi, E.-K., & Hamdaoui, O. (2016). Evaluation of potato peel as a novel adsorbent for the removal of Cu(II) from aqueous solutions: Equilibrium, kinetic, and thermodynamic studies. *Desalination and Water Treatment*, 57(23), 10677–10688. https://dx.doi.org/10.1080/19443994.2015.1038739

Hamdaoui, O., & Naffrechoux, E. (2007). Modeling of adsorption isotherms of phenol and chlorophenols onto granular activated carbon. Part I. Two-parameter models and equations allowing determination of thermodynamic parameters. *Journal of Hazardous Materials*, 147(1–2), 381–394. https://doi.org/10.1016/j.jhazmat.2007.01.021

Ho, Y.-S. (2006). Isotherms for the sorption of lead onto peat: Comparison of linear and non-linear methods. *Polish Journal of Environmental Studies*, 15(1), 81–86.

Hu, Q., Lan, R., He, L., Liu, H., & Pei, X. (2023). A critical review of adsorption isotherm models for aqueous contaminants: Curve characteristics, site energy distribution and common controversies. *Journal of Environmental Management*, 329, 117104. https://doi.org/10.1016/j.jenvman.2022.117104

Kinniburgh, D. G. (1986). General purpose adsorption isotherms. *Environmental Science & Technology*, 20(9), 895–904. https://doi.org/10.1021/es00151a008

Kumar, K. V., & Sivanesan, S. (2006). Pseudo second order kinetics and pseudo isotherms for malachite green onto activated carbon: Comparison of linear and non-linear regression methods. *Journal of Hazardous Materials*, 136(3), 721–726. https://doi.org/10.1016/j.jhazmat.2006.01.003

Langmuir, I. (1918). The adsorption of gases on plane surfaces of glass, mica and platinum. *Journal of the American Chemical Society*, 40(9), 1361–1403.

Li, X., Wang, J., Zhang, X., & Chen, C. (2015). Powdered activated carbon adsorption of two fishy odorants in water: Trans, trans-2,4-heptadienal and trans,trans-2,4-decadienal. *Journal of Environmental Sciences (China)*, 32, 15–25. https://doi.org/10.1016/j.jes.2015.01.001

Liu, Y., Chen, C., He, L., Hu, L., Ding, Z., Liao, S., & Tan, N. (2021). Preparation of a fungal-modified material linked by the monoamidoxime terminal open-chain polyether and its uranyl adsorption. *Industrial Engineering Chemistry Research*, 60(12), 4705–4713. https://dx.doi.org/10.1021/acs.iecr.0c05213

Magdy, Y., & Daifullah, A. (1998). Adsorption of a basic dye from aqueous solutions onto sugar-industry-mud in two modes of operations. *Waste Management*, 18(4), 219–226. https://doi.org/10.1016/S0956-053X(98)00022-1

Mozaffari Majd, M., Kordzadeh-Kermani, V., Ghalandari, V., Askari, A., & Sillanpää, M. (2022). Adsorption isotherm models: A comprehensive and systematic review (2010–2020). *The Science of the Total Environment*, 812, 151334. https://doi.org/10.1016/j.scitotenv.2021.151334

Marczewski, A. W. (2011). Extension of Langmuir kinetics in dilute solutions to include lateral interactions according to regular solution theory and the Kiselev association model. *Journal of Colloid and Interface Science*, 361(2), 603–611. https://doi.org/10.1016/j.jcis.2011.06.013

Melo, D. Q. et al. (2013). Adsorption equilibria of Cu^{2+}, Zn^{2+}, and Cd^{2+} on EDTA-functionalized silica spheres. *Journal of Chemical & Engineering Data*, 58(3), 798–806. https://dx.doi.org/10.1021/je3013364

Melo, D. Q. de. (2015). Use of activated lignocellulosic residues as adsorbents for the removal of toxic metal ions: Batch and column studies using experimental design [Doctoral dissertation, Federal University of Ceará]. Cearense Foundation for Scientific and Technological Development Support. [Dissertation written in Portuguese].

Millar, G., Schot, A., Couperthwaite, S., Shilling, A., Nuttall, K., & Bruyn, M. (2015). Equilibrium and column studies of iron exchange with strong acid cation resin. *Journal of Environmental Chemical Engineering*, 3, 373–385. https://dx.doi.org/10.1016/j.jece.2014.12.023

Padmavathy V. (2008). Biosorption of nickel (II) ions by baker's yeast: Kinetic, thermodynamic and desorption studies. *Bioresource Technology*, 99(8), 3100–3109. https://doi.org/10.1016/j.biortech.2007.05.070

Popoola, L. T. (2019). Characterization and adsorptive behaviour of snail shell-rice husk (SS-RH) calcined particles (CPs) towards cationic dye. *Heliyon*, 5(1), e01153. https://doi.org/10.1016/j.heliyon.2019.e01153

Preetha, B., & Viruthagiri, T. (2007). Batch and continuous biosorption of chromium (VI) by *Rhizopus arrhizus*. *Separation and Purification Technology*, 57(1), 126–133. https://dx.doi.org/10.1016/j.seppur.2007.03.015

Prihatdini, R. W., Suratman, A., & Siswanta, D. (2023). Linear and nonlinear modeling of kinetics and isotherm of malachite green dye adsorption to trimellitic-modified pineapple peel. *Materials Today: Proceedings*, 88, 33–40. https://doi.org/10.1016/j.matpr.2023.07.108

Redlich, O., & Peterson, D. L. (1959). A useful adsorption isotherm. *Journal of Physical Chemistry*, 63, 1024–1026. https://doi.org/10.1021/j150576a611

Ruthven, D. M. (1984). *Principles of Adsorption & Adsorption Process*. John Wiley & Sons, New York.

Satyanarayana, J., Murthy, G. S., & Sasidhar, P. (1999). Adsorption studies of caesium on zirconium molybdoarsenate (ZrMAs). *Waste Management*, 19, 427. https://doi.org/10.1016/S0956-053X(99)00200-7

Sousa Neto, V. d. O. (2011). Use of coconut bagasse as alternative adsorbent for separation of copper (ii) ions from aqueous solutions: Isotherms, kinetics, and thermodynamic studies. *BioResources*, 6(3), 3376–3395.

Subramanyam, B., & Das, A. (2014). Linearised and non-linearised isotherm models optimization analysis by error functions and statistical means. *Journal of Environmental Health Science & Engineering*, 12, 92. https://doi.org/10.1186/2052-336X-12-92

Suzuki, M. (1990). *Adsorption Engineering*, volume 14. Kodansha Tokyo.

Temkin, M. (1940). Kinetics of ammonia synthesis on promoted iron catalysts. *Acta Physiochimica URSS*,12, 327–356.

Vidal, C. B., Barros, A. L., Moura, C. P., de Lima, A. C., Dias, F. S., Vasconcellos, L. C., Fechine, P. B., & Nascimento, R. F. (2011). Adsorption of polycyclic aromatic hydrocarbons from aqueous solutions by modified periodic mesoporous organosilica. *Journal of Colloid and Interface Science*, 357(2), 466–473. https://doi.org/10.1016/j.jcis.2011.02.013

Wang, J., & Guo, X. (2020). Adsorption isotherm models: Classification, physical meaning, application and solving method. *Chemosphere*, 258, 127279. https://doi.org/10.1016/j.chemosphere.2020.127279

Wu, F. C., Tseng, R. L., & Juang, R. S. (2010). A review and experimental verification of using chitosan and its derivatives as adsorbents for selected heavy metals. *Journal of Environmental Management*, 91(4), 798–806. https://doi.org/10.1016/j.jenvman.2009.10.018

Yan, D. (2012). Dimensional analysis and its applications (in Chinese). *Physics and Engineering*, 22(6), 22–26. https://doi.org/10.1080/00224065.2014.11917964

Zhou, X., Maimaitiniyazi, R., & Wang, Y. (2022). Some consideration triggered by misquotation of Temkin model and the derivation of its correct form. *Arabian Journal of Chemistry*, 15(11), 104267. https://dx.doi.org/10.1016/j.arabjc.2022.104267

3 Adsorption Kinetics

Vicente de Oliveira Sousa Neto,
Carla Bastos Vidal, Ari Clecius Alves de Lima,
Diego de Quadros Melo,
Giselle Santiago Cabral Raulino, and
Ronaldo Ferreira do Nascimento

INTRODUCTION

ADSORPTION KINETICS

Adsorption technology is extensively utilized in the treatment of water and wastewater due to its cost-effectiveness and high efficiency. Kinetic models of adsorption are commonly employed to assess adsorbent performance and elucidate the mechanisms governing mass transfer during adsorption. However, the physical interpretations and solution strategies associated with these models remain inadequately defined. An accurate understanding of their physical significance and the establishment of standardized methods for solving these models are critical for their effective application.

This chapter primarily examines the underlying physical interpretations, practical applications, and solution methodologies of some adsorption kinetic models.

ADSORPTION KINETICS: A GENERAL APPROACH

Adsorption and ion exchange mechanisms are fundamental to separation technologies employed in diverse industrial applications, particularly in the purification of wastewater and potable water. These processes rely on porous solid substrates, wherein sorption is predominantly governed by diffusional mass transfer. In bidisperse porous materials—such as activated carbons and zeolites—macropores serve as primary conduits, facilitating the transport of adsorbate molecules into the particle's interior matrix. It is crucial to emphasize that there is a significant distinction between ion exchange and adsorption, particularly the latter's stoichiometric nature. As a result of the electroneutrality requirement, any ions that exit the solid phase are replaced by an equal amount of ions from the liquid phase (Zagorodni, 2006).

Pore diffusion, also known as molecular and/or Knudsen diffusion, usually dominates mass transfer in macropores. The process by which the adsorbate migrates on its surface in micropores is known as surface diffusion.

Adsorption kinetics is expressed as the rate of removal of the adsorbate in the fluid phase over time, involving the mass transfer of one or more components contained in an external liquid mass into the interior of the adsorbent particle, from where they

DOI: 10.1201/9781003654391-3

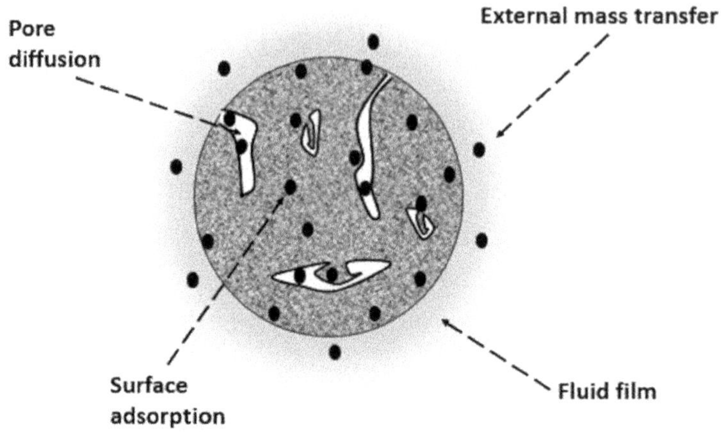

FIGURE 3.1 Adsorption mass transfer steps. (This figure by Vicente Neto is licensed under the Creative Commons Attribution 4.0 International License. https://commons.wikimedia.org/wiki/File:Surface_adsorption.jpg.)

must migrate through the macropores to the innermost regions of this particle (Vidal et al., 2013; Sarma et al., 2019).

In principle, adsorption kinetics can be driven by different processes (Inglezakis et al., 2019):

- External Mass Transfer: the process by which molecules are transported from the fluid phase to the outer surface of the adsorbent particle, facilitated by a fluid layer surrounding the particle.
- Pore Diffusion: occurring as molecules diffuse within the fluid into the interior of the pores.
- Surface Diffusion: referring to the diffusion of fully adsorbed molecules along the pore surface.

Since adsorption and ion exchange are crucial in many processes, kinetic models have been used in numerous studies for over a century. In general, the models fall into two categories: diffusion-based models and reaction kinetics-based models. These steps described above can be observed in Figure 3.1.

Diffusion-Based Models

The first stage of adsorption can be influenced by the adsorbate concentration and agitation. Therefore, an increase in solute concentration may accelerate its diffusion from the solution to the solid surface. The maximum adsorption capacity is obtained from the equilibrium adsorption isotherm. The second stage is generally considered the rate-determining step, particularly in the case of microporous adsorbents.

The adsorption rate can be affected by temperature, pH, ionic strength, initial adsorbate concentration, agitation, particle size, and pore size distribution.

Considering a homogeneous spherical adsorbent particle subjected to a concentration change of an adsorbable species at the surface, the adsorbate will diffuse into the particle according to Fick's law, also known as the homogeneous particle diffusion model (Equation 3.1) (Ruthven, 1984; Rani et al., 2018; Inglezakis et al., 2019).

$$\frac{\partial q}{\partial t} = \frac{1}{r^2}\frac{\partial}{\partial r}\left(D_e r^2 \frac{\partial q}{\partial r}\right) \tag{3.1}$$

where:

- q: adsorption capacity (mg g^{-1})
- De: effective diffusion coefficient ($m^2 s^{-1}$)
- r: pore radius (m)
- t: time (s)

In practice, it is observed that commonly used adsorbents typically exhibit a wide distribution of pore sizes within the adsorbent particles, and various diffusion mechanisms occur depending on the relative size of the adsorbate molecule compared to the pore size. Intraparticle diffusion can be of two types: macropore diffusion and micropore diffusion.

Macropore diffusion occurs easily due to the size differences between solute particles and the pores of the solid phase. The macropore diffusion rate is influenced by the following parameters: the mutual diffusivity between solute and adsorbent, tortuosity factor, pellet size, and solute concentration in the system. The solute can diffuse through several mechanisms, including molecular diffusion, Knudsen diffusion, surface diffusion, and Poiseuille flow (Ruthven, 1984).

Molecular Diffusion (D_m)

Pore diffusion is caused by the diffusion of molecules in the fluid into the pores. This occurs in very large pores, where molecules diffuse as if there were no pore walls. The effective diffusivity (D_e) in the pore depends on some properties of the porous matrix, such as porosity (ε_p) and tortuosity factor (τ_p) (Equation 3.2) (Wilke & Chang, 1955; Valderrama et al., 2008; Souza et al., 2017):

$$D_e = D_m \frac{\varepsilon_p}{\tau_p} \tag{3.2}$$

The tortuosity factor is calculated by the actual distance traveled inside the pore divided by the particle diameter. Typical values for tortuosity in macroporous adsorbents range between 2 and 6 (Ruthven, 1984). D_m ($cm^2 s^{-1}$) represents the molecular diffusion coefficient, and its value is estimated according to the Wilke-Chang correlation, Equation 3.3 (Wilke & Chang, 1955; Reid et al., 1987; Souza et al., 2019).

$$D_m = 7.4 \times 10^{-8}\left(\frac{2.6M^{1/2}T}{\mu V_m^{0.6}}\right) \tag{3.3}$$

where:

- μ: solvent viscosity (cP)
- M: solvent molar mass (g mol^{-1})
- T: absolute temperature (K)
- V_m: solute molar volume at normal boiling point ($cm^3 mol^{-1}$)

The value 2.6 is related to the solvent (dimensionless), with this value being recommended when the solvent is water.

Knudsen Diffusion (D_k)

Occurs when adsorbate collisions with pore walls are more frequent than intermolecular collisions. This typically happens when the adsorbent pore diameter is comparable to the adsorbate size. The Knudsen diffusion coefficient (cm^2s^{-1}) is pressure-independent and given by Equation 3.4 (Ruthven, 1984; Rani et al., 2018; Reshetnikov & Kurzina, 2019):

$$D_k = 9{,}700r\left(\frac{T}{M}\right)^{1/2} \tag{3.4}$$

where:

- r: mean pore radius (m);
- M: adsorbate molar mass (g mol^{-1});
- T: temperature (K).

Generally, Knudsen diffusion occurs simultaneously with molecular diffusion, such that the effective diffusivity can be expressed by Equation 3.5:

$$\frac{1}{D_s} = \frac{1}{D_k} + \frac{1}{D_m} \tag{3.5}$$

where:

- D_s is the surface diffusion (m^2s^{-1})
- D_k is the Knudsen diffusion coefficient (m^2s^{-1})
- D_m is the molecular diffusion coefficient (m^2s^{-1})

Surface Diffusion (D_s)

Surface diffusion refers to the transport of molecules through the adsorbed layer on macropore surfaces. This mechanism becomes significant only for strongly adsorbed species at sufficiently low temperatures, where a thick enough adsorbed layer forms to enable transport. Typically, surface diffusion becomes more important in small pores where Knudsen diffusion (D_k) is also significant. Its contribution to overall diffusivity depends on the adsorption equilibrium. For linear equilibrium conditions (Ruthven, 1984; Xu et al., 2018):

$$(1-\varepsilon_p)D_e = D_k + \left(\frac{1}{k_p}\right)kD_s \tag{3.6}$$

where:

- k: Henry's constant (mol kg^{-1} Pa^{-1})
- D_k: Knudsen diffusion coefficient ($cm^2 s^{-1}$)
- ε_p: porosity (dimensionless)
- D_s: surface diffusion coefficient ($cm^2 s^{-1}$)

Poiseuille Flow

In very large pores under pressure gradients, convective mass transport into adsorbent particles may contribute significantly. Under these laminar flow conditions, the resulting diffusion coefficient is given by Equation 3.7 (Ruthven, 1984):

$$D_e = \frac{Pr^2}{8\mu} \tag{3.7}$$

where:

- r: pore radius (m)
- μ: fluid viscosity (Pa·s)
- P: system pressure (Pa)

This diffusion mechanism is more common in gaseous systems and in high-pressure chromatographic processes that utilize resins with extremely wide pores. Typically, in liquid-phase systems, molecular diffusion dominates the diffusive process, whereas, for gas-phase systems, Knudsen diffusion and surface diffusion are generally more significant.

In the micropore region, resistance to diffusion is often the most important factor and is responsible for shape selectivity. In the intracrystalline region, different molecular interactions between various adsorbates and the micropore structure can lead to significant differences in the kinetic behavior of the species. In this region, steric effects and potential field interactions between the molecule and the microporous channels are dominant. The kinetics can be studied by considering a set of resistances to mass transfer:

- In the film (boundary layer);
- Outside the adsorbent particle (external diffusion);
- Inside the particle (pore diffusion).

External Diffusion

When the resistance to mass transfer is confined to a film around the particles, with a thickness sufficiently small to assume a linear profile, the solute transport across the adsorbent is generally modeled by the rate law, Equation 3.8 (Cooney, 1999; Jain et al., 2016; Guiza et al., 2019):

$$\frac{d\bar{q}}{dt} = k_f S_0 (c - c_i) \tag{3.8}$$

where:

- $d\bar{q}$: average solute concentration in the solid ($mol \cdot cm^{-3}$);
- c: solute concentration in the liquid far from the surface ($mol \cdot cm^{-3}$);
- c_i: solute concentration in the liquid at the liquid/particle interface ($mol \cdot cm^{-3}$).
- S_0: Surface area of the adsorbent particle per unit volume of the adsorbent particle (cm^{-1});
- k_f: Mass transfer coefficient in the film ($cm^2 \cdot s^{-1}$).

The value of k_f is typically estimated and depends on the liquid-solid system. The resistance to mass transfer through the external fluid film depends on the hydrodynamic conditions of the system, which affect the thickness of the laminar layer surrounding the adsorbent particle. This parameter does not show significant differences when compared to other types of mass transfer from the fluid to the adsorbent. In practice, for most real systems, the resistances to mass transfer in the pore are more significant than the external resistance to the particle.

Pore Diffusion and Film Diffusion

To investigate the contribution of the mass transfer parameter to adsorption kinetics, the pore diffusion model can be employed through mass balance in the adsorbent particle and the fluid phase. To determine these parameters, the concentration profiles obtained from kinetic tests, as well as knowledge of the adsorption isotherms, must be used. The pore diffusion model assumes the following premises:

a. Mass transfer into the adsorbent pores is governed by the boundary layer model, characterized by the mass transfer coefficient, k_f;
b. The adsorbent must consist of a porous material through which the solute can diffuse;
c. The adsorbent particles must be spherical and of uniform size;
d. Adsorption is isothermal, and the equilibrium behavior can be represented by the Langmuir equation.

This model can be represented by Equations 3.9 and 3.10 Cooney (1999).

Solid Phase Balance

$$\varepsilon_p \frac{\partial C_p}{\partial t} + \rho_{ap} \frac{\partial \dot{q}}{\partial t} = D_f \left(\frac{\partial^2 C_p}{\partial r^2} + \frac{2}{r} \frac{\partial C_p}{\partial r} \right) \tag{3.9}$$

Initial condition: $t=0$; $C_p=0$

Boundary condition: $r=0$; $\frac{\partial C_p}{\partial r} = 0$

Liquid Phase Balance

$$r = R; \ \frac{\partial C_p}{\partial r} = \frac{k_f}{D_{ef}} (C - C_p) \tag{3.10}$$

where:

C_p: Concentration of the adsorbed component ($mg \cdot L^{-1}$);
$q*$: Theoretical adsorption capacity ($mg \cdot g^{-1}$);
ε_p: Porosity of the adsorbent particle;
k_f: Film mass transfer coefficient ($cm \cdot s^{-1}$);
D_{ef}: Effective diffusivity;
ρ_{ap}: Apparent particle density.

Reaction Kinetics-Based Models

Wang and Guo (2020), when discussing the overall perspective of adsorption kinetic models, highlight the improper use of various kinetic models. According to the authors, the physical meanings and solving methods of these kinetic models have not been well established. In general, the adsorption kinetic process has been described by several adsorption kinetic models, including phenomenological mass transfer models (Blanco et al., 2017; Marin et al., 2014; Scheufele et al., 2016; Hu et al., 2019; Suzaki et al., 2017), the pseudo-first-order (PFO) model (Lagergren, 1898), the pseudo-second-order (PSO) model (Ho & McKay, 1999), the mixed-order (MO) model (Guo & Wang, 2019a), Ritchie's equation (Ritchie, 1977), the Elovich model (Elovich & Larinov, 1962), and the pseudo-first-order (PSO) model (Ho & McKay, 1999). Nevertheless, there are certain issues with using these kinetic models. The first is that the PFO and PSO models, which are most frequently used, are empirical models without specific physical meanings. We cannot use these empirical kinetic models to study mass transfer pathways. Consequently, it is necessary to establish the physical implications of these empirical kinetic models. The second issue is that while the phenomenological external/internal and adsorption in active sites models of differential kinetics have distinct physical interpretations, the procedures for addressing them are intricate. The mass transfer mechanisms using these models have not been extensively studied (Blanco et al., 2017; Marin et al., 2014; Scheufele et al., 2016; Hu et al., 2019; Suzaki et al., 2017; Guo & Wang, 2019b). The applications of these models are hampered by intricate solving techniques. Additionally, kinetic models are sometimes applied or solved improperly in certain published papers. For example, the Frusawa and Smith (F&S) model can only be used for modeling adsorption processes where the isotherm is linear (Frusawa & Smith, 1973). However, it has been applied to adsorption processes where the isotherm is Langmuir-type (Özer et al., 2005; Wang et al., 2008).

Furthermore, due to its ease of use, the linear regression approach is the most popular technique for determining model parameters (Darwish et al., 2019; Sabarinathan et al., 2019). Nevertheless, the independent and dependent variables are altered during the linearization process. This procedure may result in erroneous parameter estimations and propagate errors to the independent and dependent variables (Guo & Wang, 2019c). Model parameters can be reliably and accurately estimated using the nonlinear approach (El-Khaiary et al., 2010). Providing the nonlinear solving approach for the adsorption kinetic models is crucial.

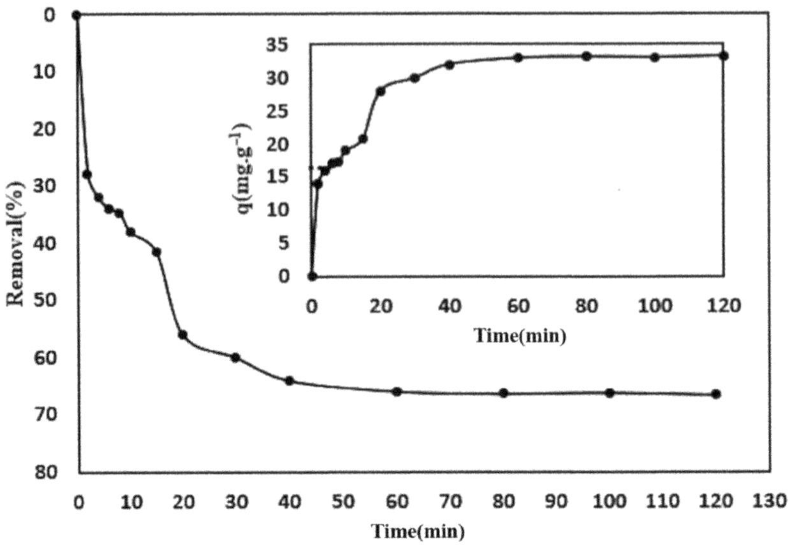

FIGURE 3.2 Adsorption of Cd^{2+} in synthetic solutions (100 mg L^{-1}) onto modified coconut bagasse as a function of time. Experimental conditions: adsorbent dosage: 2 g L^{-1}; pH: 5.5; ambient temperature: 28°C. (Adapted, reproduced with permission from Neto et al., 2012.)

Contact Time

Contact time is an important parameter as it indicates the kinetic behavior of adsorption for a given adsorbent at a specific initial adsorbate concentration.

Figure 3.2 shows the effect of contact time on the adsorption of Cd^{2+} ions onto coconut husk adsorbent in a synthetic solution at pH = 5.5 and an initial concentration of 100 mg·L^{-1}. It can be observed that removal increases rapidly during the initial stages of Cd^{2+} adsorption, and at slower removal rates, reaches equilibrium after 40 min. Within the first 17 min, approximately 50% removal was achieved. By 30 min, the removal reached about 60%. The kinetics reach equilibrium within 60 min, achieving a 66% removal efficiency.

Pseudo-First-Order (PFO) Model

The PFO model was first proposed by Lagergren (1898). The differential form of the PFO model is described by Equation 3.11 (Lagergren, 1898):

$$\frac{dq_t}{dt} = k_1 (q_e - q_t) \tag{3.11}$$

where:

k_1: Pseudo-first-order adsorption rate constant (min^{-1});
q_e and q_t: Quantities adsorbed per gram of adsorbent at equilibrium and at time t, respectively (mg g^{-1}).

After integrating Equation 3.11 and applying boundary conditions: $q_t = 0$, $t = 0$; when $q_e = q_t$, $t = t$, Equation 3.12 is obtained.

$$q_t = q_e\left(1 + e^{-k_1 t}\right) \tag{3.12a}$$

The linearized form of the PFO model is presented as follows.

$$\ln\left(q_e - q_t\right) = \ln q_e - k_1 t \tag{3.12b}$$

By plotting $\ln(q_e - q_t)$ vs. t, Equation 3.3 has been widely used to fit the kinetics data and determine the parameters q_e and k_1. However, the linearization procedure could result in imprecise parameter predictions.

Pseudo-Second-Order (PSO) Model

The pseudo-second-order model can be expressed according to Equation 3.13 (Ho & McKay, 1999):

$$\frac{dq_t}{dt} = k_2\left(q_e - q_t\right)^2 \tag{3.13}$$

where:

q_e and q_t: Quantities adsorbed per gram of adsorbent at equilibrium and at time t, respectively (mg g^{-1}).
k_2: Pseudo-second-order adsorption rate constant (g mg^{-1} min^{-1}).

Integrating Equation 3.13, similarly to Equation 3.11, we obtain Equation 3.14.

$$\frac{t}{\left(q_e - q_t\right)} = \frac{1}{q_e^2} + k_2 t \tag{3.14}$$

Linearizing Equation 3.14, we obtain Equation 3.15:

$$\frac{t}{q_t} = \frac{1}{k_2 q_e^2} + \frac{t}{q_e} \tag{3.15}$$

The values of q_e and k_2 can be obtained through the intercept and slope of the curve presented in the graph of (t/q_t) versus t. If the pseudo-second-order kinetic model is applicable, the plot of (t/q_t) versus t should present a linear relationship close to 1.

Weber and Morris Model

According to Weber and Morris (1963), if intraparticle diffusion is the rate-determining factor, the removal of the adsorbate varies with the square root of time. Thus, the intraparticle diffusion coefficient (k_d) can be defined by Equation 3.16.

$$q_t = k_d \cdot t^{0.5} + C \tag{3.16}$$

where:

- q_t: Quantity of adsorbate adsorbed in the solid phase (mg g^{-1}) at a given time.
- t: Time (min).
- k_d: Intraparticle diffusion rate constant (mg g^{-1} $min^{-0.5}$).
- C: Constant related to diffusion resistance (mg g^{-1}).

According to Yang and Al-Duri (2005), once the diffusion rate constant (k_d) has been determined, the diffusion coefficient (D) ($cm^2 \cdot min^{-1}$) can be calculated using the following Equation 3.17:

$$k_d = \left(\frac{6q_e}{R}\right)\sqrt{\frac{D}{\pi}} \quad (3.17)$$

where:

- k_d: Intraparticle diffusion rate constant (mg g^{-1} $min^{-0.5}$).
- q_e: Amount of metal adsorbed per unit of biomass ($mg \cdot g^{-1}$) at equilibrium time
- D: Intraparticle diffusivity coefficient ($cm^2 \cdot min^{-1}$)

The value of k_d can be obtained from the slope of the curve qt vs. $t^{0.5}$, while the value of C corresponds to the intercept. The C values provide an indication of the boundary layer thickness; the larger the C value, the greater the effect of the boundary layer. However, this model has the drawback of not considering important parameters such as porosity and particle radius, as well as neglecting kinetic behavior at the initial stages of the process near $t=0$, since the equation involves $t^{1/2}$.

In practice, the process is complex, involving multiple linear segments where each equation corresponds to a different adsorption stage. Equilibrium is reached when q (mg g^{-1}) remains constant over time, resulting in a horizontal line. The Weber and Morris model suggests that if the first linear segment (initial adsorption stages) has a zero intercept (i.e., the line passes through the origin), then intraparticle diffusion governs the adsorption process. Conversely, if the intercept is nonzero, the adsorption process may be controlled by film diffusion, with its thickness determined by the linear coefficient (mg g^{-1}).

Boyd Diffusion Model

In 1947, Boyd et al. published a series of scientific articles presenting theoretical models for ion exchange, simulating equilibrium, kinetics, and non-equilibrium conditions. Over the years, researchers discovered that these kinetic models also apply to adsorption systems, leading to the widespread use of Boyd diffusion models in numerous adsorption studies. To determine whether pore diffusion is the rate-limiting step in the adsorption process, the results can be analyzed using Equations 3.17–3.21.

$$F = 1 - \left(\frac{6}{\pi^2}\right)\sum_{n=1}^{\infty}\left(\frac{1}{n^2}\right)\exp(-\pi^2 Bt) \quad (3.18)$$

where:

F is obtained from Equation 3.18.

- q_e: Quantity of adsorbate adsorbed at infinite time (mg g^{-1}).
- q_t: Quantity of adsorbate adsorbed at time t (mg g^{-1}).
- Bt: Mathematical function of F.

The value of Bt is determined by the combination of the two equations developed by Reichenberg (1953). Depending on the value of F, the following equations apply:

$$F > 0.85 \quad Bt = -0.4977 - \ln(1 - F) \tag{3.19}$$

$$F < 0.85 \quad Bt = \left(\sqrt{\pi} - \sqrt{\pi - \left(\frac{\pi^2 F}{3}\right)}\right)^2 \tag{3.20}$$

The term Bt is calculated for each value of F, and the resulting values of Bt are plotted against t (Boyd plot). If the plot generates a straight line passing through the origin, it can be concluded that intraparticle diffusion is the rate-limiting step of the adsorption process within the specified time range.

The slope of the line corresponds to the Boyd constant (B). The diffusion coefficient (D, $cm^2 \cdot min^{-1}$) is related to the Boyd constant through Equation 3.21.

$$B = \frac{(\pi^2 D)}{d^2} \tag{3.21}$$

where:

- D: Diffusion coefficient ($cm^2 \cdot min^{-1}$).
- d: Mean particle diameter (cm).
- B: Boyd constant.

Elovich Equation

The Elovich equation (Roginsky & Zeldovich, 1934) was originally developed for the kinetic study of chemical adsorption of a gas onto the surface of a solid material (Ozacar & Sengil). Surprisingly, this equation has been successfully applied to the adsorption of various chemical species in liquid media.

The nonlinear equation (Equation 3.23) was derived from the following differential equation (Equation 3.22):

$$\frac{dq_t}{dt} = \propto e^{-\beta q_t} \tag{3.22}$$

Its nonlinear form is expressed as follows:

$$q_t = \frac{1}{\beta} \ln(1 + \alpha\beta t) \tag{3.23}$$

where:

- α: Initial adsorption rate (mg g^{-1} min^{-1}). – β: Desorption constant (mg g^{-1}).
- q_t: Amount of metal adsorbed per unit of biomass (mg g^{-1}) at time t.
- t: Time (min).

After integration and applying the same boundary conditions as the pseudo-first-order and pseudo-second-order equations, the equation takes the following linear form:

$$q = \frac{1}{\beta}\ln(\alpha\beta) + \frac{1}{\beta}\ln(t) \tag{3.24}$$

A plot of q versus ln(t) produces a straight line, from which the parameters α and β can be obtained based on the linear and angular coefficients, respectively.

Application of Kinetic Models

Vidal et al. (2013) conducted adsorption kinetics studies of BTEX on hydrophobically modified zeolite and applied the homogeneous particle diffusion model (Equation 3.1) to obtain the effective diffusion coefficients of the compounds, in accordance with Equation 3.1. The results are presented in Figure 3.3.

With the following initial and boundary conditions:

$$q\ (r,0) = q_0;\ q\ (r,t) = q_0;\ \left.\frac{\partial q}{\partial r}\right|_{r=0} = 0 \tag{3.25}$$

The kinetics results indicated that the reaction equilibrium time was 6h. The authors also observed that the initial adsorption process was rapid, as adsorption in this case occurs predominantly on the surface of the adsorbent material, followed by a slower phase (intraparticle adsorption).

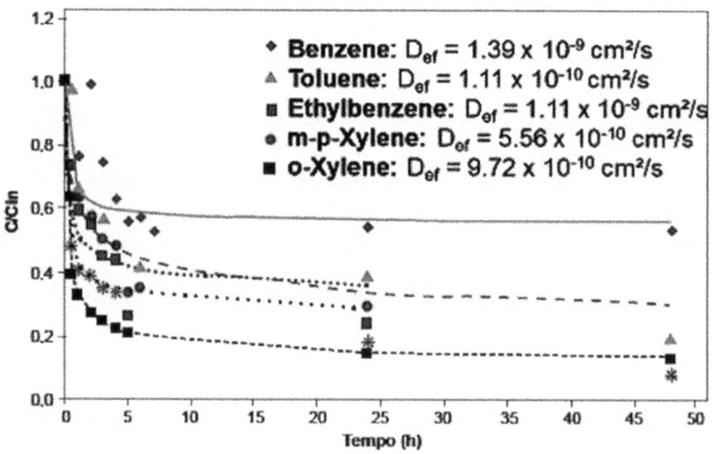

FIGURE 3.3 Multicomponent adsorption kinetics of BTEX (Co = 10 mg L^{-1}, T = 28°C, and pH = 7). (Reproduced with permission from Vidal et al., 2013.)

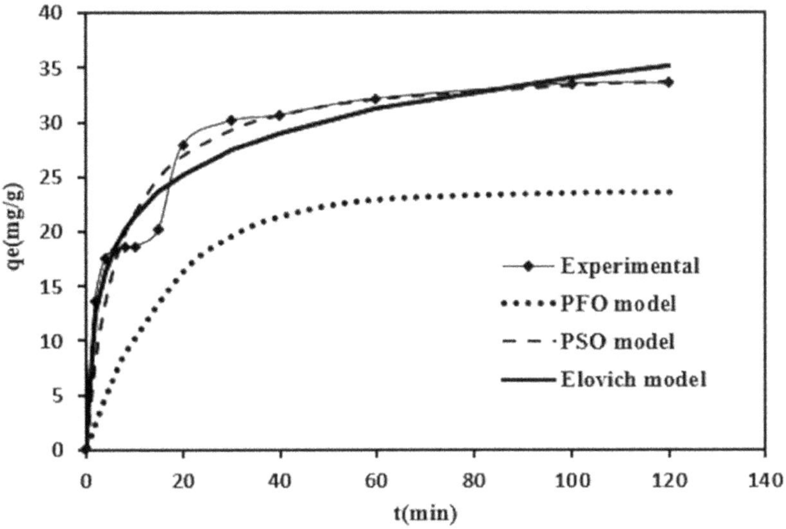

FIGURE 3.4 Adsorption kinetics of Cd^{2+} in synthetic solution (100 mg L^{-1}) on coconut bagasse treated with thiourea/ammonia solution (CBT). Linear models of Lagergreen (PFO), Ho (PSO), and Elovich. (Figure adapted, reproduced with permission from Neto et al., 2012.)

Neto et al. (2012) investigated the adsorption of Cd^{2+} ions using coconut bagasse treated with thiourea/ammonia solution (CBT) as an adsorbent, applying the linear models of pseudo-first-order (Lagergren), pseudo-second-order (Ho), Weber and Morris, Boyd, and the Elovich equation. As shown in Figure 3.4, the linear pseudo-first-order model did not accurately represent the experimental results, presenting a theoretical value significantly lower than the experimental value despite a good linear correlation. In contrast, the pseudo-second-order model demonstrated a better fit to the experimental data.

Figure 3.5 illustrates the diffusion model proposed by Weber and Morris. The graph indicates a strong agreement between theoretical and experimental data, suggesting that the model effectively describes the adsorption mechanism.

Intraparticle diffusion refers to the transport of molecules within the adsorbent particles, which can occur through different mechanisms, such as diffusion within the pore volume or surface diffusion. Intrapore diffusion, on the other hand, is more specifically related to the movement of molecules within the material's pores and can be one of the mechanisms contributing to intraparticle diffusion. Figure 3.6 illustrates the stages involved in the adsorption of Cd^{2+} onto the surface of the coconut bagasse treated with thiourea/ammonia solution (CBT). It is observed that, within the time interval of 2–20 min, intrapore diffusion was not the predominant adsorption mechanism. In Stage I, the linear coefficient is different from zero, and the value of C is approximately 8.19 mg g^{-1}. The intraparticle diffusion rate constant for this stage is $k_d = 4.12$ mg $g^{-1}t^{-1/2}$. In Stage II, a reduction in the intraparticle diffusion rate constant is observed, with a value of $k_d = 0.67$ mg $g^{-1}t^{-1/2}$ and C equal to 26.55 mg g^{-1}. The data regarding the application of the Weber-Morris model by Neto et al. (2012)

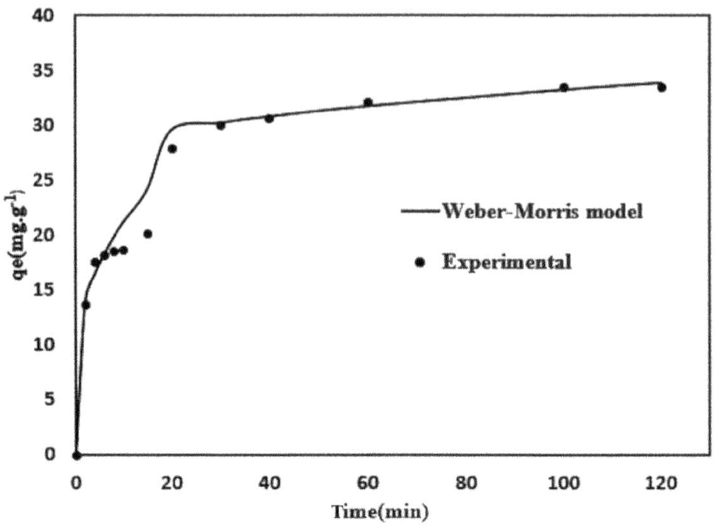

FIGURE 3.5 Weber-Morris intraparticle diffusion model applied to the adsorption of Cd^{2+} in synthetic solutions (100 mg L^{-1}) onto coconut bagasse treated with thiourea/ammonia solution (CBT). Experimental conditions: pH = 5.5. (Figure adapted, reproduced with permission from Neto et al., 2012.)

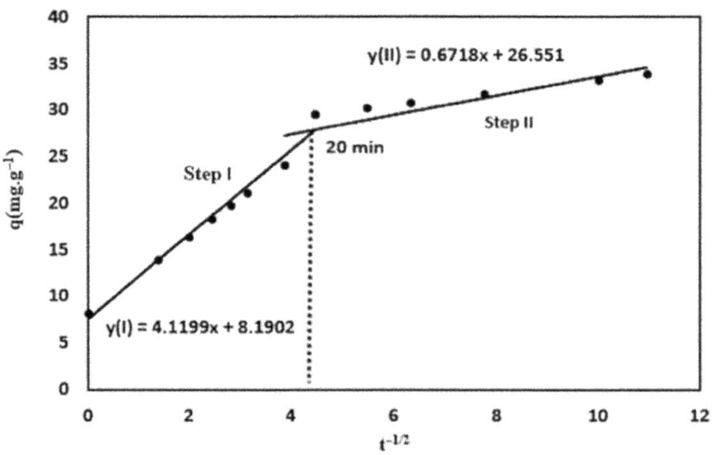

FIGURE 3.6 Weber and Morris intraparticle diffusion model applied to the adsorption of Cd^{2+} in a synthetic solution (100 mg L^{-1}) onto coconut shell. (Figure adapted, reproduced with permission from Neto et al., 2012.)

to the adsorption of Cd^{2+} onto the surface of coconut bagasse treated with a thiourea/ ammonia solution (CBT) are presented in Table 3.1.

Figure 3.7 presents the diffusion graph based on the Boyd model, illustrating how intraporous diffusion governs the initial stages of adsorption, as indicated by the

TABLE 3.1
Results for the Sorption of Cadmium Ions onto Coconut Bagasse Treated with Thiourea/Ammonia Solution (CBT)

	Morris-Weber Pore Diffusion Model			
Step	**k_d (mg·g^{-1}·min$^{-1/2}$)**	**C (mg·g^{-1})**	**D (cm^2·s^{-1})**	**R^2**
I	4.12	8.19	9.74×10^{-8}	0.976
II	0.672	26.55	2.59×10^{-9}	0.961
	Boyd Diffusion Model			
Step	**B**		**D (cm^2·s^{-1})**	**R^2**
I	0.0587		1.05×10^{-4}	0.9838
II	0.032		5.40×10^{-5}	0.9858

Data Obtained by the Weber's Pore-Diffusion Model and Boyd's Diffusion Model Using the Linear Method

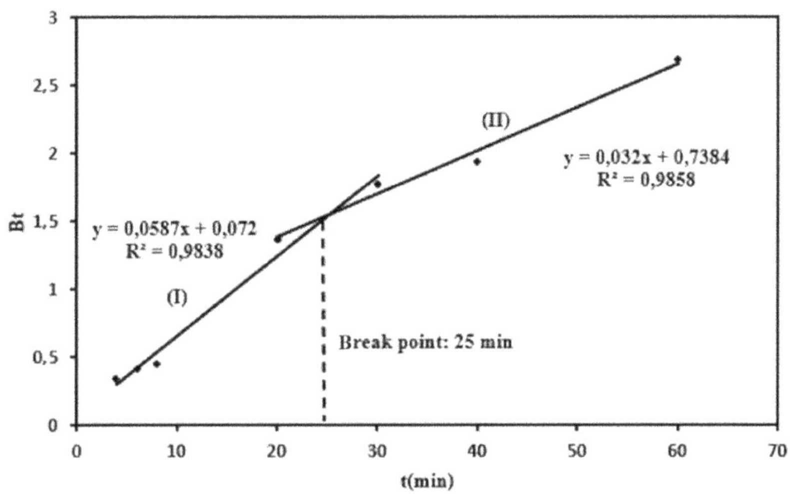

FIGURE 3.7 Boyd intraparticle diffusion model applied to the adsorption of Cu^{2+} in synthetic solutions (100 mg L^{-1}) onto coconut bagasse treated with thiourea/ammonia solution (CBT). (Figure adapted, reproduced with permission from Neto et al., 2012.)

line crossing the graph's origin. It is important to note that the Weber model has limitations in assessing the diffusion mechanism at very short time intervals due to the presence of the $t^{1/2}$ variable. In practice, determining q_e within extremely short periods is prone to significant experimental errors, as separation procedures require a minimum duration to ensure reliable results. Therefore, we strongly advise against using times shorter than 1 min in kinetic adsorption studies.

SUMMARY

Adsorption is widely applied in water treatment due to its efficiency and economic viability. To evaluate adsorbent performance and understand mass transfer mechanisms, adsorption kinetic models are essential. However, their physical interpretation and standardized solution methods remain inadequately defined. This chapter explored key kinetic models, distinguishing between ion exchange and adsorption, emphasizing that adsorption involves stoichiometric exchange due to electroneutrality. Adsorption kinetics refers to the rate at which adsorbates are removed from the fluid phase. This process is influenced by mechanisms such as external mass transfer, pore diffusion, and surface diffusion. Diffusion-based models, especially for porous materials like activated carbon, consider both macropore and micropore transport processes. Effective diffusion within these materials is governed by parameters such as porosity, tortuosity, molecular size, and temperature. The models incorporate different types of diffusion—molecular, Knudsen, surface, and Poiseuille flow—each described by specific mathematical expressions. Accurate modeling requires integrating these mechanisms to estimate real adsorptive behavior and optimize system performance under varying operational conditions.

CHAPTER 3 EXERCISES LIST

Objective Questions on Adsorption Kinetics

1. **What is the primary role of macropores in bidisperse porous materials?**
   a. To store adsorbate molecules permanently
   b. To facilitate the transport of adsorbate molecules into the particle's interior
   c. To restrict the movement of adsorbate molecules
   d. To promote chemical bonding between the adsorbent and the adsorbate
2. **What distinguishes ion exchange from adsorption?**
   a. Ion exchange follows a stoichiometric principle due to electroneutrality requirements
   b. Adsorption involves permanent bonding, while ion exchange does not
   c. Adsorption is irreversible, whereas ion exchange can only occur in gases
   d. Ion exchange depends solely on surface diffusion
3. **Which mechanism dominates mass transfer in macropores?**
   a. External mass transfer
   b. Surface diffusion
   c. Pore diffusion
   d. Chemical reaction kinetics
4. **How is adsorption kinetics commonly expressed?**
   a. As the total amount of adsorbate present in the fluid phase
   b. As the molecular weight of the adsorbate
   c. As the chemical reactivity of the adsorbent material
   d. As the rate of adsorbate removal from the fluid phase over time

5. **What does external mass transfer refer to in adsorption kinetics?**
   a. Adsorbate movement between micropores and macropores
   b. The transport of molecules from the fluid phase to the outer surface of the adsorbent particle
   c. The interaction of adsorbate molecules with the innermost regions of the adsorbent
   d. The conversion of adsorbate molecules into gaseous compounds

SUBJECTIVE QUESTIONS: CONTACT TIME

Refer to Figure 3.2 to Answer the Questions

1. What conclusions can be drawn about the efficiency of the adsorption process over time?
2. How does the rapid decline in removal percentage impact the interpretation of the system's performance?
3. What factors might explain the stabilization of adsorption capacity after 40 min?
4. Considering the trend observed, how could this process be optimized for higher adsorption efficiency?
5. What practical applications could these findings have in industrial or environmental settings?

Subjective Questions: Homogeneous Particle Diffusion Model

1. Explain the homogeneous particle diffusion model used by Vidal et al. (2013) for the adsorption of BTEX onto hydrophobically modified zeolite.
2. What is the significance of the effective diffusion coefficients (D_{eff}) of BTEX compounds presented in the figure, and how do they influence adsorption?
3. Describe the phases of the adsorption process observed by Vidal et al. (2013) and their implications for BTEX removal.
4. What factors may influence the rate of BTEX adsorption onto hydrophobically modified zeolite?
5. How can the findings of Vidal et al. (2013) contribute to the treatment of water contaminated with BTEX?

Objective Questions

1. **What was the adsorption reaction equilibrium time observed by Vidal et al. (2013)?**
   a. 4 h
   b. 6 h
   c. 8 h
   d. 10 h
   Answer: (b) 6 h

2. **Which of the following adsorption steps occurs first, as described by Vidal et al. (2013)?**
   a. Intraparticle diffusion
   b. Chemical equilibrium
   c. Surface adsorption
   d. Complex formation
   Answer: (c) Surface adsorption
3. **Which compound exhibited the highest effective diffusion coefficient (D_eff) in the hydrophobically modified zeolite?**
   a. Toluene
   b. m-p-Xylene
   c. Benzene
   d. o-Xylene
   Answer: (c) Benzene
4. **The model used by Vidal et al. (2013) to determine the diffusion coefficients of BTEX was:**
   a. Freundlich surface model
   b. Homogeneous particle diffusion model
   c. Langmuir-Hinshelwood model
   d. Convective flow model
   Answer: (b) Homogeneous particle diffusion model
5. **Intraparticle adsorption of BTEX compounds occurs:**
   a. Rapidly, like surface adsorption
   b. Slowly, after the initial adsorption phase
   c. Simultaneously with surface adsorption
   d. Only at high temperatures
   Answer: (b) Slowly, after the initial adsorption phase

Objective Questions: Boyd Model

1. According to Boyd et al. (1947a,b), what was the primary purpose of their theoretical models?
   a. To analyze equilibrium conditions
   b. To simulate ion exchange kinetics
   c. To determine adsorption efficiency
   d. To study surface diffusion mechanisms
   Correct Answer: (b) To simulate ion exchange kinetics
2. What condition must be met for the Boyd plot to confirm that intraparticle diffusion is the rate-limiting step?
   a. The plot must be nonlinear
   b. The plot must pass through the origin
   c. The plot must have a negative slope
   d. The plot must show exponential growth
   Correct Answer: (b) The plot must pass through the origin
3. In Equation 3.17, what does the term qt represent?
   a. The Boyd constant
   b. The diffusion coefficient

c. The quantity of adsorbate adsorbed at time t
d. The mean particle diameter
Correct Answer: (c) The quantity of adsorbate adsorbed at time t

4. What mathematical function is used to determine Bt when F > 0.85?
a. $Bt = -0.4977 - \ln(1 - F)$
b. $Bt = \left(\sqrt{\pi} - \sqrt{\pi - \left(\frac{\pi^2 F}{3}\right)}\right)^2$
c. $Bt = \pi^2 D/d^2$
d. $Bt = 6/\pi^2 \sum(1/n^2)\exp(-\pi^2 Bt)$

5. What role do macropores play in adsorption kinetics according to the provided figure?
a. They serve as primary conduits for adsorbate transport
b. They limit the adsorption rate
c. They prevent surface diffusion
d. They increase the Boyd constant
Correct Answer: (a) They serve as primary conduits for adsorbate transport

Subjective Questions

1. How does the Boyd Diffusion Model contribute to understanding adsorption kinetics in porous materials?
2. What implications does the distinction between ion exchange and adsorption have for industrial applications?
3. Considering the provided figure, what conclusions can be drawn about the adsorption process over time?
4. What challenges arise when applying the Boyd Diffusion Model to real-world adsorption systems?

Questions: Elovich Equation

1. What is the main difference between the nonlinear and linear forms of the Elovich equation in adsorption kinetics modeling?
2. How can the parameters α (initial adsorption rate) and β (desorption constant) be determined from the graph of the Elovich equation?
3. How does the differential equation of Elovich (Equation 3.22) describe the adsorption rate over time?

Applied Questions

1. What does the fit of the Elovich equation to experimental data indicate about the adsorption mechanism of the metal onto biomass?

   Based on Figure 3.8, answer the following questions.
2. Based on the linear regression equations provided in Figure 3.8, determine the alpha (α) and beta (β) parameters for the Elovich equation regarding the adsorption of bichromate anions using the mod.1 sorbent.

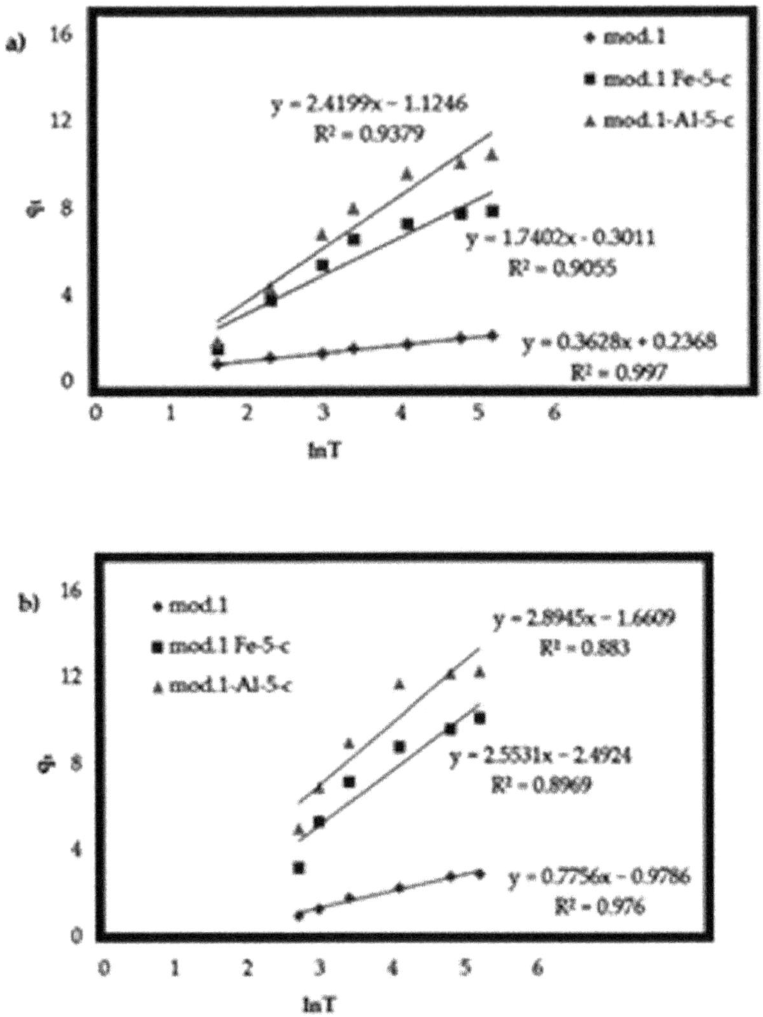

FIGURE 3.8 Elovich model: adsorption of bichromate and arsenate anions by bentonite and modified iron (III) and aluminum polyhydroxocations by bentonite-based sorbents: (a) bichromate anion (b) arsenate anion. (Figure adapted, reproduced with permission from Kussainova et al., 2024.) This article is an open-access article distributed under the terms and conditions of the Creative Commons Attribution (CC BY) license (https://creativecommons.org/licenses/by/4.0/).

3. Analyze the linear regression equations in Figure 3.8 and calculate the alpha (α) and beta (β) parameters for the Elovich equation in the adsorption of arsenate anions using the mod.1 Fe-5-c sorbent.
4. *Using* the data presented in Figure 3.8, derive the alpha (α) and beta (β) parameters for the Elovich equation applied to the adsorption of bichromate anions using the mod.1-Al-5-c sorbent.

5. Examine the linear regression equations in Figure 3.8 and determine the alpha (α) and beta (β) parameters for the Elovich equation associated with the adsorption of arsenate anions using the mod.1 sorbent.

REFERENCES

Blanco, S. P. D. M., Scheufele, F. B., Módenes, A. N., Espinoza-Quiñones, F. R., Marin, P., Kroumov, A. D., & Borba, C. E. (2017). Kinetic, equilibrium and thermodynamic phenomenological modeling of reactive dye adsorption onto polymeric adsorbent. *Chemical Engineering Journal*, 307, 466–475. https://dx.doi.org/10.1016/j.cej.2016.08.104

Boyd, G. E., Adamson, A. W., Myers Jr., L. S. (1947b). The exchange adsorption of ions from aqueous solutions by organic zeolites. II: Kinetics. *Journal of the American Chemical Society*, 69(11), 2836–2848. https://doi.org/10.1021/ja01203a066

Boyd, G. E., Schubert, J., & Adamson, A. W. (1947a). The exchange adsorption of ions from aqueous solutions by organic zeolites. I. Ion-exchange equilibria. *Journal of the American Chemical Society*, 69(11), 2818–2829, 2818–2829. https://doi.org/10.1021/ja01203a064

Cooney, D. O. (1999). Adsorption design for wastewater treatment (1st ed.). CRC Press, Boca Raton.

Darwish, A. A. A., Rashad, M., & AL-Aoh, H. A. (2019). Methyl orange adsorption comparison on nanoparticles: Isotherm, kinetics, and thermodynamic studies. *Dyes Pigments*, 160, 563–571. https://doi.org/10.1016/j.dyepig.2018.08.045

El-Khaiary, M. I., Malash, G. F., & Ho, Y. S. (2010). On the use of linearized pseudo-secondorder kinetic equations for modeling adsorption systems. *Desalination,* 257, 93–101. https://doi.org/10.1016/j.desal.2010.02.041

Elovich, S. Y., Larinov, O. G. (1962). Theory of adsorption from solutions of non-electrolytes on solid (I) equation adsorption from solutions and the analysis of its simplest form, (II) verification of the equation of adsorption isotherm from solutions. *Izvestiya Akademii Nauk SSSR — Seriya Khimicheskaya*, 2, 209–216.

Frusawa, T., & Smith, J. M., 1973. Fluid-particle and intraparticle mass transport rates in slurries. *Industrial & Engineering Chemistry Research*, 12, 197–203. https://doi.org/10.1021/i160046a009

Guiza, S., Hajji, H., & Bagante, M. (2019). External mass transport process during the adsorption of fluoride from aqueous solution by activated clay. *Comptes Rendus Chimie*, 22(2–3), 161–168.

Guo, X., & Wang, J. L. (2019a). A general kinetic model for adsorption: Theoretical analysis and modeling. *Journal of Molecular Liquids*, 288, 111100. https://doi.org/10.1016/j.molliq.2019.111100

Guo, X., & Wang, J. L. (2019b). Sorption of antibiotics onto aged microplastics in freshwater and seawater. *Marine Pollution Bulletin*, 149, 110511. https://doi.org/10.1016/j.marpolbul.2019.110511

Guo, X., & Wang, J. L. (2019c). Comparison of linearization methods for modeling the Langmuir adsorption isotherm. *Journal of Molecular Liquids*, 296, 111850. https://doi.org/10.1016/j.molliq.2019.111850

Ho, Y. S., & Mckay, G. A. (1999). Kinetic study of dye sorption by biosorbent waste product pith. *Resources, Conservation and Recycling, Kowloon*, 25(3–4), 171–193. https://doi.org/10.1016/S0921-3449(98)00053-6

Hu, Y. M., Guo, X., Chen, C., Wang, J. L., 2019. Algal sorbent derived from Sargassum horneri for adsorption of cesium and strontium ions: Equilibrium, kinetics, and mass transfer. *Applied Microbiology and Biotechnology*, 103, 2833–2843. https://doi.org/10.1007/s00253-019-09619-z

Inglezakis, V. J., Fyrillas, M. M., & Park, J. (2019). Variable diffusivity homogeneous surface diffusion model and analysis of merits and fallacies of simplified adsorption kinetics equations. *Journal of Hazardous Materials*, 367(5), 224–245.

Jain, R., Dominic, D., Jordan, N., Rene, E. R., Weiss, S., van Hullebusch, E. D., Hübner, R., & Lens, P. N. L. (2016). Higher Cd adsorption on biogenic elemental selenium nanoparticles. *Environmental Chemistry Letters*, 14, 381–386. https://doi.org/10.1007/s10311-016-0560-8

Kussainova, B., Tazhkenova, G., Kazarinov, I., Burashnikova, M., Nurlybayeva, A., Seitbekova, G., Kantarbayeva, S., Murzakasymova, N., Baibazarova, E., Altynbekova, D., Shinibekova, A., & Bazarkhankyzy, A. (2024). Adsorption of bichromate and arsenate anions by a sorbent based on bentonite clay modified with polyhydroxocations of iron and aluminum by the "Co-precipitation" method. *Molecules*, 29(15), 3709. https://doi.org/10.3390/molecules29153709

Lagergren, S. (1898). About the theory of so-called adsorption of soluble substances. *Kungliga Svenska Vetenskapsakademiens Handlingar*, 24, 1–39.

Marin, P., Borba, C. E., Módenes, A. N., Espinoza-Quiñones, F. R., Oliveira, S. P. D., & Kroumov, A. D. (2014). Determination of the mass transfer limiting step of dye adsorption onto commercial adsorbent by using mathematical models. *Environmental Technology*, 35, 2356–2364. https://doi.org/10.1080/09593330.2014.904445

Neto, V. D. O. S., Carvalho, T. V., Honorato, S. B., Gomes, C. L., Barros, F. C. F., Araújo-Silva, M. A., Freire, P. T. C., & Nascimento, R. F. (2012). Coconut bagasse treated by thiourea/ammonia solution for cadmium removal: Kinetics and adsorption equilibrium. *BioResources*, 7(2), 1504–1524.

Özer, A., Akkaya, G., & Turabik, M. (2005). The biosorption of Acid Red 337 and Acid Blue 324 on *Enteromorpha prolifera*: The application of nonlinear regression analysis to dye biosorption. *Chemical Engineering Journal*, 112, 181–190. https://doi.org/10.1016/j.cej.2005.07.007

Rani, S., Prusty, B. K., & Pal, S. K. (2018). Adsorption kinetics and diffusion modeling of CH_4 and CO_2 in Indian shales. *Fuel*, 216, 61–70.

Reichenberg, D. (1953). Properties of ion exchange resins in relation to their structure. III. Kinetics of exchange. *Journal of the American Chemical Society*, 75, 589. https://doi.org/10.1021/ja01099a022

Reid, B. E., Prausnitz, J. M., & Poling, R. C. (1987). *The Properties of Gases and Liquids*. McGraw-Hill, New York.

Reshetnikov, S., & Kurzina, I. (2019). Investigation of adsorption of water vapor on porous aluminium oxide material. *Materials Science and Engineering*, 597(1), 012011.

Ritchie, A. G. (1977). Alternative to the Elovich equation for the kinetics of adsorption of gases on solids. *Journal of the Chemical Society, Faraday Transactions 1: Physical Chemistry in Condensed Phases*, 73, 1650–1653. https://doi.org/10.1039/F19777301650

Roginsky, S. Z., & Zeldovich, J. (1934). Die katalytische oxydation von kohlenmonoxyd auf mangandioxyd. *Acta Physicochima USSR*, 1, 554.

Ruthven, D. M. (1984). *Principles of Adsorption & Adsorption Process*. John Wiley & Sons, New York.

Sabarinathan, G., Karuppasamy, P., Vijayakumar, C. T., & Arumuganathan, T. (2019). Development of methylene blue removal methodology by adsorption using molecular polyoxometalate: Kinetics, thermodynamics and mechanistic. *Microchemical Journal*, 146, 315–326. https://doi.org/10.1016/j.microc.2019.01.015

Sarma, G. K., Gupta, S. S., & Bhattacharyya, K. G. (2019). Nanomaterials as versatile adsorbents for heavy metal ions in water: A review. *Environmental Science and Pollution Research*, 26(7), 6245–6278.

Scheufele, F. B., Módenes, A. N., Borba, C. E., Ribeiro, C., Spinoza-Quiñones, F. R., Bergamasco, R., & Pereira, N. C. (2016). Monolayer-multilayer adsorption phenomenological model: Kinetics, equilibrium and thermodynamics. *Chemical Engineering Journal*, 284, 1328–1341. https://doi.org/10.1016/j.cej.2015.09.085

Souza, P. R., Dotto, G. L., & Salau, N. P. G. (2017). Detailed numerical solution of pore volume and surface diffusion model in adsorption systems. *Chemical Engineering Research and Design*, 122, 298–307.

Souza, P. R., Dotto, G. L., & Salau, N. P. G. (2019). Experimental and mathematical modeling of hindered diffusion effect of cationic dye in the adsorption onto bentonite. *Journal of Environmental Chemical Engineering*, 7(1), 102891.

Suzaki, P. Y. R., Munaro, M. T., Triques, C. C., Kleinübing, S. J., Klen, M. R. F., Bergamasco, R., & Jorge, L. M. D. M. (2017). Biosorption of binary heavy metal systems: Phenomenological mathematical modeling. *Chemical Engineering Journal*, 313, 364–373. https://doi.org/10.1016/j.cej.2016.12.082

Valderrama, C., Gamisans, X., de las Heras, X., Farrán, A., & Cortina, J. L. (2008). Sorption kinetics of polycyclic aromatic hydrocarbons removal using granular activated carbon: Intraparticle diffusion coefficients. *Journal of Hazardous Materials*, 157(2–3), 386–396.

Vidal, C. B., Raulino, G. S. C., da Luz, A. D., da Luz, Cleuzir, do Nascimento, R. F., & de Keukeleire, D. (2013). Experimental and theoretical approach to multi-component adsorption of selected aromatics on hydrophobically modified zeolite. *Chemical Engineering Journal, Belgica*, 59(2), 282—288. https://dx.doi.org/10.1021/je400780f

Wang, B. E., Hu, Y. Y., Xie, L., & Peng, K. (2008). Biosorption behavior of azo dye by inactive CMC immobilized *Aspergillus fumigatus* beads. *Bioresource Technology*, 99, 794–800. https://doi.org/10.1016/j.biortech.2007.01.043

Wang, J., & Guo, X. (2020). Adsorption kinetic models: Physical meanings, applications, and solving methods. *Journal of Hazardous Materials*, 390, 122156. https://doi.org/10.1016/j.jhazmat.2020.122156

Weber, W.J., & Morris, J. C. (1963). Kinetics of adsorption on carbon from solutions. *Journal of the Sanitary Engineering Division, American Society of Civil Engineers*, 89, 31–60. https://doi.org/10.1061/JSEDAI.0000430

Wilke, C. R., & Chang, P. (1955). Correlation of diffusion coefficients in dilute solutions. *AIChE Journal*, 1, 264–270.

Xu, G., Guan, W., Shi, S., & Blersch, D. (2018). Adsorption model development for mass transport characteristics of mfep structure by physisorption method. *Chemical Engineering Journal*, 354(15), 922–931. https://doi.org/10.1016/j.cej.2018.08.080

Yang, X., & Al-Duri, B. (2005). Kinetic modeling of liquid-phase adsorption of reactive dyes on activated carbon. *Journal of Colloid and Interface Science*, 287(1), 25–34. https://doi.org/10.1016/j.jcis.2005.01.093

Zagorodni, A. A. (2006). *Ion Exchange Materials: Properties and Applications*. Elsevier, Amsterdam.

4 Adsorption Thermodynamics

Ari Clecius Alves de Lima, Cícero Wellington Brito Bezerra, Iara Jennifer Moura Duarte, Antonia Mayza de Morais França, Carla Bastos Vidal, Diego de Quadros Melo, Giselle Santiago Cabral Raulino, and Ronaldo Ferreira do Nascimento

Several physical quantities, such as energy, entropy, and equilibrium constants, undergo variations in adsorption processes. Thermodynamics is a branch of physical science that examines such variations and therefore has a broad scope and application. The content enables comprehension of the spontaneity, interaction strength, adsorbent capacity, and the distinction between physical and chemical adsorption, a key research area in this field. The primary thermodynamic parameters most employed in adsorption analysis are Gibbs free energy ($\Delta G°$), a measure of the spontaneity, or lack thereof, of a process; adsorption enthalpy ($\Delta H°$), which indicates whether the process is exothermic or endothermic; and entropy ($\Delta S°$), a measure of the variation of disorder in the system during adsorption.

These observations lend themselves to the phenomenon of adsorption (Equation 4.1), which is physicochemical. In adsorption, attractive forces act in such a way as to allow the coating of a given surface (adsorbent, $\equiv$S) by similar species (adsorbate, $A_{(aq)}$), so that there is always a release of energy.

$$\equiv S_{(s)} + A_{(aq)} \leftrightarrow \equiv S - A_{(s)} + \text{Energy} \tag{4.1}$$

The following sections provide additional information to help the target audience understand, in greater depth, the theoretical and practical aspects related to the kinetic and thermodynamic models associated with recent case studies (Patil et al., 2024; Abdulrahman et al., 2023; Chakraborty et al., 2021; Ebelegi et al., 2020; Singh et al., 2020).

EQUILIBRIUM CONSTANT (K_E)

Isothermal models are used to obtain the equilibrium constant (K_e) of the adsorption process, which is then used in thermodynamic equations. The adsorption equilibrium constant (K_{ads}) is a quantitative measure of the affinity between the adsorbate

 DOI: 10.1201/9781003654391-4

(molecule or ion) and the adsorbent surface. It reflects the intensity of the interaction and is fundamental to describing the retention and selection of the adsorbent in industrial and environmental processes. The equilibrium constant can be defined in different ways, depending on the adsorption model employed. The term "K_e" is calculated using adsorption isotherms (e.g., Langmuir, Freundlich, and Temkin). However, it is often calculated by the direct method, using experimental data, called the "distribution coefficient" (K_{ads}), using Equation 4.2:

$$K_{ads} = q_e / C_e \quad (4.2)$$

where:

q_e = amount of adsorbate adsorbed per unit mass of the adsorbent at equilibrium (mg g^{-1}),
C_e = equilibrium concentration of adsorbate (mg L^{-1}).

PHYSICAL INTERPRETATION OF K_{ADS}

K_{ads} >1: adsorption is spontaneous and favored ($\Delta G < 0$);
K_{ads} < 1: Adsorption is non-spontaneous and less favored ($\Delta G > 0$);
High K_{ads}: Strong interaction between adsorbate and adsorbent, typical of chemisorption processes;
Low K_{ads}: Weak interactions, typical of physisorption.

GIBBS ENERGY CHANGE ($\Delta G°$)

The energy flow between the system and its surroundings can be used as a criterion of spontaneity. Those in which the energy leaves the system are called exergonic ($\Delta G° < 0$) and, therefore, spontaneous. Conversely, when the energy enters the system ($\Delta G° > 0$), the process is said to be endergonic; consequently, it is not spontaneous (Chakraborty et al., 2021). Thus, whenever at a given temperature, the variation of the Gibbs energy relative to the adsorption process is negative, the process will occur spontaneously, although nothing can be predicted about the time required for it to occur. For a given temperature, $\Delta G°$ can be expressed as the free energy change in an adsorption process (ΔG^{o}_{ads}).

In adsorption, the equilibrium constant (K_{ads}) in the Gibbs energy equation is used to describe the affinity between the adsorbate (adsorbed substance) and the adsorbent (solid surface). This constant is important to understand the efficiency of the adsorption process and can be expressed in Equation 4.3:

$$\Delta G^{o}_{ads} = -RT \ln K_{ads} \quad (4.3)$$

where:

R = gas constant (8.314 kJ mol K^{-1}), T = absolute temperature (K), and K_{ads} = adsorption equilibrium constant (dimensionless or in units of concentration depending on the model).

PHYSICAL INTERPRETATION OF ΔG^o_{ads}

Physisorption is weak, and its value is less than 40 KJ mol^{-1}, and chemisorption involves strong forces compared to physisorption and possesses a higher magnitude (Singh et al., 2020). If $\Delta G^o_{ads} < 0$, the process is spontaneous, while if $\Delta G^o_{ads} > 0$, it is non-spontaneous. When the system reaches the state of thermodynamic equilibrium, that is, the absence of a tendency of the system to move toward reactants or products, it assumes the following condition: $\Delta G_{ads} = 0$.

As stated in Chapter 1, when the forces involved in the process are weak, such as van der Waals forces, the adsorption is referred to as physical or physisorption, and the energy released by the system is small, on the order of magnitude comparable to condensation. In this case, the adsorbed molecule is not fixed at a specific site on the surface and can move across it (Atkins, 1999). Conversely, when chemical bonds are involved, the process is designated as chemisorption, or chemical adsorption (Atkins, 1999).

ENTHALPY CHANGE (ΔH^o_{ads})

The adsorption enthalpy (ΔH_{ads}) is defined as the amount of energy released or absorbed when an adsorbate (a molecule or ion) binds to the surface of an adsorbent. This fundamental thermodynamic property is pivotal in comprehending the strength of interactions between molecules and the surface and can be utilized to predict the efficiency and stability of the adsorption process. In addition to the difference in enthalpy between the processes of physisorption and chemisorption, other characteristics can be thought of for these systems. For example, in the case of interaction of a covalent nature between the participants, there will also be a directional dependence due to the need to overlap the orbitals involved; surface coverage tends to occur in a monolayer; there will be greater dependence of adsorption on temperature, and the process tends to be irreversible, because if energy has been released to form a bond, it also needs energy for its disruption and desorption of the adsorbate.

PHYSICAL INTERPRETATION OF ΔH_{ADS}

The enthalpy of adsorption can be measured by determining the temperature variation of a sample. ΔH_{ads} values less negative than 25 kJ mol^{-1} are typically associated with physisorption.

When the enthalpy of the process assumes sufficient values to indicate breakage and formation of chemical bonds, i.e., more negative than 40 kJ mol^{-1}, the adsorption can be classified as chemisorption (Singh et al., 2020).

In practice, this criterion is not absolute, although the tendency is valid that the greater the energy released, the greater the propensity for the formation of chemical bonds between adsorbent and adsorbate.

$\Delta H_{ads} < 0$ (Negative): Exothermic process, releasing energy to the environment. Adsorption is favored at lower temperatures.

$\Delta H_{ads} > 0$ (Positive): Endothermic process, absorbing energy. Adsorption is favored at higher temperatures (less common, usually associated with major structural changes in the adsorbent).

ENTROPY CHANGE ΔS^o_{ads}

The entropy of adsorption, denoted by ΔS^o_{ads}, is defined as the change in the degree of molecular randomness or freedom in the system upon the binding of an adsorbate to the surface of an adsorbent. It has been demonstrated that this parameter is indicative of alterations in the freedom of movement and distribution of molecules and thus represents a critical component in comprehending the spontaneity and affinity of the adsorption process.

PHYSICAL INTERPRETATION OF ΔS_{ADS}

If $\Delta S^o_{ads} > 0$: increased disorder in the system (e.g., physical processes). Although adsorption often restricts the degrees of freedom of adsorbates, thus leading to negative $\Delta S^0{}_{ads}$ values, positive values can arise. This occurs when a larger number of solvent molecules are displaced from both the adsorbent surface and the adsorbate's hydration sphere to accommodate fewer adsorbate molecules, resulting in a net increase of entropy.

$\Delta S^o_{ads} < 0$: greater organization at the solid-liquid interface.

RELATIONSHIP TO ENTHALPY, ENTROPY, AND FREE ENERGY

For a given temperature and known equilibrium constant (K_{ads}), the relationship between enthalpy and entropy can be expressed by the nonlinear form of the Van't Hoff Equation 4.4. Once the values of $\Delta H°$ and $\Delta S°$ are known, $\Delta G°$ is then calculated by Equation 4.5, which relates the three parameters.

$$K_{ads} = e^{\left(\frac{\Delta S_{ads}}{R}\right)} e^{\left(\frac{-\Delta H_{ads}}{RT}\right)} = e^{\left(\frac{T\Delta S_{ads} - \Delta H_{ads}}{RT}\right)} \tag{4.4}$$

$$\Delta G^o_{ads} = \Delta H^o_{ads} - T\Delta S^o_{ads} \tag{4.5}$$

where:

K_{ads} = equilibrium constant of the adsorption process (dimensionless or in L mol^{-1}), $\Delta H°$ = standard enthalpy change (J mol^{-1}), $\Delta S°$ = standard entropy change (J mol^{-1} K^{-1}), R = gas constant (8.314 J mol^{-1} K^{-1}), T = absolute temperature (K).

ISOSTERIC HEAT OF ADSORPTION (ΔH_x)

The isosteric heat of adsorption (ΔH_x) is defined as the amount of energy released when a molecule of adsorbate adheres to the surface of an adsorbent, thereby maintaining the amount adsorbed constant. This parameter is fundamental for describing the affinity of the adsorbate for the adsorbent and the heterogeneity of the surface. To calculate ΔH_x, one usually plots $\ln C_e$ versus 1/T at different surface coverages (q_e) and determines the slope of the line.

ΔH_x is usually calculated using the Clausius-Clapeyron Equation 4.6:

$$\frac{d(\ln C_e)}{dT} = \frac{\Delta H_x}{RT^2} \tag{4.6}$$

where:

C_e = equilibrium concentration of adsorbate (mg L^{-1}), ΔH_x = isosteric heat of adsorption (J mol^{-1}), R = gas constant (8.314 J mol^{-1} K^{-1}), T = absolute temperature (K).

PHYSICAL INTERPRETATION OF ΔH_X

For an adsorption system, if ΔH_x is less than 80 kJ mol^{-1}, it may be dictated by a physical adsorption type. On the other hand, if ΔH_x ranges between 80 and 400 kJ mol^{-1}, it may be dominated by chemical adsorption (Patil et al., 2024).

$\Delta H_x < 0$ (exothermic): The process releases energy, indicating a strong interaction between adsorbate and adsorbent. $\Delta H_x > 0$ (endothermic, rare): The process absorbs energy, typically associated with chemical reactions on the surface.

ACTIVATION ENERGY (E_A)

The Swedish chemist Svante August Arrhenius (1859–1927) observed that the rate of chemical reactions increases with temperature, a phenomenon that can be expressed by Equation 4.7. The activation energy (E_a) in adsorption is defined as the minimum amount of energy required for a molecule to be adsorbed onto the surface of an adsorbent. This parameter is critical to understanding the speed and mechanism of the adsorption process, thus differentiating between physisorption (physical adsorption) and chemisorption (chemical adsorption). The close relationship between kinetic studies at varying temperatures provides an estimate of the activation energy of the adsorption process, which can be determined by using the Arrhenius Equation 4.7:

$$k_n = A \cdot e^{\frac{-Ea}{RT}} \tag{4.7}$$

where:

k_n = rate constant velocity for the reaction of order n (i.e., k_2, $mg^{-1}min^{-1}$); A = Arrhenius factor (or pre-exponential factor); E_a = activation energy (kJ mol^{-1}); R = constant of gases (8.314 J mol^{-1} K^{-1}); T= absolute temperature (Kelvin).

For practical purposes, Equation 4.7 can be linearized by taking the natural logarithm of both sides (Equation 4.8). A graph of $\ln k_n$ versus 1/T provides a straight line from which the E_a can be calculated from the slope of the line (slope) multiplied by R.

$$\ln(k_n) = \ln(k_o) - \frac{Ea}{RT} \tag{4.8}$$

PHYSICAL INTERPRETATION OF E_A

For sorption processes, if $E_a < 0$, this suggests that a low temperature favors an exothermic sorption process, and if$E_a > 0$, it is an indication that the system requires energy (high temperature), which favors an endothermic sorption process (Ebelegi et al., 2020).

Low Ea: Usually physisorption, a fast and reversible process with weak interactions (Van der Waals forces). Typically between 5 and 40 kJ mol^{-1}.

High E_a: Usually chemisorption involves the formation of stronger chemical bonds. Typically between 40 and 400 kJ mol^{-1}.

ADSORPTION THERMODYNAMIC PARAMETERS

The estimation of the values for the thermodynamic parameters of adsorption is of great utility and importance, as it allows, among other points, the following aspects:

- Determine the feasibility (spontaneity) of the process under specific temperature and pressure conditions;
- Verify whether the process is predominantly governed by enthalpic or entropic contributions;
- Estimate the maximum adsorption capacity of the adsorbent;
- Provide information regarding the heterogeneity of the adsorbent surface;
- Indicate the nature of the process (physisorption or chemisorption), in addition to favoring the understanding of the mechanistic study.

For this last point, the interpretation of thermodynamics and kinetics aspects needs to be considered (Ebelegi et al., 2020). All these parameters can be estimated from equilibrium or isothermal data, as long as the measurements are carried out at different temperatures, with a minimum of three points. The adsorption isotherm, the basis of thermodynamic calculations, is a graph that provides the adsorbed quantity of the component per mass of the adsorbent as a function of the equilibrium concentration of this component (McQuarrie & Simon, 1997). Originally, the mathematical expressions for most adsorption isothermal models were empirically derived, but they can be thought of from thermodynamic arguments (Ticianelli & Gonzalez, 1998). It is an equilibrium condition that for species A (Equation 4.2), its chemical potentials are equal in the two phases (Equation 4.9):

$$\mu_A^{\equiv S} = \mu_A^{(aq)} \tag{4.9}$$

If: $\mu_A = \mu_A^0 + RT\ln a_A$, where R is the constant of the gases (8.314 J K^{-1}mol^{-1}) and a_A is the activity of species A, it follows that (Equation 4.10):

$$\mu_A^{0,\equiv S} + RT\ln a_A^{\equiv S} = \mu_A^{0,\equiv S} + RT\ln\big(f(\theta)\big) = \mu_A^{0,(aq)} + RT\ln a_A^{(aq)} \tag{4.10}$$

where:

$a_A^{\equiv S}$: It can be better represented as a function of the degree of coverage (θ) of the adsorbent surface: $f(\theta)$. Setting the standard Gibbs energy to (Equation 4.11):

$$\Delta G^0_{ads} = \mu_A^{0,\equiv S} - \mu_A^{0,(aq)} \tag{4.11}$$

And applying the exponential in Equation 4.10, the generic thermodynamic expression for the adsorption isotherm is (Ticianelli & Gonzalez, 1998):

$$a_{A(s)} = a_{A(aq)} e^{\frac{-\Delta G^0_{Ads}}{RT}} \tag{4.12}$$

The different models developed for the calculation of the parameters of the adsorbate/adsorbent interactions start to consider the term of the constant exponential and to express the activity of the adsorbate on the surface of the adsorbent in different ways. Another way to consider the approach is to substitute, in Equation 4.10, the chemical potentials for their respective terms, defining the equilibrium constant for the adsorption process (Equation 4.13) (Ramesh et al., 2005):

$$\Delta G^0_{ads} = -RT[\ln\left(\frac{f(\theta)}{a_A^{(aq)}}\right) = -RT\ln K_{ads} \tag{4.13}$$

Among the diversity of adsorption isotherms, the Langmuir and Freundlich models can be mentioned as having two adjustable parameters; the Redlich–Peterson, Sips, Langmuir–Freundlich, Fritz–Schlunder, Radke–Prasnitz, Tóth, and Jossens models as having three parameters; the Weber–van Vliet, Fritz–Schlunder, and Baudu models as having four parameters; and the Fritz–Schlunder model as having five parameters (Hamdaoui & Naffrechoux, 2007; Cestari et al., 2009).

As the Langmuir and Freundlich models are the most commonly employed, therefore, they warrant more detailed discussion.

Langmuir Isotherm

Considering the adsorption equilibrium (Equation 4.2) and the site occupancy relation presented by Equation 4.14, the expression for Langmuir's adsorption constant (Langmuir, 1918) becomes:

$$K_{ads} = \frac{f(\theta)}{a_A^{(aq)}} = \frac{\theta}{(1-\theta)a_A^{(aq)}} \tag{4.14}$$

where:

$$\theta = \frac{\text{Species adsorbed per gram of adsorbent}}{\text{Species required for formation of a monolayer}} = \frac{q_e}{q_{max}} \tag{4.15}$$

Thus:

$$\theta = \frac{q_e}{q_{max}} = \frac{K_{ads} a_A^{(aq)}}{1 + K_{ads} a_A^{(aq)}} \tag{4.16}$$

Rearranging:

$$q_e = \frac{q_{max} K_{ads} a_A^{(aq)}}{1 + K_{ads} a_A^{(aq)}} \tag{4.17}$$

Or in linearized form:

$$\frac{a_A^{(aq)}}{q_e} = \frac{1}{q_{max} K_{ads}} + \frac{a_A^{(aq)}}{q_{max}} \tag{4.18}$$

The activity of species A in solution ($a_A^{(aq)}$) is often approximated by its equilibrium concentration, C_{eq}, particularly in dilute solutions. Therefore, the analysis of the $C_{eq}/q_e \times C_{eq}$ graph will provide the parameters q_{max} and K_{ads} from the slope ($1/q_{max}$) and intercept ($1/q_{max}K_{ads}$), respectively. This model admits, in general terms, that:

i. the surface of the solid contains a defined number of adsorption sites, and each site may adsorb only one molecule;
ii. there is the formation of a monolayer on the surface, which establishes the value for q_{max};
iii. there is no interaction between neighboring molecules;
iv. all adsorption sites are equivalent, meaning they have the same energy, and no site is preferred over another.

Freundlich Isotherm

Freundlich applied and popularized an empirical equation, for which he provided no theoretical justification (Freundlich, 1906; McBain, 1932; Dąbrowski, 2001). Although this equation was not originally proposed by Freundlich, it became known as Freundlich's equation due to his extensive use and dissemination (Equations 4.19 and 4.20).

$$q_e = K_F C_{eq}^{1/n} \tag{4.19}$$

$$\ln q_e = \ln K_F + \frac{1}{n} \ln C_{eq} \tag{4.20}$$

Since K_F and n are constant for a given adsorption system at a constant temperature, the other terms have been previously defined. Using the linearized Equation 4.19, these two parameters can be determined. This model reflects an empirical logarithmic relationship between the amount adsorbed and the concentration of the solute (adsorbate). Usually in an adsorption process, the adsorption energy tends to decrease (Asnin et al., 2000). Empirically, this decrease is attributed to surface heterogeneity, making the Freundlich equation suitable for modeling adsorption on non-uniform surfaces.

MODELING OF EXPERIMENTAL DATA

Isothermal models have varying degrees of congruence with experimental data, ranging from a higher to a lower degree of congruence. In the context of correlation analysis, the most frequently used parameters are the coefficients of determination (R^2) and the chi-square (χ^2).

The percentage error variance (ΔE) is a relative measure that quantifies the discrepancy (percentage) between the model and the experimental data. R^2 indicates the proportion of the variance in the experimental data that is explained by the model. This term is used to analyze how well the models fit the experimental data. Thus, the χ^2 value is important for determining the most suitable model for data derived from adsorption procedures, which evaluates the quality of the fit between the model and the experimental data.

The mathematical expressions can be found below (Equations 4.21–4.23). It is also possible to compare the experimental and theoretical qmax values (predicted by the model), depending on the model used.

$$R^2(\%) = 100x\frac{\sum\left(q_{teo} - \bar{q}_{exp}\right)^2}{\sum\left(q_{teo} - \bar{q}_{exp}\right)^2 + \sum\left(q_{teo} - q_{exp}\right)^2} \tag{4.21}$$

$$\chi^2 = \sum\frac{\left(q_{exp} - q_{teo}\right)^2}{q_{teo}} \tag{4.22}$$

$$\Delta E(\%) = 100x\sqrt{\frac{\sum\left[\left(\Gamma_E - \Gamma_T\right)/\Gamma_E\right]^2}{n-1}} \tag{4.23}$$

where:

q_{exp}: Amount experimentally adsorbed (mg g^{-1}); q_{teo}: Theoretical amount predicted by the models (mg g^{-1}); $\bar{q}_{exp}$: Mean of the experimental values (mg g^{-1});

A better fit is indicated by low χ^2, q_{exp} values close to q_{teo}, and R^2 values of 100%. These calculations can be performed using the resources of statistical programs (e.g., Excel, Origin, etc.). Excellent works on the subject can be found in the references (Lima et al., 2012; Santos et al., 2025; Vasconcelos Da Silva et al., 2023; Romão et al., 2025; Yusoff et al., 2025).

THE DETERMINATION OF THE THERMODYNAMIC PARAMETERS OF ADSORPTION

When isotherms are available at different temperatures, thermodynamic parameters, such as Gibbs free energy change (ΔG_{ads}), enthalpy change (ΔH_{ads}), entropy change (ΔS_{ads}), and isosteric heat of adsorption (ΔH_x) can be estimated using graphical methods.

One of the main challenges is the accurate estimation of K_{ads} (Equation 4.14). This equilibrium constant can be rewritten as follows (Equation 4.24):

$$K_{ads} = \frac{q_e}{(q_{max} - q_e)} \frac{1}{C_{eq}} \tag{4.24}$$

Where, according to Equation 4.14, q_{max} is the number of sites available in the solid, or its maximum adsorption capacity given in mol g^{-1} or mg g^{-1}. The problem is that this parameter is not always properly considered in this analysis, and the maximum values observed in the isotherms under study are usually used, as if all the active sites of the solid were, in fact occupied for that specific case. When the Langmuir model fits well with the experimental data, its adsorption constant can be reliably used to estimate thermodynamic parameters.

APPLICATION OF THE VAN'T HOFF LINEAR EQUATION IN ADSORPTION

The Van't Hoff Equation 4.4 in its linearized form is widely used to analyze the effects of temperature on adsorption processes, allowing the determination of the important thermodynamic parameters, enthalpy ($\Delta H°$) and entropy ($\Delta S°$), and as a consequence the Gibbs free energy ($\Delta G°$). The Van't Hoff linear equation can be written in the following expression (Equation 4.25):

$$\ln(K_{ads}) = -\frac{\Delta H^{\circ}_{ads}}{RT} + \frac{\Delta S^{\circ}_{ads}}{R} \tag{4.25}$$

From the estimated value for K_{ads}, it is possible to make the graph $\ln K_{ads}$ versus 1/T, whose linear behavior can allow the direct estimation of the thermodynamic parameters (ΔS°_{ads}) and (ΔH°_{ads}), from the respective intercept and slope coefficients. However, in some cases, it is possible that the Van't Hoff equation does not assume a linear behavior but a second-order polynomial equation (Equations 4.26 and 4.27) (Cestari et al., 2009; Zubieta et al., 2008):

$$\ln K_{ads} = A + B \cdot T^{-1} + C \cdot \left(T^{-1}\right)^2 \tag{4.26}$$

$$\frac{\partial(\ln K_{ads})}{\partial\left(T^{-1}\right)} = B + 2C \cdot T^{-1} \tag{4.27}$$

where:

A, B, and C are the coefficients of the second-order polynomial regression.

Once the values of (ΔS°_{ads}, ΔH°_{ads}), and the constant of adsorption (K_{ads}) are known, the ΔG°_{ads} can be calculated (see Equations 4.3 and 4.5).

The Clausius-Clapeyron relationship (Equations 4.6 and 4.28) also allows the estimation of the value of the enthalpy of adsorption, ΔH_x (Isosteric heat of adsorption) which is essential in the characterization of adsorption processes (Ebelegi et al., 2020). ΔH_x is calculated from the slope of the linear relationship between $\ln C_e$ and 1/T, under the condition of constant q_e:

$$\ln Ce = \frac{-\Delta H_x}{RT} + \text{Constant} \tag{4.28}$$

The equilibrium concentration (C_e) is obtained from the adsorption isotherms, with a value established for q_e and for each temperature. If ΔH_x varies as a function of q_e, it is assumed that the surface of the adsorbent is heterogeneous (Patil et al., 2024).

As previously mentioned, negative values for ΔG°_{ads} indicate that the adsorption process is spontaneous and that the adsorbent has an affinity for the adsorbate. Furthermore, positive values for ΔS°_{ads} suggest that the spontaneity is entropically driven, often due to the desorption of a larger number of solvent molecules from the solid/liquid interface, leading to a net increase in the system's disorder.

Next, to illustrate the application of the aforementioned thermodynamic parameters associated with kinetics, case studies will be presented demonstrating their application in the adsorption of ions and molecules.

Sousa Neto et al. (2011) reported a thermodynamic study applied to the adsorption of Cu^{2+} on the surface of lignocellulosic adsorbent (coconut bagasse). The Gibbs energy of adsorption was estimated using Equations 4.26–4.28. Table 4.1 shows the values of the thermodynamic parameters ΔG°_{ads} (kJ mol^{-1}), ΔH°_{ads} (kJ mol^{-1}), ΔS°_{ads} (kJ mol^{-1} K^{-1}) for Cu^{2+} adsorption. Method I and method II refer to the use, respectively, of the constant K_L (Langmuir) and K_T (Temkin) obtained from the isotherms at temperatures of 301, 308, 313, and 318K, using Equation 4.27 (free energy).

The lnK versus 1/T ln graph (in Figure 4.1) provided a straight line, where the values of the ΔH°_{ads} and ΔS°_{ads} could be obtained from the slope and intercept, respectively (Ticianelli & Gonzalez, 1998; Hamdaoui & Naffrechoux, 2007).

Employing method I, Gibbs' free energy of Cu^{2+} adsorption was -16.34 kJ mol^{-1} (301K); –18.00 kJ mol^{-1} (308K), –19.06 kJ mol^{-1} (313K), and –22.44 kJ mol^{-1} (318K).

TABLE 4.1
Thermodynamic Parameters for the Adsorption of Cu^{2+} in Coconut Bagasse

		Method I			Method II		
Adsorbent	**T (K)**	**ΔG^{o} (kJ mol^{-1})**	**ΔH^{o} (kJ mol^{-1})**	**ΔS^{o} (kJ mol^{-1} K^{-1})**	**ΔG^{o} (J mol^{-1})**	**ΔH^{o} (kJ mol^{-1})**	**ΔS^{o} (J mol^{-1} K^{-1})**
Coconut	301	16.34	86.20	0.34	–22.14	85.05	0.36
bagasse	308	18.00			–25.25		
	313	19.06			–25.97		
	318	22.44			–28.53		

Source: author

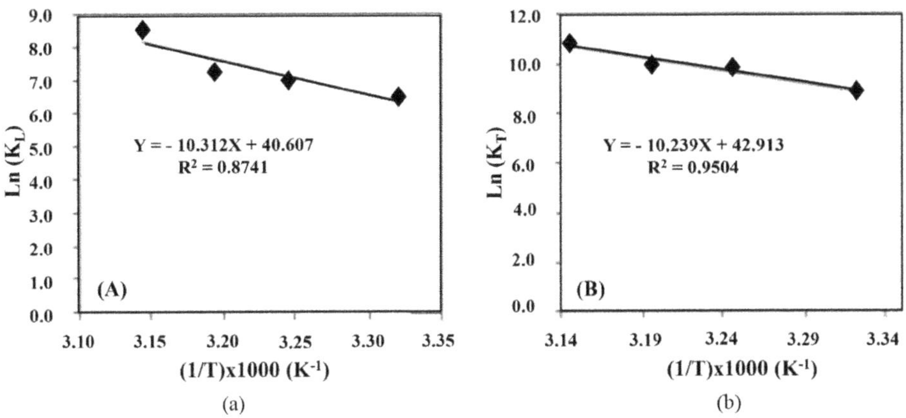

FIGURE 4.1 Thermodynamic study of Cu^{2+} adsorption onto coconut bagasse adsorbent: (a) method I using Langmuir's constant (K_L) and (b) method II using Temkin's constant (K_T). (Reproduced with permission from Sousa Neto et al., 2011.)

Using method II, the ΔG°_{ads} value were –22.14 kJ mol^{-1} (301K), –25.25 kJ mol^{-1} (308K), –25.976 kJ mol^{-1} (313K) and –28.53 J mol^{-1} (318K), as shown in Table 4.1. It is observed that the values of Langmuir's constant (K_L) increased with the increase in temperature, confirming the endothermic nature of the adsorption. It is verified that the Temkin constants (K_T) follow the same trend.

Recent studies were carried out by França (2024) involving glass fibers coated predominantly with zeolite A (called FB2–95) in order to adsorb Cu^{2+}, Cd^{2+}, and Pb^{2+}. In this study, the Freundlich and Langmuir isotherm models were analyzed.

The thermodynamic parameters for the adsorption of Cu^{2+}, Cd^{2+}, and Pb^{2+} on FB2–95 were calculated using the Van't Hoff graph (see Figure 4.2) at temperatures of 298, 308, and 318 K. The calculated values of entropy, enthalpy, and Gibbs free energy are presented in Table 4.2. A predominantly endothermic process is observed, which requires the absorption of energy to occur effectively (Bai et al., 2020; Khamis et al., 2023).

The adsorption equilibrium constants were calculated using Equation 4.2, using the average value of K(eq) for each isotherm (Cestari et al., 2009). The enthalpy values obtained for Cu^{2+} and Pb^{2+} are within the physisorption range (5–40 kJ mol^{-1}), indicating a predominance of weak interactions. Cd^{2+} showed an adsorption process in the chemisorption range (40–800 kJ mol^{-1}). This observation indicates the possibility of electron exchange or sharing between the adsorbate molecules and the surface of the adsorbent, resulting in more intense binding interaction compared to physisorption (Agarwala and Mulky, 2023; Cheu et al., 2016; He et al., 2024; Wu et al., 2017).

The positive values of the entropy changes (ΔS) that were observed indicate an increase in randomness at the solid/solution interface, accompanied by some structural changes in the adsorbents and adsorbates, demonstrating their affinity for Cu^{2+},

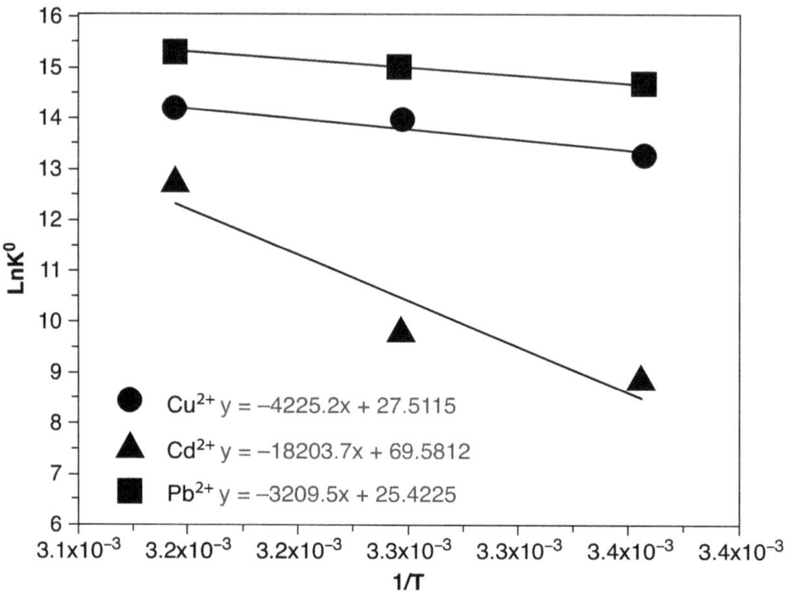

FIGURE 4.2 Van't Hoff Chart of Multi-Component System for Adsorption of Cu^{2+}, Cd^{2+}, and Pb^{2+} in FB2-95 (Experimental data of the authors).

TABLE 4.2
Thermodynamic Adsorption Parameters of Cu^{2+}, Cd^{2+}, and Pb^{2+}on FB2-95.

FB2-95					
			ΔG (kJ mmol⁻¹)		
Metal Ion	**ΔS (J mol⁻¹ K⁻¹)**	**ΔH (kJ mol⁻¹)**	**298 K**	**308 K**	**318 K**
Cu^{2+}	228.726	35.128	−33.032	−35.319	−37.606
Cd^{2+}	578.496	151.348	−21.043	−26.828	−32.613
Pb^{2+}	211.358	26.684	−36.301	−38.414	−40.528

Source: Experimental data of the authors.

Cd^{2+}, and Pb^{2+}. As demonstrated by Abdulkareem et al. (2023), Shaker (2014), and Srivastava et al. (2007), an increase in entropy also results in an increase in the degree of freedom of adsorbed species. The free energy (ΔG) exhibited negative values for all ionic systems examined. Furthermore, an increase in temperature was found to result in even more negative ΔG values for the ions onto FB2–95. The intensification of the negativity of ΔG appears to indicate greater viability and spontaneity of the adsorption process for these systems at higher temperatures (Mo et al., 2023; Rassaei, 2023).

The estimated entropy values demonstrate a direct correlation with temperature. The entropy of adsorption is indicative of alterations in the structural and energetic organization of the system. The increase in the value of this parameter is typically ascribed to the fact that the solvent molecules, which surround the active sites and the adsorbate, become freer in the medium. In accordance with the Gibbs energy values, the phenomenon under investigation was spontaneous in all cases, thus indicating an exergonic process. It is evident that the adsorptive processes were endothermic in nature. Consequently, the driving force behind the spontaneity was attributed to the entropic contribution.

Duarte et al. (2019) conducted an important study involving the removal of nitrate (NO_3^-) with nano-functionalized clay (MAO) and A520E ion exchange resins (RAC). The parameters (ΔH°) and (ΔS°) were determined using the graph of Ln $K_L \times 1/T$, being K_L of Langmuir (Equation 4.23). Equation 4.2 was then applied to determine the Gibbs free energy (ΔG°). Therefore, by applying the Van't Hoff Equation 4.24 to the linearized fit of the data illustrated in Figure 4.3, which was obtained through the modeling of the equilibrium data employing the Langmuir isotherm modeling, with the K_L values in L mol^{-1}, the thermodynamic properties presented in Table 4.3 are derived.

The Gibbs free energy of NO_3^- adsorption (see Table 4.2) was determined to be −16.24 kJ mol^{-1} at 25°C; −17.45 kJ mol^{-1} at 35°C; −18.38 kJ mol^{-1} at 45°C for MAO

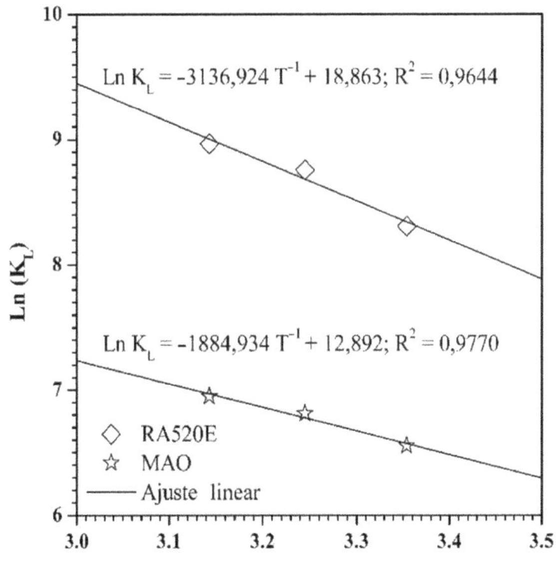

FIGURE 4.3 Langmuir parameters as a function of thermodynamic temperature for the nitrate adsorption on functionalized nanoclay (MAO) and commercial anion exchange resin A520E (RAC) adsorbents. (Reproduced with permission from Duarte et al., 2019.)

TABLE 4.3
Adsorption Thermodynamic Parameters

Solid	T (°C)	T^{-1} (K^{-1})	k′ (L mol^{-1})	ΔG° (kJ mol^{-1})	ΔH° (kJ mol^{-1})	ΔS° (J mol^{-1} K^{-1})
MAO	25	298	701.22	−16.24	15.67	107.18
	35	308	907.06	−17.45		
	45	318	1042.22	−18.38		
RA520E	25	298	4047.36	−20.59	26.08	156.82
	35	308	6349.42	−22.43		
	45	318	7823.16	−23.71		

Source: Adapted with permission from Duarte et al. (2019).

and −20.59 kJ mol^{-1} at 25°C; −22.43 kJ mol^{-1} at 35°C; −23.71 kJ mol^{-1} at 45°C for RA520E.

In addition, negative values were observed for ΔG°, indicating that the nitrate adsorption process is thermodynamically viable and spontaneous. As shown, the enthalpy variation (ΔG°) decreased with increasing temperature, indicating that the process is exothermic. According to Table 4.3, the ΔH values for MAO and RA520E were 15.67 kJ mol^{-1} and 26.08 kJ mol^{-1}, respectively. As both values are lower than 40 kJ mol^{-1}, it can be concluded that the adsorption process is physical in nature, with weak Van der Waals bonds between the nitrate (NO_3^-) in the clay and the resin.

When ΔH_{ads} values are positive, the process is designated as endothermic, while negative values indicate an exothermic process. According to the positive values shown in Table 4.3, it can be concluded that the affinity of these adsorbents for nitrate is significant. This conclusion is in line with the results obtained by Bhardwaj et al. (2012). The entropy values recorded were 107.18 kJ mol^{-1} and 156.82 kJ mol^{-1}. As mentioned, positive ΔS values indicate changes in the structural and energetic arrangement.

Thermodynamic findings by Patil et al. (2024) on the adsorption of phenolic compounds using high surface area activated carbon derived from biomass at different temperatures (283–323) were obtained on the adsorption of phenolic compounds (phenol and 2–4dichlophenol, 2–4DCP) using high surface area activated carbon derived from biomass. The Van Hoff equation was applied (using logK vs 1/T), and the values of (ΔH°), (ΔS°), and (ΔG°) were calculated based on Figure 4.4. The values of the ΔG° ranged from 0.673 to 3.653 kJ mol^{-1}, indicating a spontaneous process (see Table 4.4). An increase in the ΔG° was also observed in response to an increase in temperature, accompanied by a reduction in the efficiency of the adsorption process of phenolic compounds. The enthalpy value of adsorption, as determined through data analysis, was found to be less than 40 kJ mol^{-1}, thereby indicating a physical adsorption process for both compounds. In the case of a negative enthalpy of adsorption, the process is classified as exothermic (Wang et al., 2021).

The thermodynamic parameters obtained by the authors are listed in Table 4.4. The ΔG° values found ranged from −0.673 to -3.653 kJ mol^{-1}, indicating a spontaneous

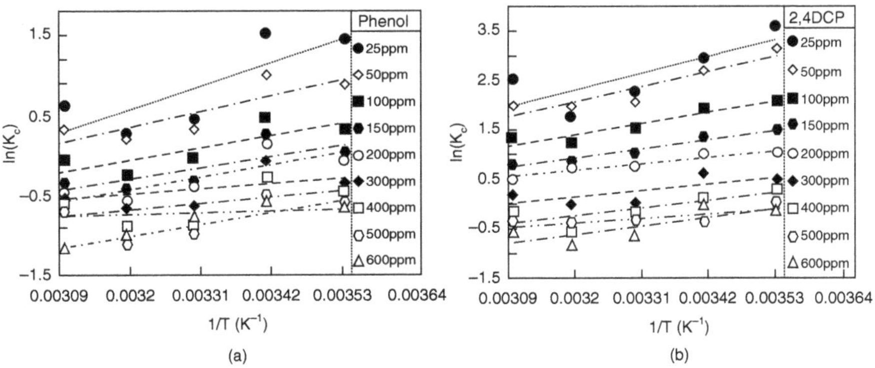

FIGURE 4.4 Enhanced adsorption of phenolic compounds using biomass-derived high surface area activated carbon: isotherms, kinetics, and thermodynamics. (Reproduced with permission from Praveengouda Patil et al., 2024.)

TABLE 4.4
Thermodynamic Parameters for Adsorption of Phenol and 2,4-DCP on Biomass-Derived High Surface Area Activated Carbon (CFPAC)

Pollutant	C_o (mg L^{-1})	ΔG (kJ mol^{-1})					ΔH (kJ mol^{-1})	ΔS (kJ mol^{-1} K^{-1})
		283 K	293 K	303 K	313 K	323 K		
Phenol	25	−3.395	−3.653	−1.117	−0.673	−1.685	−22.161	−0.066
2,4-DCP	50	−7.060	−6.489	−5.918	−5.347	−4.776	−23.22	−0.057

Source: Adapted with permission from Praveengouda Patil et al. (2024).

adsorption process (see Table 4.4). It was also observed that an increase in Gibbs free energy (ΔG°) values resulted in a decrease in adsorption as the temperature increased.

The negative value of the adsorption enthalpy (ΔH°) indicated an exothermic process, and a ΔH° value below 40 kJ mol^{-1} indicated a physical adsorption of phenol and 2,4-DCP on the adsorbent (Sriramoju et al., 2021). The decrease in randomness between the adsorbent and the adsorbate (phenol and 2,4-DCP) upon adsorption is indicated by $\Delta S° < 0$ (Wang et al., 2021).

LINEAR ARRHENIUS EQUATION APPLIED TO ADSORPTION

One of the primary applications of the method is the calculation of the E_a (the minimum energy required for the reaction to progress) of a reaction, based on experimental kinetic data at varying temperatures. The most efficacious method of arriving at this determination is through graphical representation. To facilitate experimental analysis, the linearized Equation 4.8 is applied.

$$\ln(k_n) = \ln(A) - \frac{E_a}{RT}$$

where:

k_n = rate constant of reaction of order n, A = pre-exponential factor independent of temperature (Arrhenius factor).
E_a = Activation energy, expressed in kJ mol^{-1}; R = Gas constant (8.314 J mol^{-1} K^{-1}); T = Absolute temperature in Kelvin.

When applied to the adsorption study, the k_n value is obtained from the kinetic studies of first- or second-order adsorption as described in Chapter 3.

As posited by Khattri and Singh (1999), an increase in temperature has the capacity to affect the solubility and chemical potential of adsorption. Consequently, fluctuations in temperature during a process result in alterations to the adsorption capacity. It is imperative to acknowledge the pivotal role that temperature plays in the adsorption process, given its dual impact on the system (Abdulrahman et al., 2023).

As demonstrated, an increase in temperature can lead to a pore enlargement in the structure of the adsorbent, thus allowing the penetration of larger molecules of the adsorbate (Atkins, 1999). It is evident that temperature exerts a significant influence on the velocity of diverse physicochemical processes, as evidenced by monitoring of adsorption over time at varying temperatures resulting in the determination of adsorption rate constants (McQuarrie & Simon, 1997).

A notable illustration of the application of the Arrhenius equation to ascertain thermodynamic parameters, particularly activation energy (E_a) and heat of adsorption, was documented by Patil et al. (2024) in the kinetic study of phenolic compounds (phenol and 2,4-dichlorophenol (2,4 DCP)) utilizing biomass-derived activated carbon at varying temperatures (20°C–50°C).

The calculation of E_a (see Figure 4.5) estimates whether the energy required for the process involves physical adsorption (in the range of 5–40 kJ mol^{-1}) or whether it is indicative of chemisorption (in the range of 40–800 kJ mol^{-1}).

The results were obtained from the kinetics graph at different temperatures as shown in Figure 4.5. In that study, the activation energy (E_a) values for phenol and 2,4-DCP adsorption were 35.62 and 8.5 kJ mol^{-1}, respectively, indicating a physical adsorption process for both phenolic compounds. It has been pointed out that physical adsorption typically occurs at lower temperatures compared to temperatures associated with chemical adsorption (Patil et al., 2024).

Considering that the calculation of the heat of reaction (ΔH_x) is relevant to propose the predominant adsorption mechanism, Patil et al. (2024) determined ΔH_x by the Clausius-Clapeyron Equation 4.27, using the graph of ln (C_e) versus 1/T (Equation 4.3), as shown in Figure 4.6.

The slope (inclination) of the plot was used to calculate ΔH_x (kJ mol^{-1}). It is clear that physical adsorption prevails for $\Delta H_x < 80$ kJ mol^{-1}. Chemical adsorption is the dominant process if ΔH_x ranges between 80 and 400 kJ mol^{-1}. The authors found for phenol adsorption, ΔH_x, between 0.59 and 15.95 kJ mol^{-1}. For 2,4-DCP, the range is 3.12–24.14 kJ mol^{-1}. It is clear that $\Delta H_x < 80$ kJ mol^{-1} for both pollutants. This means that the adsorption of phenol and 2,4-DCP is dominated by physical adsorption. This corroborates the thermodynamic results (Iheanacho et al., 2023). The relationships discussed above are illustrated in the plots below.

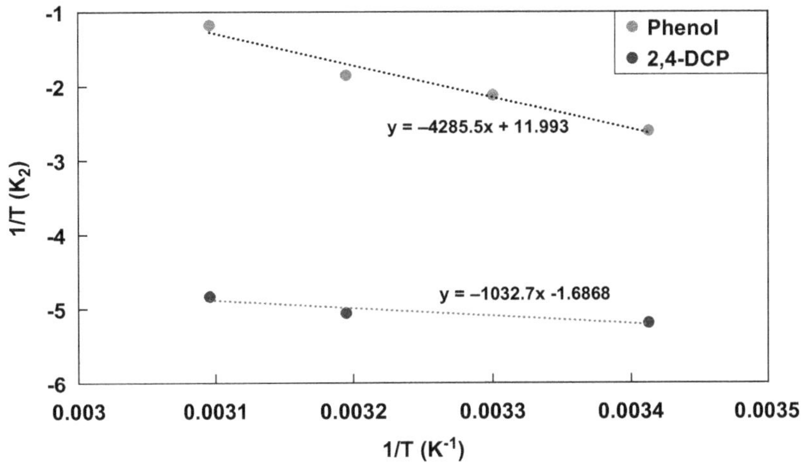

FIGURE 4.5 Arrhenius Plot ($\ln k_2$ vs 1/T) for activation energy determination. (Reproduced with permission from Praveengouda Patil et al., 2024.)

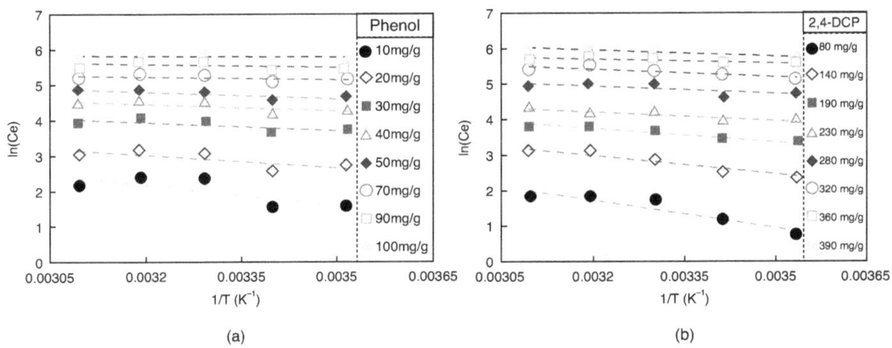

FIGURE 4.6 Isosteric heat of adsorption for (a) phenol and (b) 2,4-DCP on CFPAC. (Reproduced with permission from Praveengouda Patil et al., 2024.)

ERROR IN ESTIMATING THERMODYNAMIC PARAMETERS IN ADSORPTION

The thermodynamic parameters offer insights into the spontaneity and energetics of the adsorption process. In order to comprehend the application of varying temperatures and ascertain whether the adsorption process is spontaneous, thermodynamic parameters such as the Gibbs free energy change (ΔG°), enthalpy change (ΔH°), and entropy change (ΔS°) are employed (Özsin et al., 2019).

As demonstrated previously, it is indeed possible to determine the thermodynamic parameters (ΔG°, ΔH°, and ΔS°) from the variation of the equilibrium constant with temperature, in accordance with the Van't Hoff linear Equation 4.25 (Lima et al., 2019).

The linear graph of lnK versus 1/T is employed to determine ΔH° and ΔS°. Utilizing the equation of the line, the angular coefficient and linear coefficient are thus determined as ΔH°/R and ΔS°/R, respectively (Hameed et al., 2007).

The Langmuir and Freundlich isotherm equations have been extensively utilized to calculate thermodynamic parameters, particularly in the determination of variations in free energy, i.e., ΔG° (Anastopoulos & Kyzas, 2016). The isothermal data is typically modeled using a linear equation. As demonstrated in the works, this provides the equilibrium constant at a given temperature (Lima et al. 2019; Ćurković et al., 2019; Hong et al., 2009; Saucier et al., 2017).

The equilibrium constants, calculated from the isotherm equations, are dimensional in nature and, under ideal conditions, should be calculated in the unit of L mol^{-1}. However, they are usually obtained in different units (e.g., L mg^{-1}, L g^{-1}, and L $mmol^{-1}$) for application in the Van't Hoff equation. Consequently, several studies have reported serious errors associated with the determination of ΔG°, mainly using the equilibrium constant K_L of the Langmuir isotherm, in the equation $\Delta G° = -RT \ln Ke$.

These errors are usually associated with the incorrect use of K_L to determine ΔG°, which is expressed in concentration units.

In a strictly theoretical sense, the calculation method in question is incorrect, since K in the Langmuir isotherm is not dimensionless. In a strictly theoretical sense, the calculation method in question is incorrect, since K in the Langmuir isotherm is not dimensionless. As established by the International Union of Pure and Applied Chemistry (IUPAC), the standard equilibrium constant (K) for the calculation of ΔG° must be a dimensionless parameter. However, several scientific studies in the literature do not consider the dimensionality of the constants acquired through isotherm models, resulting in an incorrect identification of the value of the thermodynamic equilibrium constant (Zhou & Zhou, 2014; Castellini et al., 2005; Gong et al., 2008; Zhou et al., 2011). Therefore, it is necessary to analyze the aspects associated with the dimensionality of K, since both the sign and the magnitude of the thermodynamic parameters (ΔG°, ΔH°, and ΔS°) can lead to conclusions inconsistent with the real data (Tran et al., 2016).

Therefore, the calculation is performed by converting the units of the best-fitted isotherm model, initially expressed, for example, in L mg^{-1} to L mol^{-1} (Liu, 2009; Anastopoulos & Kyzas, 2016). This is achieved by multiplying the K_e value (L mg^{-1}) by 1,000, converting it into L g^{-1}, and then multiplying this result by the molecular weight of the adsorbate (g mol^{-1}) multiplied by the standard unit concentration of the adsorbate (1 mol L^{-1}) and dividing by the activity coefficient (dimensionless) (Chang & Thoman, 2014; Tran et al., 2017).

Considering that all concentrations of adsorbent, adsorbate, and adsorbent-adsorbate are also expressed in mol L^{-1}, the equilibrium constant will be dimensionless (Levine, 2009; Chang & Thoman, 2014).

Some studies use expressions that are also addressed for the calculation of the equilibrium constant (Equations 4.29 and 4.30), where the most common according to Lima et al. (2019), are:

$$K_d = \frac{q_e}{C_e}\frac{m}{V} \tag{4.29}$$

$$K_c = \frac{C_s}{C_e} = \frac{C_0 - C_e}{C_e} \tag{4.30}$$

In these equations, K_d (distribution constant in L g^{-1}) or K_c (dimensionless equilibrium constant) are applied in their dimensionless form; however, they do not lead to an exact value of the equilibrium constant. These K_c and K_d coefficients do not represent an equilibrium constant, as mentioned in the literature (Tran et al., 2017; Tran et al., 2016). Thus, it should be borne in mind that this strategy is incorrect and does not follow the principles of chemical equilibrium. This is just a strategy of leaving K dimensionless, disregarding whether the equilibrium constant obtained has physical meaning (Lima et al., 2019).

APPROPRIATE PROCEDURE FOR CALCULATING THERMODYNAMIC PARAMETERS

Depending on the unit of measurement of the isotherm constant applied, some correction factors will be indicated. These can be used to make the unit of the Ke constant dimensionless (see Table 4.4), thus facilitating the calculation of the thermodynamic parameters. This issue has been identified as a frequent occurrence; therefore, it is proposed that an analysis of the issue be conducted in order to alert readers to the need to avoid similar errors.

The use of the following correction factors (Table 4.4) takes into account that the adsorption process takes place in solution. The factor of 55.5 mol L^{-1} is obtained by dividing 1,000 g L^{-1} of H_2O by its molar weight (18 g mol^{-1}). In another situation, it is proposed that the molecular weight of the adsorbent (M*) be applied (Lima et al., 2019). However, in these examples, the type of adsorbent applied, as well as the charge (whether weakly charged, neutral, or charged adsorbents) needs to be investigated so that the calculation of ΔG° and successively the estimation of ΔH° and ΔS° do not show errors.

The Langmuir equation will be used as an example, since it has become popular for its applications in summarizing adsorption data and estimating adsorption equilibrium constants. Thus, the following approaches have been proposed to make the unit of the thermodynamic constant (K_e) dimensionless (Table 4.5). The parameter K_L in this equation is usually referred to as the adsorption coefficient or the adsorption equilibrium constant. In addition, K_L values at equal temperatures were used to estimate the standard free energy changes ΔG°. For a better understanding, we will examine four theoretical examples (reflecting practical cases) involving the incorrect and correct calculation of ΔG° (Table 4.6).

In order to illustrate the concepts discussed above, Table 4.5 shows fictitious theoretical K_L values derived from the best Langmuir isotherm fit. The ΔG° values obtained using the $\Delta G° = -RT\ln K°$ equation were found to be unstable, since the

TABLE 4.5
Correction Factors of the Units of Measurement of Langmuir's Constant

Unit of Measurement (K_L)	Correction Factor	References
$dm^3\ mol^{-1}$	$K° = K \times 55.5$	Zhou and Zhou (2014)
$L\ mmol^{1}$	$K° = K \times 10^3 \times 55.5$	Zhou and Zhou (2014)
$L\ g^{1}$	$K° = K \times M_* \times 55.5$	Zhou and Zhou (2014)
$L\ mg^{1}$	$K° = K \times M_* \times 10^3 \times 55.5$	Zhou and Zhou (2014)

Source: Adapted from Silva (2018) with permission.

TABLE 4.6
Theoretical Values Calculated for ΔG° Considering the Dimensionality of the K_L Constant

					Dimensional		Dimensionless	
Adsorbate	T (C°)	T (K)	Constant (K_L)	Unit	Formula	ΔG° (J mol⁻¹)	Formula	ΔG° (J mol⁻¹)
A	25	298.15	3.236×10^6	$dm^3\ mol^{-1}$	$\Delta G° = -RT\ln K°$	−37157	$K° = 55.5K$	−47,113
B	25	298.15	16.8	$L\ mmol^{1}$		−6,994	$K° = K \times 10^3 \times 55.5$	−34,073
C	25	298.15	33.45	$L\ g^{1}$		−8,701	$K° = K \times M^* \times 55.5$	−33,182
D	25	298.15	35.85	$L\ mg^{1}$		−8,873	$K° = K \times M_* \times 10^3 \times 55.5$	−54,353

Source: Authors.
Note: M* is the molar mass of the solute (g mol⁻¹), where the fictitious masses are MC = 350.63 and MD = 1,675, corresponding to items C and D, respectively.

dimensional and dimensionless K_L values are significantly different. However, it is well known that for the calculation of ΔG° based on the K parameter, it must be dimensionless, according to IUPAC. Therefore, calculations that do not consider the dimensionless analysis do not correctly represent the ΔG° constant. As an illustration, we can consider the dimensional case of the adsorbate (A) in $dm^3\ mol^{-1}$, shown in Table 4.6, which is calculated as follows:

$$\Delta G° = -RT\ln K° = -\left(8.314\ J\ mol^{-1}K^{-1}\right) \times \left(298.15\ K\right) \times \ln\left(3.236 \times 10^6\ dm^3 mol^{-1}\right)$$
$$= -37{,}157\ \left(J\ mol^{-1}\right)$$

After analyzing the example presented, it can be seen that the result obtained is not compatible with the determinations described by IUPAC. IUPAC states that the constant K should be dimensionless, while the units for ΔG°, R, and T should be J mol^{-1}, J mol^{-1} K^{-1}, and K, respectively. In the K_L example, the magnitude of the constant was expressed in dm^3 mol^{-1}. Consequently, the example presented should not be used to indicate the value of ΔG°. As stated by Zhou and Zhou (2014), a dimensional parameter cannot be used in a logarithmic expression, as such a procedure would result in a lack of consistency. Therefore, the precise calculation of ΔG° for the dimensionless adsorbate (A) requires the use of the correction factor (Table 4.4) corresponding to the unit dm^3 mol^{-1} for the appropriate conversion of units.

$$\Delta G° = -RT\ln K° = -\left(8.314\ \text{J mol}^{-1}\text{K}^{-1}\right)\times(298.15\ \text{K})\times\ln(55.5\times \text{K})$$

$$= -\left(8.314\ \text{J mol}^{-1}\text{K}^{-1}\right)\times(298.15\ \text{K})\times\ln\left(55.5\times(3,236\times10^{6}\ \text{dm}^3\text{mol}^{-1}\right)$$

$$= -47,113\left(\text{J mol}^{-1}\right)$$

In this calculation, the dimensionless unit was taken into account in the conversion, with K° obtained through the equation, thus obtaining the value in J mol^{-1}, as recommended by IUPAC, which is the correct value for this parameter.

The calculation involving dimensional adsorbate (B, Lmmol^{-1}), Table 4.6, follows the same principle as described for adsorbate (A). Thus, we obtain the correct value taking into account the dimensionless fit by substituting the dimensionally calculated K° value, according to the equation $K°=K\times M^*\times 55.5$. Therefore, to stipulate the correct value of ΔG°, the following calculation is used:

$$\Delta G° = -RT\ln K° = -\left(8.314\ \text{J mol}^{-1}\text{K}^{-1}\right)\times(298.15\ \text{K})\times\ln\left(\text{K}\times 10^3\times 55.5\right)$$

$$= -\left(8.314\ \text{J mol}^{-1}\text{K}^{-1}\right)\times(288\ \text{K})\times\ln\left(16.8\ \text{L mmol}^{-1}\times10^3\times55.5\ \right)$$

$$= -34,073\ \left(\text{J mol}^{-1}\right)$$

In the case of adsorbate (C), the dimension of K_L is L·g^{-1}. As previously described, the calculation below is incorrect:

$$\Delta G° = -RT\ln K° = -\left(8.314\ \text{J mol}^{-1}\text{K}^{-1}\right)\times(298.15\ \text{K})\times\ln\left(33.45\ \text{L g}^{-1}\right)$$

$$= -8,701\ \left(\text{J mol}^{-1}\right)$$

Therefore, it should be corrected as follows:

$$\Delta G° = -RT\ln K° = -\left(8.314\ \text{J mol}^{-1}\text{K}^{-1}\right)\times(288\ \text{K})\times\ln\left(\text{K}\times M^*\times 55.5\right)$$

$$= -\left(8.314\ \text{J mol}^{-1}\text{K}^{-1}\right)\times(298.15\ \text{K})\times\ln(33.45\times350.63\times55.5\ )$$

$$= -33,182\ \left(\text{J mol}^{-1}\right)$$

In this calculation, the multiplication of the variable (55.5) by the K_L value is considered, as well as the multiplication of this value by the molar mass of the adsorbate (M*). The same principle is applied for the calculation of dimensionless adsorbate (D), described in Table 4.5, as shown below:

$$\Delta G = -RT\ln K = -\left(8.314\ \text{J mol}^{-1}\text{K}^{-1}\right)\times(298.15\ \text{K})\times\ln\left(K\times M_{*}\times 10^{3}\times 55.5\right)$$

$$= -\left(8.314\ \text{J mol}^{-1}\text{K}^{-1}\right)\times(298.15\ \text{K})\times\ln\left(35.85\times 1{,}675\times 10^{3}\times 55.5\right)$$

$$= -37{,}230\ \left(\text{J mol}^{-1}\right)$$

The calculation of the thermodynamic equilibrium constant, in aqueous systems, can be performed using the modified Langmuir equation. To see the full deduction of the equation, see the reference Zhou and Zhou (2014). According to the authors using the modified Langmuir equation (Equation 4.31), the Langmuir constant (K) becomes dimensionless.

$$\theta = \frac{K\dfrac{C_e}{C^{\circ}}}{1+K\dfrac{C_e}{C^{\circ}}} \tag{4.31}$$

where θ is the fraction of the covered area at equilibrium; C_0 and C_e are, respectively, the molar concentrations of the standard reference solution (1 mol L^{-1}) and the molar concentration of the adsorbate in the solution at equilibrium (mol L^{-1}).

SOLVED PROBLEMS ON ADSORPTION THERMODYNAMICS IN VOCS

Exercise 1: Free Energy of Adsorption of Toluene

An adsorption experiment of toluene on activated carbon was conducted at 298 K. The experimentally determined adsorption equilibrium constant (K) was 2.5×10^3. Calculate the standard free energy of adsorption (ΔG°).

Data:

- $R = 8.314$ J mol-K^{-1}
- $T = 298$ K
- $K = 2.5\times 10^3$

Solution:

$\Delta G^{\circ} = -RT \ln K$
$\Delta G^{\circ} = -(8.314)(298) \ln(2{,}500)$

$\Delta G° \approx -2477 \times 7.824 \approx -19{,}378\ J\ mol^{-1} = -19.4\ kJ\ mol^{-1}$
Answer: $\Delta G° = -19.4\ kJ\ mol^{-1}$

Exercise 2: Determining ΔH° and ΔS° with the Van't Hoff Equation

The equilibrium constant of acetone adsorption on a solid adsorbent was determined at two temperatures:

- $K_1 = 1.2 \times 10^2$ at $T_1 = 298$ K
- $K_2 = 2.3 \times 10^1$ at $T_2 = 318$ K

Calculate the standard enthalpy of adsorption (ΔH°) and the standard entropy (ΔS°).

Solution:

$\ln(K_2/K_1) = -\Delta H°/R \times (1/T_2 - 1/T_1)$
$\text{lnc}(23/120) = -\Delta H°/8.314 \times (1/318 - 1/298)$
$\text{lnc}(0.1917) = -\Delta H°/8.314 \times (-2.115 \times 10^{-4})$
$-1.650 = \Delta H°/8.314 \times 2.115 \times 10^{-4}$
$\Delta H° \approx -64.79\ kJ\ mol^{-1}$

For ΔS°:

$\Delta G° = -RT \ln K = -(8.314)(298) \ln(120) \approx -13.76\ kJ\ mol^{-1}$
$\Delta S° = (\Delta H° - \Delta G°)/T = (-64.79 - (-13.76))/298 \approx -171.2\ J\ mol\text{-}K^{-1}$

Answers:

- $\Delta H° \approx --64.8\ kJ\ mol^{-1}$
- $\Delta S° \approx -171\ J\ mol\text{-}K^{-1}$

Exercise 3: Benzene Adsorption Energy on MOF Using Van't Hoff

The following line equation was obtained by plotting ln K vs. 1/T for benzene adsorption on MOF:

$\ln K = -8{,}200(1/T) + 20$

Determine the standard enthalpy and entropy of adsorption.

Solution:

Comparing with $\ln K = -\Delta H°/R - (1/T) + \Delta S°/R$
- Slope $= -\Delta H°/R = -8{,}200 \rightarrow \Delta H° = 8{,}200 \times 8.314 \approx 68.17\ kJ\ mol^{-1}$
- Intercept $= \Delta S°/R = 20 \rightarrow \Delta S° = 20 \times 8.314 \approx 166.28\ J\ mol\text{-}K^{-1}$

Answers:

$-\Delta H° \approx -68.2\,kJ\ mol^{-1}$
$-\Delta S° \approx 166\,J\ mol\text{-}K^{-1}$

Exercise 4: Adsorption of Formaldehyde on Activated Carbon

The equilibrium constant of formaldehyde adsorption on activated carbon was measured as 1.8×10^3 at 298 K. Calculate the standard free energy of adsorption ($\Delta G°$) and estimate the standard entropy, considering that the standard enthalpy ($\Delta H°$) is $-45\,kJ\ mol^{-1}$.

Data:

$-R = 8.314\,J\ mol\text{-}K^{-1}$
$-T = 298\ K$
$-K = 1.8 \times 10^3$

Solution:

$\Delta G° = -RT\ \ln\ K = -(8.314)(298)\ \ln(1{,}800) \approx -2477 \times 7.495 \approx -18{,}572\,J\ mol^{-1} \approx -18.6\,kJ\ mol^{-1}$
$\Delta S° = (\Delta H° - \Delta G°)/T = (-45 - (-18.6))/298 \approx -88.6\,J\ mol\text{-}K^{-1}$

Answers:

$-\Delta G° \approx -18.6\,kJ\ mol^{-1}$
$-\Delta S° \approx -88.6\,J\ mol\text{-}K^{-1}$

Exercise 5: Ethanol Adsorption on Zeolite

The Van't Hoff equation for ethanol adsorption on a zeolite was given by:

$$\ln K = -7{,}500(1/T) + 17.5$$

Determine the values of $\Delta H°$ and $\Delta S°$.

Solution:

$\Delta H° = 7{,}500 \times 8.314 \approx 62.36\,kJ\ mol^{-1}$
$\Delta S° = 17.5 \times 8.314 \approx 145.5\,J\ mol\text{-}K^{-1}$

Answers:

$-\Delta H° \approx -62.4\,kJ\ mol^{-1}$
$-\Delta S° \approx 145.5\,J\ mol\text{-}K^{-1}$

Exercise 6: Adsorption of n-Hexane on Metal-Organic Frameworks (MOF)

An experiment at 303 K showed that the equilibrium constant for the adsorption of n-hexane on MOF was 3.1×10^2. Given that the adsorption enthalpy is -70 kJ mol^{-1}, determine $\Delta G°$ and $\Delta S°$.

Solution:

$\Delta G° = -RT \ \ln \ K = -(8.314)(303) \ \ln(310) \approx -2{,}518 \times 5.737 \approx -14{,}449 \text{ J} \ mol^{-1} = -14.4 \text{ kJ } mol^{-1}$

$\Delta S° = (\Delta H° - \Delta G°)/T = (-70 - (-14.4))/303 \approx -183 \text{ J mol-K}^{-1}$

Answers:

–$\Delta G° \approx -14.4 \text{ kJ } mol^{-1}$
–$\Delta S° \approx -183 \text{ J mol-K}^{-1}$

Exercise 7: Adsorption Capacity as a Function of Temperature – n-Hexane

The adsorption of n-hexane on an MOF-type material follows the equation:

$q(T) = q_0 \cdot e^{-\Delta H°/RT}$

Data:

–$q_0 = 2.5 \text{ mmol } g^{-1}$
–$\Delta H° = -70{,}000 \text{ J } mol^{-1}$
–$R = 8.314 \text{ J } mol^{-1} \ K^{-1}$

a. Calculate q(T) for:
   –T = 298 K
   –T = 308 K
   –T = 318 K
   –T = 328 K
b. Generate the graph of ln K vs. 1/T with this data.
c. Is the adsorption process exothermic or endothermic?

Answers:

a)

Temperature (K)	q(T) (mmol/g)
298	2.462
308	2.435
318	2.407
328	2.379

b. Van't Hoff Chart – Acetone

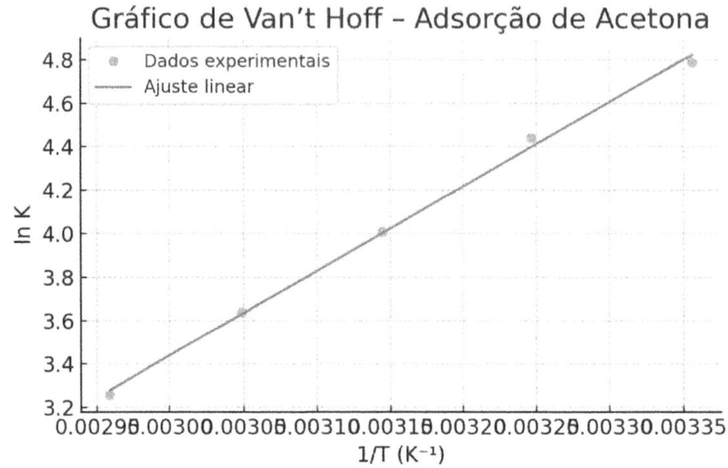

c. As expected for exothermic adsorption, the adsorptive capacity decreases with increasing temperature, due to the unfavorability of the process in more energetic thermal environments.

SOLVED EXERCISES ON ACTIVATION ENERGY IN AIR POLLUTANTS

Exercise 8: Calculating Activation Energy using the Arrhenius Equation

Data:

An atmospheric pollutant removal process obeys first-order kinetics. The speed of the reaction was measured at two different temperatures:

Temperature (K)	Velocity constant (k) (s^{-1})
298	1.5×10^{-3}
318	6.0×10^{-3}

Calculate the activation energy (E_a) of the reaction in kJ mol^{-1}.

Resolution:

The Arrhenius equation is:

$$k = A - \exp(-E_a/(R - T))$$

Using the linearized form between two temperatures:

$$\ln(k_2/k_1) = E_a/R - (1/T_1 - 1/T_2)$$

Data:

$k_1 = 1.5 \times 10^{-3}\ s^{-1}$ at $T_1 = 298$ K
$k_2 = 6.0 \times 10^{-3}\ s^{-1}$ at $T_2 = 318$ K
$R = 8.314$ J (mol-K)$^{-1}$

Calculating the left-hand side:

$$\ln(k_2/k_1) = \ln(6.0 \times 10^{-3}/1.5 \times 10^{-3}) = \ln(4) = 1.386$$

Calculating the right-hand side:

$$1/T_1 - 1/T_2 = 1/298 - 1/318 = 0.003356 - 0.003145 = 0.000211\ K^{-1}$$

Isolating E_a:

$$E_a = (R \times \ln(k_2/k_1))/(1/T_1 - 1/T_2) = (8.314 \times 1.386)/0.000211 = 54{,}598\ J\ mol^{-1} = 54.6\ kJ\ mol^{-1}$$

Exercise 9: Determination of Activation Energy by Graphical Method

Data:

The rate constant for the degradation of a VOC was measured at different temperatures:

Temperature (K)	Constant k (s^{-1})
290	1.0×10^{-4}
300	3.0×10^{-4}
310	7.5×10^{-4}
320	1.8×10^{-3}

Using this data, determine the activation energy E_a of the reaction by plotting ln(k) versus 1/T and calculating the slope.

Resolution:

1. Calculate ln(k) and 1/T for each temperature:

T (K)	k (s^{-1})	ln(k)	1/T (K^{-1})
290	1.0×10^{-4}	–9.210	0.003448
300	3.0×10^{-4}	–8.111	0.003333
310	7.5×10^{-4}	–7.194	0.003226
320	1.8×10^{-3}	–6.321	0.003125

2. Plotting ln(k) versus 1/T, we get a straight line with slope $m = -E_a/R$.

3. Calculating the slope between the first two points:
   $m = (\ln(k_2) - \ln(k_1))/(1/T_2 - 1/T_1) = (-8.111 + 9.210)/(0.003333 - 0.003448)$
   $= 1.099/(-0.000115) = -9{,}565\ K$
4. Therefore, $E_a = -m \times R = 9{,}565 \times 8.314 = 79{,}561\ J\ mol^{-1} = 79.6\ kJ\ mol^{-1}$

Exercise 10: Calculating the Velocity Constant from E_A and Temperature

Data:

The activation energy for the degradation of a pollutant is 65 kJ mol^{-1}. The pre-exponential constant (A) is $2.0 \times 10^{13}\ s^{-1}$.

Calculate the rate constant k at 350 K.

Resolution:

Using the Arrhenius equation:

$$k = A - \exp(-E_a/(R - T))$$

Converting E_a to J mol^{-1}:

$$E_a = 65 \times 10^3 = 65{,}000\ J\ mol^{-1}$$

Substituting:

$$k = 2.0 \times 10^{13} - \exp(-65{,}000/(8.314 \times 350))$$
$$k = 2.0 \times 10^{13} - \exp(-22.33) = 2.0 \times 10^{13} - 1.99 \times 10^{-10} = 3{,}980\ s^{-1}$$

PROPOSED PROBLEMS

4.1 At 298 K and for a specific adsorption system, the standard enthalpy change ($\Delta H°$) and standard entropy change ($\Delta S°$) were determined to be −23 kJ mol^{-1} and −8 J $mol^{-1}\ K^{-1}$. Based on this, indicate whether the process is spontaneous and whether it can be better classified as physisorption or chemisorption.

4.2 Two systems, A and B, present the following values for $\Delta H°$ and $\Delta S°$, respectively: 15 kJ mol^{-1}; 90 J $mol^{-1}\ K^{-1}$, and −22 kJ mol^{-1}; −50 J $mol^{-1}\ K^{-1}$. Compare both systems at 298 K in terms of spontaneity and entropic contribution. Which system is more entropically driven?

4.3 Adsorption is fundamentally a spontaneous process. However, the spontaneity and favorability of adsorption can change significantly with temperature, influenced by the interplay of enthalpy and entropy. Consider an adsorption system where the following standard thermodynamic parameters were experimentally determined:

System A: $\Delta H° = -15\ kJ\ mol^{-1}$; $\Delta S° = -40\ J\ mol^{-1}\ K^{-1}$;
System B: $\Delta H° = 20\ kJ\ mol^{-1}$; $\Delta S° = 70\ J\ mol^{-1}\ K^{-1}$;

a. For system A, calculate ΔG° at two temperatures: 273 and 300 K. Discuss how temperature affects the spontaneity of this adsorption process. What type of adsorption is more likely for system A: physisorption or chemisorption, and why?
b. For system B, calculate ΔG° at the same two temperatures (273 and 300 K) and discuss how temperature affects the spontaneity of this adsorption process. What is the main driving force for spontaneity in system B at higher temperatures?

4.4 A research team is investigating the equilibrium adsorption of a pharmaceutical compound (paracetamol) from aqueous solutions onto a novel activated carbon adsorbent. To understand the adsorption behavior at different concentrations, batch adsorption experiments were conducted at a constant temperature of 303 K.

In each experiment, 0.05 g of the activated carbon was added to 100 mL of paracetamol solutions with varying initial concentrations. After allowing sufficient time for equilibrium to be reached (determined from kinetic studies), the equilibrium concentration of paracetamol in the solution (C_e) was measured.

a. Calculate the equilibrium adsorption capacity (q_e, mg g^{-1}) for each experimental point.
b. Construct the isothermal graph (q_e as a function of C_e).
c. Check the applicability of the linear Langmuir and Freundlich isothermal models to the experimental data.

4.5 Consider that, at a temperature of 298 K, the Langmuir constant, K_L, for a specific adsorption process in the liquid phase was estimated to be 2.3 L mol^{-1}.

a. Convert this K_L constant to a dimensionless equilibrium constant (K_{ads}).
b. Calculate the corresponding standard Gibbs free energy change (ΔG°) in kJ mol^{-1}.
c. Based on the calculated ΔG° value, determine if the adsorption process is spontaneous at 298 K. Justify your answer.

4.6 The Freundlich isotherm is an empirical model commonly used to describe adsorption on heterogeneous surfaces. Its constants, K_F and n, are determined from experimental data. Unlike the Langmuir constant, Freundlich constants have complex units and an empirical nature.

Consider a given adsorption system at 298 K where the Freundlich constants were $K_F = 5.2$ (mg g^{-1})/(L mg^{-1})$^{1/n}$, and $n = 2.5$.

a. Discuss how these Freundlich parameters influence the spontaneity of the adsorption process.
b. Comment if these constants can be directly used in Van't Hoff calculations. Justify your answer.

4.7 The isosteric heat of adsorption (ΔH_x), also known as the differential heat of adsorption, is a crucial thermodynamic parameter that provides insight into the energy change occurring during the adsorption process at a constant surface loading (q_e). Its value can vary with the amount adsorbed, reflecting the heterogeneity of the adsorbent surface and the nature of the adsorbate-adsorbent interactions. Consider the following equilibrium concentrations (C_e) for a specific adsorbate at a constant equilibrium adsorbed amount (q_e) of 50 mg g^{-1}, obtained at different temperatures, and determine the isosteric heat of adsorption.

Hypothetical experimental data

T (K)	C_e (mg L^{-1})
298	5.0
308	3.0
318	1.8

4.8 Below are the values for the adsorption equilibrium constant (K_{ads}) for a specific and hypothetical adsorbent-adsorbate system at different temperatures. Calculate and interpret the following standard thermodynamic parameters for this system: ΔH°, ΔS°, and ΔG° (at 298 K).

T (K)	K_{ads} (L mol^{-1})
298	1.5
308	2.0
318	2.7

4.9 The Van't Hoff equation, in its linearized form, assumes that the standard enthalpy (ΔH°) and entropy (ΔS°) of a process are constant with temperature within the studied range. However, in some adsorption systems, a plot of $\ln K_{ads}$ versus 1/T may exhibit significant curvature, indicating that this assumption is not valid. In such a case, a higher-order polynomial equation is often used to model the relationship (see: Dos Santos et al., 2018).

Consider that, for a given adsorption system, the relationship between $\ln K_{ads}$ and T was best fitted by the following equation: $\ln K_{ads} = A + BT^{-1} + CT^{-2}$.

a. Suggest a physical interpretation for the curvature in the plot $\ln K_{ads} \times T^{-1}$;
b. Discuss what a significant C term implies about the adsorption process.

4.10 Consider an adsorption system at equilibrium, where the standard enthalpy of adsorption is –55 kJ mol^{-1} and the standard entropy adsorption is – 120 J mol^{-1} K^{-1}.

a. Is the adsorption process spontaneous at 298 K?
b. Calculate the standard enthalpy and entropy of desorption ($\Delta H°_{des}$, $\Delta S°_{des}$). How do these relate to the values for adsorption ($\Delta H°_{ads}$ and $\Delta S°_{ads}$)?
c. Discuss how increasing the temperature affects the equilibrium position.

REFERENCES

Abdulkareem, A. S., Hamzat, W. A., Tijani, J. O., Egbosiuba, T. C., et al. (2023). Isotherm, kinetics, thermodynamics and mechanism of metal ions adsorption from electroplating wastewater using treated and functionalized carbon nanotubes. *Journal of Environmental Chemical Engineering*, 11(1), 09180.

Abdulrahman, M. S., et al. (2023). Activated carbon from sugarcane as an efficient adsorbent for phenol from petroleum refinery wastewater: equilibrium, kinetic, and thermodynamic study. *Open Engineering*, 13(1), 20220442.
Agarwala, R., & Mulky, L. (2023). Adsorption of dyes from wastewater: A comprehensive review. *ChemBioEng Reviews*, 10(3), 326–335.
Anastopoulos, I., & Kyzas, G. Z. (2016). Are the thermodynamic parameters correctly estimated in liquid-phase adsorption phenomena? *Journal of Molecular Liquids*, 218, 174–185.
Asnin, L. D., Fedorov, A. A., & Chekryshkin, Yu S. (2000). Thermodynamic quantities of adsorption described by Freundlich isotherm. *Russian Chemical Bulletin*, 49(1), 178–180.
Atkins, P. W. (1999). *Físico-química* (Vol. 3). 6ª. ed. LTC-Livros Técnicos e Científicos, Rio de Janeiro.
Bai, C., Wang, L., & Zhu, Z. (2020). Adsorption of Cr (III) and Pb (II) by graphene oxide/alginate hydrogel membrane: Characterization, adsorption kinetics, isotherm and thermodynamics studies. *International Journal of Biological Macromolecules*, 147, 898–910.
Bhardwaj, D., Sharma, M., Sharma, P., & Tomar, R. (2012). Synthesis and surfactant modification of clinoptilolite and montmorillonite for the removal of nitrate and preparation of slow release nitrogen fertilizer. *Journal of Hazardous Materials*, 227, 292–300.
Castellini, E., Lusvardi, G., Malavasi, G., & Menabue, L. (2005). Thermodynamic aspects of the adsorption of hexametaphosphate on kaolinite. *Journal of Colloid and Interface Science*, 292(2), 322–329.
Cestari, A. R., Vieira, E. F., Vieira, G. S., Da Costa, L. P., Tavares, A. M., Loh, W., Airoldi, C. (2009). The removal of reactive dyes from aqueous solutions using chemically modified mesoporous silica in the presence of anionic surfactant—The temperature dependence and a thermodynamic multivariate analysis. *Journal of Hazardous Materials*, 161(1), 307–316.
Chakraborty, S., Mukherjee, A., Das, S., Maddela, N. R., Iram, S, & Das, P. (2021). Study on isotherm, kinetics, and thermodynamics of adsorption of crystal violet dye by calcium oxide modified fly ash. *Environmental Engineering Research*, 26(1), 190372.
Chang, R., & Thoman Jr. J. W. (2014). *Physical Chemistry for the Chemical Sciences*. University Science Books. MIT Press, Cambridge, Massachusetts.
Cheu, S. C., Kong, H., Song, S. T., Saman, N., et al. (2016). High removal performance of dissolved oil from aqueous solution by sorption using fatty acid esterified pineapple leaves as novel sorbents. *RSC Advances*, 6(17), 13710–13722.
rković, L., Ašperger, D., Babić, S., & Župan, J. (2019). Adsorption of enrofloxacin onto natural zeolite: Kinetics, thermodynamics, isotherms and error analysis. *Indian Journal of Chemical Technology (IJCT)*, 25(6), 565–571.
Dąbrowski, A. (2001). Adsorption—from theory to practice. *Advances in Colloid and Interface Science*, 93(1–3), 135–224.
DosSantos, C. C., Cestari, A. R., Lopes, E. C. N., Barreto, L. S., Lazaro, G. S., & Almeida, L. E. (2018). Chitosan-edible oil based materials as upgraded adsorbents for textile dye. *Carbohydrate Polymers*, 180, 182–191.
Duarte, I. J. M., Soares, A. J. R., Carvalho, J. R., & Buarque, H. L. D. B. (2019). Avaliação do uso de argilas montmorilonitas modificadas na adsorção de nitrato em meio aquoso. *Engenharia Sanitaria Ambiental*, 24(1), 21–31.
Ebelegi, A. N., Ayawei, N., & Wankasi, D. (2020). Interpretation of adsorption thermodynamics and kinetics. *Open Journal of Physical Chemistry*, 10(3), 166–182.
França, A. M. M. (2024). Development of hierarchical structure Zeolite A as a versatile adsorbent for the removal of Cu^{2+}, Cd^{2+}, and Pb^{2+} ions in aqueous medium (Doctoral thesis, Universidade Federal do Ceará, 209 pages). (Original work written in Portuguese).
Freundlich, H. M. F. (1906). Over the adsorption in solution. *The Journal of Physical Chemitry*, 57(1), 385–470.

Gong, R., Zhu, S., Zhang, D., Chen, J., Ni, S., & Guan, R. (2008). Adsorption behavior of cationic dyes on citric acid esterifying wheat straw: Kinetic and thermodynamic profile. *Desalination*, 230(1–3), 220–228.

Hamdaoui, O., & Naffrechoux, E. (2007). Modeling of adsorption isotherms of phenol and chlorophenols onto granular activated carbon: Part II. Models with more than two parameters. *Journal of Hazardous Materials*, 147(1–2), 401–411.

Hameed, B. H., Ahmad, A. A., & Aziz, N. (2007). Isotherms, kinetics and thermodynamics of acid dye adsorption on activated palm ash. *Chemical Engineering Journal*, 133(1), 195–203.

Hong, S., Wen, C., He, J., Gan, F., & Ho, Y. S. (2009). Adsorption thermodynamics of methylene blue onto bentonite. *Journal of Hazardous Materials*, 167(1–3), 630–633.

He, Y., Yin, L., Yuan, N., & Zhang, G. (2024). Adsorption and activation, active site and reaction pathway of photocatalytic CO_2 reduction: A review. *Chemical Engineering Journal*, 481(9), 48754.

Iheanacho, O. C., Nwabanne, J. T., Obi, C. C., Igwegbe, C. A., Onu, C. E., & Dahlan, I. (2023). Adsorptive dephenolization of aqueous solutions using thermally modified corn cob: mechanisms, point of zero charge, and isosteric heat studies. *Adsorption Science & Technology*, 2023, 1–14.

Khamis, F., Hegab, H. M., Banat, F., Arafat, H. A., et al. (2023). Comprehensive review on pH and temperature-responsive polymeric adsorbents: Mechanisms, equilibrium, kinetics, and thermodynamics of adsorption processes for heavy metals and organic dyes. *Chemosphere* 349, 140801.

Khattri, S. D., & Singh, M. K. (1999). Colour removal from dye wastewater using sugar cane dust as an adsorbent. *Adsorption Science & Technology*, 17(4), 269–282.

Langmuir, I. (1918). The adsorption of gases on plane surfaces of glass, mica and platinum. *Journal of the American Chemical Society*, 40(9), 1361–1403.

Levine, I. N. (2009). *Physical Chemistry*, 6th ed. Mc Graw Hill, New York.

Lima, A. C. A., Nascimento, R. F., Sousa, F. F., Filho, J. M., & Oliveira, A. C. (2012). Modified coconut shell fibers: A green and economical sorbent for the removal of anions from aqueous solutions. *Chemical Engineering Journal*, 185–186, 274–284.

Lima, E. C., Hosseini-Bandegharaei, A., Moreno-Piraján, J. C., & Anastopoulos, I. (2019). A critical review of the estimation of the thermodynamic parameters on adsorption equilibria. Wrong use of equilibrium constant in the Van't Hoof equation for calculation of thermodynamic parameters of adsorption. *Journal of Molecular Liquids*, 273, 425–434.

Liu, Y. (2009). Is the free energy change of adsorption correctly calculated? *Journal of Chemical & Engineering Data*, 54(7), 1981–1985.

Mcbain, J. W. (1932). *The Sorption of Gases and Vapours by Solids*. George Routledge and Sons, Londres.

McquarriE, D. A., & Simon, J. D. (1997). *Physical Chemistry: A Molecular Approach*. University Science Book, Califórnia.

Mo, Z., Zhang, H., Shahab, A., Chen, J., et al. (2023). Functionalized metal-organic framework UIO-66 nanocomposites with ultra-high stability for efficient adsorption of heavy metals: Kinetics, thermodynamics, and isothermal adsorption. *Journal of the Taiwan Institute of Chemical Engineers*, 146, 104778.

sin, G., Kiliç, M., Apaydin-Varol, E., & Pütün, A. E. (2019). Chemically activated carbon production from agricultural waste of chickpea and its application for heavy metal adsorption: Equilibrium, kinetic, and thermodynamic studies. *Applied Water Science*, 9(3), 56.

Patil, P., Jeppu, G., Vallabha, M. S., et al. (2024). Enhanced adsorption of phenolic compounds using biomass-derived high surface area activated carbon: Isotherms, kinetics and thermodynamics. *Environmental Science and Pollution Research*, 31, 67442–67460.

Ramesh, A., Lee, D. J., & Wong, J. W. C. (2005). Thermodynamic parameters for adsorption equilibrium of heavy metals and dyes from wastewater with low-cost adsorbents. *Journal of Colloid and Interface Science*, 291(2), 588–592.

Rassaei, F. (2023). Adsorption kinetics and isotherm modeling of lead in calcareous soils: Insights into thermodynamics, desorption, and soil properties. *Communications in Soil Science Plant Analysis*, 54(15), 2059–2076.

Romão, A. L. E, Souza Borba, R. N., Martins Carrilho, E. N. V., Nascimento, R. F., & Alves, C. R. (2025). Green coconut (*Coco nucifera*) biowaste: Its conversion into a sustainable magnetic nanomodified biosorbent for dye removal. *Biomass Conversion and Biorefinery*.

Santos, J. A. P., Rocha Pires, N., Loiola, A. R., & Nascimento, R. F. (2025). New insights into the removal of indigo blue dye from aqueous effluent using a low-cost adsorbent. *Colloids and Surfaces A: Physicochemical and Engineering Aspects*, 711, 136338.

Saucier, C., Karthickeyan, P., Ranjithkumar, V., Lima, E. C., Dos Reis, G. S., & De Brum, I. A. (2017). Efficient removal of amoxicillin and paracetamol from aqueous solutions using magnetic activated carbon. *Environmental Science and Pollution Research*, 24(6), 5918–5932.

Shaker, M. A. (2014). Dynamics and thermodynamics of toxic metals adsorption onto soil-extracted humic acid. *Chemosphere*, 111, 587–595.

Silva, F. N. (2018). *Investigação da Seletividade e da Termodinâmica de Adsorção de Cr(VI) de Águas Residuárias por Nanoadsorventes Magnéticos do Tipo* $CoFe_2O_4$*@(-*Fe_2O^3*. Dissertação (Mestrado em Ciência de Materiais)*. Universidade de Brasília, Planaltina.

Singh, A., Sonal, S., Kumar, R., & Mishra, B. K. (2020). Adsorption of chlorhexidine digluconate on acid-modified fly ash: Kinetics, isotherms and influencing factors. *Environmental Engineering Research,* 25(2), 205–211.

SousaNeto, V. O., Oliveira, A. G., Teixeira, R. N., Silva, M. A., Freire, P. C., De Keukeleire, D., & Nascimento, R. F. (2011). Use of coconut bagasse as alternative adsorbent for separation of copper (II) ions from aqueous solutions: Isotherms, kinetics, and thermodynamic studies. *BioResources*, 6(3), 3376–3395.

Sriramoju, S. K., Dash, P. S., & Majumdar, S. (2021). Meso-porous activated carbon from lignite waste and its application in methylene blue adsorption and coke plant effluent treatment. *Journal of Environmental Chemical Engineering*, 9(1), 104784.

Srivastava, V. C., Mall, I. D., & Mishra, I. M. (2007). Adsorption thermodynamics and isosteric heat of adsorption of toxic metal ions onto bagasse fly ash (BFA) and rice husk ash (RHA). *Chemical Engineering Journal*, 132(1–3), 267–278.

Ticianelli, E. A., Gonzalez, E. R. (1998). *Eletroquímica.* Ed. Da Universidade de São Paulo, São Paulo.

Tran, H. N., You, S. J., Chao, H. P. (2016). Thermodynamic parameters of cadmium adsorption onto orange peel calculated from various methods: A comparison study. *Journal of Environmental Chemical Engineering*, 4(3), 2671–2682.

Tran, H. N. You, S. J., Hosseini-Bandegharaei, A., & Chao, H. P. (2017). Mistakes and inconsistencies regarding adsorption of contaminants from aqueous solutions: A critical review. *Water Research*, 120, 88–116.

VasconcelosDa Silva, L. T., Freitas, A. D. L., Martins, T. G., França, A. M. M., Loiola, A. R., Nascimento, R. F. (2023). Comparative study on the performance of two synthetic zeolite 4A on the removal of heavy metals from aqueous solution—Effect of coal fly ash as Al and Si source. *Desalination and Water Treatment*, 310, 157–166.

Wang, W., Wang, Z., Li, K., Liu, Y., Xie, D., Shan, S., Mei, Y. (2021). Enhanced adsorption of aqueous chlorinated aromatic compounds by nitrogen auto-doped biochar produced through pyrolysis of rubber-seed shell. *Environmental Technology,* 44(5), 631–646.

Wu, Y., Hu, E., Dai, W., Li, Z., et al. (2017). A facile sacrificial template method to synthesize one-dimensional porous CdO/$CdFe_2O_4$ hybrid nanoneedles with superior adsorption performance. *RSC Advances*, 7(9), 5093–5100.

Wu, Z., Xie, J., Liu, H., Chen, T., et al. (2017). Preparation, characterization, and performance of 4A zeolite based on opal waste rock for removal of ammonium ion. *Adsorption Science Technology*, 36(9–10), 1700–1715.

Yusoff, N. A., Abdullah, L. C., Md. Jamil, S. N. A., Ting, T. M., & ZainuddiN, H. S. (2025). High-performance fixed-bed adsorption using active-site-controlled amine-functionalized adsorbent for naproxen removal: Insights into isotherms, kinetics, and regeneration. *Journal of Water Process Engineering*, 74, 107802.

Zhou, X., Liu, N., GAO, J., Zhang, M., & Cao, F. (2011). Adsorption thermodynamics of toxin (65 kDa) and protoxin (130 kDa) from *Bacillus thuringiensis* by several minerals. *International Journal of Chemical Reactor Engineering*, 9(1), 1–18.

Zhou, X., & Zhou, X. (2014). The unit problem in the thermodynamic calculation of adsorption using the Langmuir equation. *Chemical Engineering Communications*, 201(11), 1459–1467.

Zubieta, C. E., Messina, P. V., Luengo, C., Dennehy, M. Pieroni, O., & Schulz, P. C. (2008). Reactive dyes remotion by porous TiO_2-chitosan materials. *Journal of Hazardous Materials*, 152(2), 765–777.

5 Common Errors in Fitting Theoretical Models

Ari Clecius Alves de Lima,
Antonia Mayza de Morais França,
Luiz Thiago Vasconcelos da Silva,
Carla Bastos Vidal, Iara Jennifer Moura Duarte,
Vicente de Oliveira Sousa Neto, and
Ronaldo Ferreira do Nascimento

The most common error in analyzing data from adsorption studies is misusing the linearization technique when applied to adsorption models, such as isotherms and kinetics. The linearization method has been used for a few decades, and its introduction occurred at a time when the availability of both computers and statistical software was limited (Carlos P. Bergmann, 2015; González-López et al., 2022). Currently, the use of resources for linearization of isothermal and kinetics equations for experimental data analysis continues to be in full use, due to the ease of obtaining parameters of the models studied (Bolster & Hornberger, 2008; González-López et al., 2022). One of the reasons for the use of linear equations is the possibility of performing a smaller number of experimental points to define a straight line or disregarding some points to increase the values of R^2, which is considered the biggest problem of using nonlinear equations, as they require a greater number of experimental points to define a curve. The main problem is that some points are discarded, without any previous statistical analysis, aiming only at obtaining more satisfactory R^2 values, not considering, also, the values of the parameters inherent to the equilibrium and adsorption kinetics models (Carlos P. Bergmann, 2015; Tran et al., 2017).

ADJUSTMENTS TO ADSORPTION ISOTHERM MODELS

Several models of nonlinear isotherms have been proposed to describe the relationship between the amount of adsorbate adsorbed on the solid surface and the amount in remaining solution, among which can be mentioned the models of Langmuir, Freundlich, Sips, Toth, Redlich–Peterson, Dubinin–Radushkevich, among others (Anako Opotu et al., 2022; Wang et al., 2024). While adjusting the models to the experimental data, errors can occur, mainly due to linearizing equations that are not originally linear. This is due to the implicit assumptions of nonlinear isothermal models, which indicate that dependent variables, for example, Q_e, are not linearly dependent on the independent variable, for example, C_e. However, linear regression and the exclusive use of the coefficient of determination (R^2) continue to be the most

DOI: 10.1201/9781003654391-5

used method for determining best fit isotherms in adsorption experiments (Shikuku et al., 2018a; Wang et al., 2024).

In the field of adsorption research, some errors in data modeling analysis constantly appear, and in this sense, several authors address the importance of the correct interpretation of experimental data (Bolster & Hornberger, 2008; Shikuku et al., 2018a; Tran et al., 2017). Considering the application of isothermal models, it is known that although linearized and nonlinearized equations are mathematically equivalent, several limitations and errors are embedded in using linearized equations (Bolster & Hornberger, 2008; López-Luna et al., 2019).

A key limitation of linear models is that transforming experimental data into a linear form may result in modifications to the error functions, introducing errors in the independent variable, and altering the weight assigned to each experimental point. Consequently, this can lead to disparities in the values of the calculated parameters when comparing linear and nonlinear models (Bolster & Hornberger, 2008; Tran et al., 2017). These disparities may be attributed to errors arising from the computational process (Subramanyam & Das, 2014). Moreover, it is important to acknowledge that the units of measurement are subject to alteration when the parameters are modified, which frequently goes unconsidered by numerous authors (Anako Opotu et al., 2022; Carlos P. Bergmann, 2015; González-López et al., 2022).

The linear regression method approximates the dispersion of points around the line, following a Gaussian distribution and standard deviation at each value of C_e (González-López et al., 2022; Nagy et al., 2017). However, this behavior is impossible with isothermal equilibrium models. The nonlinear regression method avoids these errors, making it the most suitable technique for estimating the parameters of the isotherm model studied (Nagy et al., 2017; Subramanyam & Das, 2014).

The linearization of nonlinear isothermal models implies, in several aspects, including (Bolster & Hornberger, 2008; Guo & Wang, 2019; Shikuku et al., 2018a; Tran et al., 2017):

- Nonlinear models have a theoretical basis, and the parameters of the models provide greater knowledge about the phenomenon under study than linear models;
- Nonlinear models generally provide a good fit, with fewer parameters than linear models;
- The transformation of a nonlinear model into a linear model, if on the one hand facilitates the adjustment process, implies making unrealistic assumptions about the errors (normal distribution with constant variance), losing information about the standard errors of the original parameters;
- There are intrinsically nonlinear models; they cannot be linearized by transformation.

Linearization of the Langmuir Equation

The Langmuir isotherm model (Langmuir, 1919), Equation 5.1, is widely used in adsorption studies to describe adsorptive phenomena. Four different linearized versions of the Langmuir model can be used to obtain the parameters directly from the model by solving the normal equations, that is, by linear regression (Guo & Wang, 2019; Subramanyam & Das, 2014; Wang et al., 2024). Nonetheless, it is important to acknowledge that the linearization process can introduce discrepancies in the values of the parameters, given that the original equation is inherently nonlinear (Bolster & Hornberger, 2008; Guo & Wang, 2019). The following methods have been proposed as a means of deriving linear equations from Langmuir's nonlinear equation (Equations 5.2–5.13):

$$q_e = \frac{q_{máx} K_L C_e}{1 + K_L C_e} \tag{5.1}$$

Type I Linear Equation: Inversion of Langmuir's nonlinear equation, rearrangement of the equation, and cancellation of equal constants present in the numerator and denominator of the fraction (Equations 5.2 and 5.3):

$$\frac{1}{q_e} = \frac{1 + K_L C_e}{q_{máx} K_L C_e} \tag{5.2}$$

$$\frac{1}{q_e} = \frac{1}{C_e}\left(\frac{1}{q_{máx} K_L}\right) + \frac{1}{q_{máx}} \tag{5.3}$$

Type II Linear Equation: Multiplication on both sides of Equation 5.1 by C_e and cancellation of constants present in the numerator and denominator of the fraction (Equations 5.4 and 5.5):

$$\frac{C_e}{q_e} = \frac{C_e}{q_{max} K_L C_e} + \frac{C_e}{q_{max}} \tag{5.4}$$

$$\frac{C_e}{q_e} = \frac{1}{q_{max} K_L} + C_e\left(\frac{1}{q_{max}}\right) \tag{5.5}$$

Type III Linear Equation: Rearrange the nonlinear equation to isolate q_e (Equations 5.6–5.11)

$$q_e (1 + K_L C_e) = q_{max} K_L C_e \tag{5.6}$$

$$q_e + q_e K_L C_e = q_{max} K_L C_e \tag{5.7}$$

$$q_e = C_e (q_{max} K_L - q_e K_L) \tag{5.8}$$

$$\frac{q_e}{C_e} = K_L (q_{max} - q_e) \tag{5.9}$$

$$q_{máx} - q_e = \frac{q_e}{C_e K_L} \tag{5.10}$$

TABLE 5.1
Equations of Langmuir's Linear Models

Type	Equation	Plot	Parameters
I	$\frac{1}{q_e}=\frac{1}{K_L q_{max} C_e}+\frac{1}{q_{max}}$	$1/q_e$ versus $1/C_e$	$q_{max}=1/LC$ $K_L=LC/AC$
II	$\frac{C_e}{q_e}=\frac{1}{q_{max}}C_e+\frac{1}{q_{max}K_L}$	C_e/q_e versus C_e	$q_{max}=1/AC$ $K_L=AC/LC$
III	$q_e=-\frac{q_e}{K_L C_e}+q_{max}$	q_e versus q_e/C_e	$q_{max}=LC$ $K_L=1/-AC$
IV	$\frac{q_e}{C_e}=-K_L q_e+K_L q_{max}$	q_e/C_e versus q_e	$q_{max}=-LC/AC$ $K_L=-AC$

Source: Adapted with permission from Shikuku et al. (2018a).

$$q_e = q_{max}-\left(\frac{1}{K_L}\right)\frac{q_e}{C_e} \tag{5.11}$$

Linear Equation of Type IV: Isolate q_e/C_e in Equation 5.8 (Equation 5.12):

$$\frac{q_e}{C_e} = q_{max}K_L - q_e K_L \tag{5.12}$$

Table 5.1 presents the four linearized Langmuir equations, followed by the horizontal (abscissas) and vertical (ordinate) coordinates for plotting the graphs. The parameters (q_{max} and K_L) are calculated by applying each linearized form of the Langmuir equation using the angular coefficients (AC) and linear coefficients (LC) (Santos et al., 2022).

Effects of Linearization of the Langmuir Equation

The first effect is a discrepancy between the values of the constants obtained through the linear and nonlinear equations. Furthermore, the different linear versions of the Langmuir model (Table 5.1) frequently result in variations between the values of the parameters that are adjusted to each other (Guo & Wang, 2019; Karri et al., 2017; Hamzaoui et al., 2018; Shikuku et al., 2018b).

This divergence was confirmed in a study of the variation of the parameters obtained from linear and nonlinear isotherms in the adsorption of sulfadimethoxine on kaolinite-type clay, which resulted in different q_{max} and K_L values between the linear Langmuir models (Shikuku et al., 2018b). The maximum adsorption capacity (q_{max}) ranged from 0.693 to 4.117 mg g^{-1}, and Langmuir's constant (K_L) values ranged from 0.195 to 2.827 L g^{-1}. The values of q_{max} and K_L for the Langmuir nonlinear model were 4.590 mg g^{-1} and 0.171 L g^{-1}, respectively.

TABLE 5.2
Empirical Data of Pb^{2+} Adsorption in Zeolite A Magnetic Composite

C_e (mmol L^{-1})	q_e (mmol g^{-1})	$1/C_e$ (L $mmol^{-1}$)	$1/q_e$ (g $mmol^{-1}$)	C_e/q_e (g L^{-1})	q_e/C_e (L g^{-1})
0.0	0.0	—	—	—	—
0.03	0.08	12.2	35.1	0.347	2.88
0.06	0.16	6.32	16.2	0.390	2.56
0.17	0.32	3.14	5.92	0.529	1.89
0.26	0.47	2.14	3.81	0.562	1.78
0.45	0.53	1.88	2.19	0.857	1.17
0.88	0.64	1.55	1.15	1.35	0.740
1.19	0.71	1.40	0.837	1.67	0.598
1.34	0.73	1.36	0.744	1.83	0.547
1.95	0.81	1.23	0.512	2.40	0.416
2.12	0.85	1.18	0.472	2.49	0.401

Source: Adapted from (Santos et al., 2022). Reproduced with permission from Santos et al. (2022).

In other recent studies, adjustments of linear and nonlinear Langmuir isotherms of empirical data on the adsorption of Pb^{2+}, Cd^{2+}, Zn^{2+}, and Cu^{2+} on the surface of a zeolite A magnetic composite were compared (Santos et al., 2022). This study highlighted the importance of using nonlinear isotherms in determining constants from the equations. Considering the experimental data presented in Table 5.2 of this study for Pb^{2+} adsorption, we can evaluate this effect (Santos et al., 2022). As illustrated in Figure 5.1, the four types of models are presented by the plotting instructions outlined in Table 5.1. Figure 5.1 also includes the equations of the lines, from which the q_{max} and K_L were calculated (Santos et al., 2022).

Therefore, when the acquisition of new points is evaluated through the linearization of the data, a change in the weight of each point is observed. This, in turn, results in the derivation of different values of the parameters when linear and nonlinear equations are compared (Bolster & Hornberger, 2008; González-López et al., 2022). Tables 5.3 and 5.4 show the discrepancy between the values of the parameters obtained from the linear and nonlinear models for Pb^{2+} adsorption (Santos et al., 2022). In Table 5.3, the point (0.0) was utilized to denote the initial stage of the adsorption process. Consequently, the linear forms of Langmuir's types III and IV were not considered, as the slopes of the lines are negative. This results in profiles that are contrary to those of forms I and II. The presence of (0.0) at this stage could be considered abusive, as this point would be significantly distant from the others (outlier) (Santos et al., 2022).

In this sense, as shown in Table 5.3, the type II linear model presented a q_{max} (1.118 mmol g^{-1}) closer to the value of this same parameter in the nonlinear model (0.922 mmol g^{-1}). However, considering the type II linear model, the K_L obtained presented a negative value. Nevertheless, it is observed that the nonlinear model presented R^2 closer to the unit and lower values of the sum of the square of errors (SSE), evidencing better quality in the fit and greater reliability in the representation of experimental data.

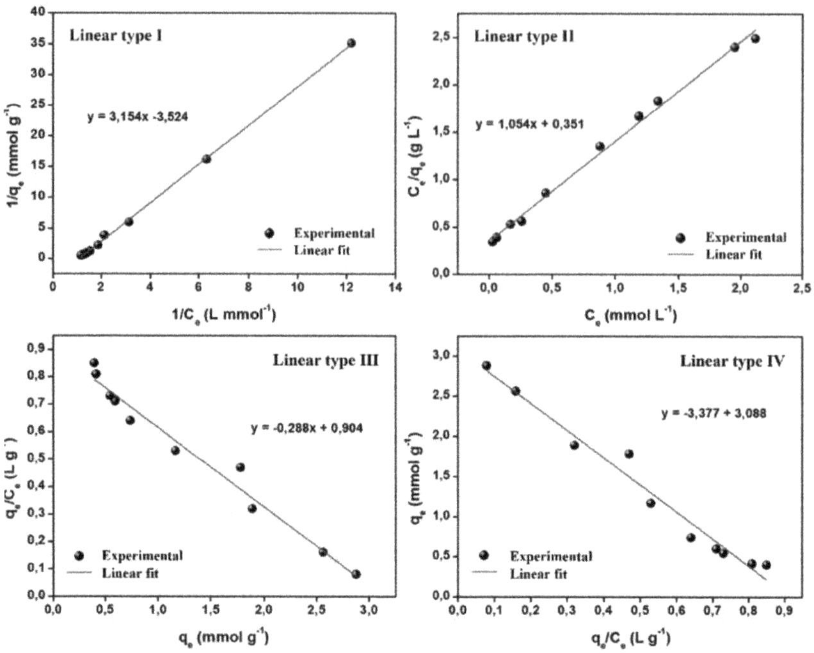

FIGURE 5.1 Adjustment of the experimental data to the linearized forms of the Langmuir equation, excluding the point (0.0). Conditions: pH: 5.0; Temperature: 25°C ± 2°C; Adsorbent mass: 50 mg; Stirring speed: 250 rpm. (Reproduced with permission from Santos et al., 2022.)

TABLE 5.3
Comparison between Linear and Nonlinear Langmuir Models Fitted to the Adsorption Isotherm Data for the Metal Cations Studied (Including the (0,0) Point)

		Parameters			
Metal	**Models**	**$q_{máx}$ (mmol g^{-1})**	**K_L (L mmol^{-1})**	**R^2**	**SSE**
Pb^{2+}	Nonlinear	0.922	3.237	0.9906	0.0077
	Linear Type I	−0.339	−0.962	0.9891	11.21
	Linear Type II	1.118	−3.694	0.9807	0.1059

Source: Adapted from (Santos et al., 2022). Reproduced with permission from Santos et al. (2022).

It has been observed that another effect and aspect of the limitations of Langmuir's linearized models is the presence or absence of the point (0.0), which can increase the SSE values and reduce the R^2 value (Santos et al., 2022). As demonstrated in Table 5.4, an analysis of the results obtained for the nonlinear model reveals that the exclusion of the point (0.0) does not result in any alteration to the values of q_{max} and K_L when compared to the fit of the nonlinear model to the experimental data using

TABLE 5.4
Comparison between Linear and Nonlinear Langmuir Models Fitted to the Adsorption Isotherm Data for the Metal Cations Studied (Excluding the (0,0) Point)

	Nonlinear	0.928	3.173	0.9888	0.0065
	Linear Type I	−0.282	−1.117	0.9991	0.813
	Linear Type II	0.948	2.999	0.9945	0.0297
	Linear Type III	0.904	−3.472	0.9707	0.017
Pb^{2+}	Linear Type IV	3.139	−0.288	0.9707	0.017

Source: Adapted from (Santos et al., 2022). Reproduced with permission from Santos et al. (2022).

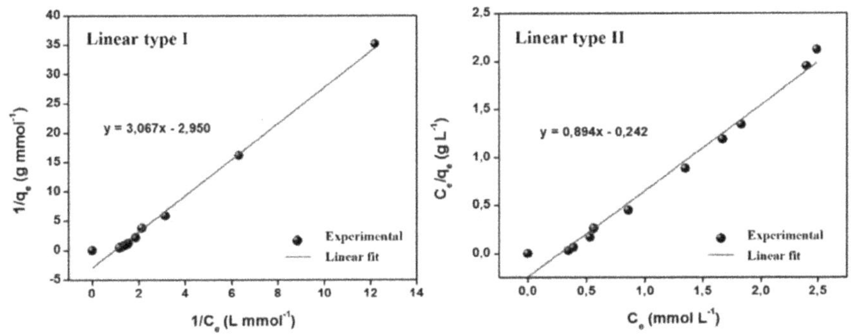

FIGURE 5.2 Adjustment of the experimental data to the linearized forms of the Langmuir equation, including the point (0,0). Conditions: pH: 5.0; Temperature: 25°C ± 2°C; Adsorbent mass: 50 mg; Stirring speed: 250 rpm. (Reproduced with permission from Santos et al., 2022.)

the point (0.0) (Table 5.3). However, a minor decrease in the value of the R^2 was observed in the absence of the point (0.0) (Santos et al., 2022). In the case of type I and II linear models, the effect differed, as the absence of the point (0.0) increased the value of R^2 and decreased the value of SSE (Santos et al., 2022). Figure 5.2 shows this effect for the type I and II models, showing that the point (0,0) affects the linear fit when comparing the fit without this point (Figure 5.1). This suggests that linear models may disregard this critical point to enhance the fit.

Certain studies have demonstrated that the point (0.0) adjusts the Langmuir model's linear model of type II. Nevertheless, these studies fail to provide evidence for the data obtained after adjustment, instead asserting solely the elevated value of R^2 (0.999) (Fang et al., 2018; López-Luna et al., 2019; Shalaby et al., 2018). Furthermore, certain studies have been observed to exhibit an absence of the point (0.0) in their graphical representations of the linearized forms of the Langmuir model, while simultaneously omitting a comparison with the nonlinear form (Qiu et al., 2018; Shalaby et al., 2018).

As a consequence of applying linear models, negative parameters such as q_{max} and K_L can be obtained, as seen in Tables 5.3 and 5.4 (Santos et al., 2022). It is evident from the regression equations presented in Figures 5.1 and 5.2 that negative values have been obtained for the angular or LC. This signifies that the regression equations are physically incoherent and devoid of rational meaning.

Certain studies have demonstrated that negative values of the constants q_{max} and K_L are presented when Langmuir's linearized models are applied. Therefore, these models do not represent any coherent physical phenomenon and are unacceptable (Gonen & Biçer, 2017; Hussain et al., 2003; Igwe & Abia, 2007). As demonstrated by Shikuku et al. (2018b), the adjustment of the linearized Langmuir model yielded negative values for the constants q_{max} and K_L in the context of the adsorption of sulfachloropyridazine (SCP) and sulfadimethoxine (SDM) in iron-phase modified crude clay. An investigation was conducted into the adsorption of clomazone in synthetic faujasite X (FAU-X) and natural zeolite from Kenya (NZ-K). Values obtained for q_{max} and K_L were less than zero after the type I linear model was adjusted to the adsorption data in FAU-X. This finding indicates that the linear model is inadequate in explaining the adsorption phenomenon in this case (Shikuku et al., 2018a). Shyam et al. (2013) discovered negative values for Langmuir's parameters (q_{max} and K_L) through their investigation of the adsorption of Cu^{2+}, Ni^{2+}, Zn^{2+}, Cd^{2+}, and Pb^{2+} using fly ash as an adsorbent. In this study, the authors did not provide evidence of the type of linear model used for the adjustment obtained, nor did they present the errors in this linearization process.

Consequently, during the linearization process, inaccuracies in parameter calculations may cause the variation in the values of maximum adsorption capacity (q_{max}) and Langmuir's constant (K_L). It can be concluded that transforming the nonlinear model to a linear one implicitly alters the error functions. Consequently, the limitations of this particular linearization render it unsuitable for this application (Bolster & Hornberger, 2008; González-López et al., 2022). Therefore, upon comparison of the two methods, it is evident that nonlinear regression is the most viable method for estimating the parameters of the isotherm model, as it is the original equation (Nagy et al., 2017).

In this sense, when plotted, Langmuir's linear models present characteristics such as grouping points close to the origin (Linear type I), as can be observed in Figures 5.1 and 5.2. In addition, the models are sensitive to variabilities at low values of q_e (high values of $1/q_e$). Conversely, these models are overestimated due to collinearity among the variables in the relationship q_e versus q_e/C_e. This results in spurious correlations, reversal of relative weights of the variables, and discrepancies in the error distributions between C_e and q_e and their values (Anako Opotu et al., 2022; Bolster & Hornberger, 2008; Tran et al., 2017).

In addition to applying linear models to experimental data, several studies do not discuss the errors inherent to fitting linear and nonlinear models, such as SSE. Furthermore, the selection of models is primarily informed by the value of R^2 rather than a comprehensive evaluation of model performance (Shikuku et al., 2018a; Tellinghuisen & Bolster, 2011). Silva et al. (2023), in their study of the adsorptive characteristics of zeolite 4A produced from steel residues for metal adsorption (Pb^{2+}, Cd^{2+}, Zn^{2+}, and Cu^{2+}), compared the fit of Langmuir and Freundlich nonlinear isotherm models (Figure 5.3). In addition to the R^2 value, they used the SSE values to confirm the

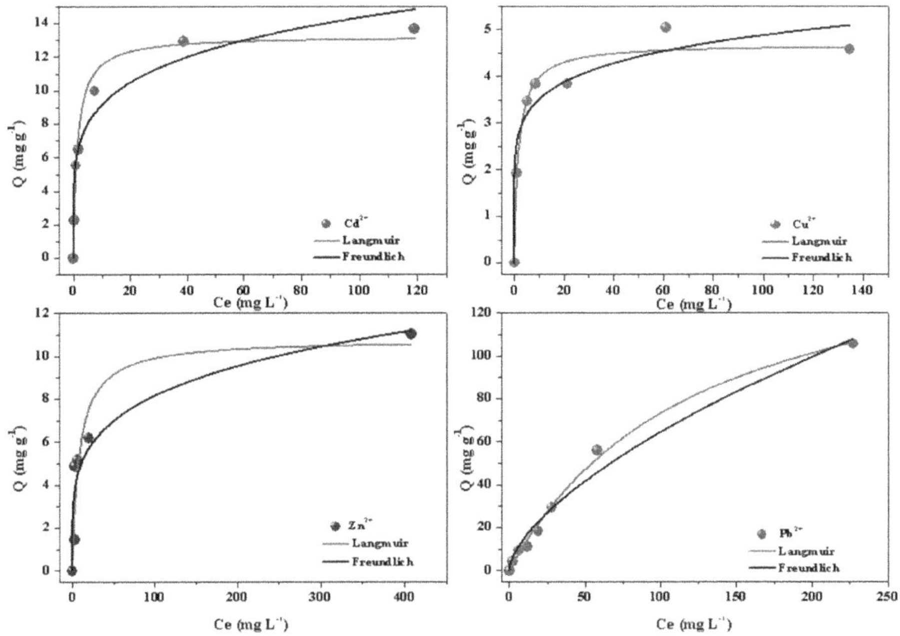

FIGURE 5.3 Multielement adsorption isotherm for zeolite 4A of metal ions (a) Cd^{2+}, (b) Cu^{2+}, (c) Zn^{2+}, and (d) Pb^{2+} (pH = 5.0; temperature, 28°C ± 2°C; stirring speed, 200 rpm; volume, 25 mL; dosage, 4 g L^{-1}). (Reproduced with permission from Silva et al., 2023.)

TABLE 5.5
Parameters of the Ion Adsorption Isotherm Models (Pb^{2+}, Cd^{2+}, Zn^{2+}, and Cu^{2+}) in the Zeolite 4A, Multielement System

Isotherms Models	Parameters	Pb^{2+}	Cd^{2+}	Zn^{2+}	Cu^{2+}
Langmuir	Q_{exp} (mg g^{-1})	105.51	13.687	11.04	4.578
	$Q_{máx}$ (mg g^{-1})	164.82	13.27	10.76	4.683
	K_L	0.0080	0.6851	0.1139	0.558
	R^2	0.9943	0.9685	0.8518	0.9668
	SSE	7.2491	0.8494	2.2343	0.0998
Freundlich	n	1.59	5.16	4.46	7.004
	K_f	3.55	5.87	2.904	2.53
	R^2	0.9772	0.9521	0.8710	0.9227
	SSE	28.87	1.2911	1.9446	0.2326

Source: Adapted from (Silva et al., 2023). Reproduced with permission from Silva et al. (2023).

best model. According to the results presented in Table 5.5, the Langmuir model was the best fit for Pb^{2+}, Cd^{2+}, and Cu^{2+}, and the Freundlich model was the best fit for Zn^{2+}.

The nonlinear regression method minimizes the error distribution between the experimental data and the predicted isotherm (González-López et al., 2022; Nebaghe et al., 2016). Nevertheless, several studies have been conducted that compare Langmuir's linear and nonlinear models using various error functions to determine which model provides a more suitable representation of the experimental adsorption data (Ayawei et al., 2017; Mallakpour & Tabesh, 2019; Nebaghe et al., 2016). Certain studies also posited that the chi-square should be utilized to select the most suitable nonlinear model. This is because the chi-square values close to zero indicate a greater proximity between the experimental values and those calculated by the model (Chakraborty et al., 2019; González-López et al., 2022).

Although the section has already covered the evidence on the errors embedded in the linearization of nonlinear equations, a study conducted by Mallakpour and Tabesh (2019) on the comparison of linear and nonlinear forms of the isotherm and kinetics models in methylene blue adsorption in tragacanth gum-based hydrogel nanocomposites found that Langmuir's linear type II form better represented the experimental adsorption data when compared to the nonlinear model. It is important to note that the results were based on the application of several statistical functions (R^2, SSE, X^2, RMSE, ARE, HYBRID, and MPSD), in which lower error values were obtained for the Langmuir type II linear model.

ADJUSTMENTS TO ADSORPTION KINETIC MODELS

Linearization of the Pseudo-First-Order Model

In the study of adsorption kinetics, it is common to use linearized forms of the equations to obtain the constants more easily. However, as with the isotherm models, this linearization process has an inherent error (Anako Opotu et al., 2022; Chakraborty et al., 2019; Ozcan et al., 2018). The most commonly used models in kinetic adsorption studies are the pseudo-first-order (PFO) and pseudo-second-order (PSO) models (Wang et al., 2024). In this sense, to exemplify the effects of linearization, the study by López-Luna et al. (2019) compared the PFO model in its nonlinear (Equation 5.13) and linear (Equation 5.14) forms in the study of arsenic adsorption with manganese ferrite nanoparticles. As demonstrated in Figure 5.4, applying the model to different concentrations of arsenic reveals that the nonlinear model provides a superior fit to the data. The data relating to the fit and the constants calculated with the two forms of the PFO model are presented in Table 5.6. It is imperative to exercise particular attention when employing the logarithm for linearization, given the occurrence of an overweight at points close to the limits of the logarithm (Anako Opotu et al., 2022; Guo & Wang, 2019; López-Luna et al., 2019). It is evident that the constant k_1 was negative, and the calculated values of q_e were significantly different from the theoretical q_e in the linear model. Additionally, the values of χ^2 were high. Conversely, in the nonlinear model, the values of k_1 were more realistic, and the calculated and theoretical q_e were close, as evidenced by the low χ^2 (Chakraborty et al., 2019; López-Luna et al., 2019).

$$q_t = \left(q_e - e^{-t k_1}\right) \tag{5.13}$$

$$\ln\left(q_e - q_t\right) = \ln q_e - k_1 t \tag{5.14}$$

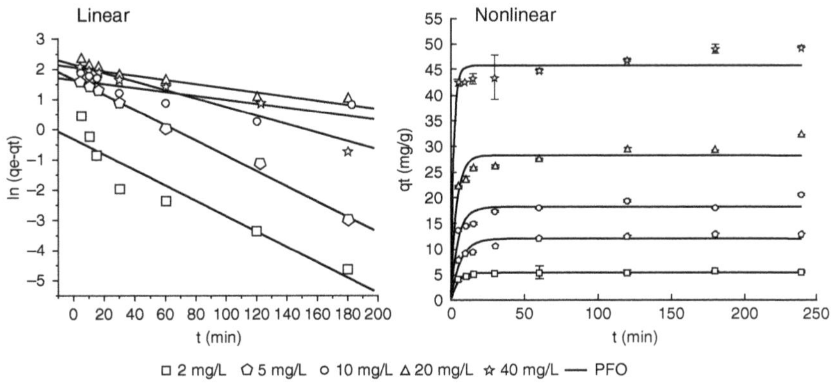

FIGURE 5.4 Linear and nonlinear kinetic PFO model for arsenic removal by manganese ferrite nanoparticles. (Adapted from López-Luna et al., 2019.)

TABLE 5.6
Kinetic Parameters of the PFO Model for Arsenic Removal by Manganese Ferrite Nanoparticles

Linear PFO Model

Arsenic	k_1 (s^{-1})	R^2	$q_{e,\,exp}$ (mg g^{-1})	$q_{e,\,cal}$ (mg/g)	χ^2
2 mg L^{-1}	−0.0253	0.89	5.39	0.72	30.20
5 mg L^{-1}	−0.0253	0.99	12.85	5.29	10.80
10 mg L^{-1}	−0.0066	0.55	20.54	5.26	44.34
20 mg L^{-1}	−0.0069	0.84	32.05	8.02	72.04
40 mg L^{-1}	−0.0140	0.92	49.53	8.72	190.91

Nonlinear PFO Model

Arsenic	k_1 (s^{-1})	MPSD	$q_{e,\,exp}$ (mg g^{-1})	$q_{e,\,cal}$ (mg g^{-1})	χ^2
2 mg L^{-1}	0.2350	0.0056	5.39	5.29	0.0019
5 mg L^{-1}	0.1596	0.0986	12.85	11.84	0.0694
10 mg L^{-1}	0.2073	0.0732	20.54	18.14	0.3175
20 mg L^{-1}	0.2507	0.0477	32.05	28.27	0.5054
40 mg L^{-1}	0.4952	0.0217	49.53	45.80	0.3038

Source: Adapted from López-Luna et al. (2019).

Interpretation of the Intraparticle Diffusion Models

Diffusion models are extensively employed in kinetic adsorption studies to identify reaction steps and mechanisms and predict the limiting step of the adsorption process (Tran et al., 2017; Wang et al., 2024). This is because models such as the pseudo-first- and PSO models are empirical and incapable of individually describing the steps of the adsorption process (Viegas et al., 2014).

As Mohseni-Bandpi et al. (2016) have demonstrated, a linear regression of the fit of the intraparticle diffusion model to the experimental data is generally expected to emerge. However, it is imperative to interpret the results differently, given that the points obtained generally form multiple linear segments. It is therefore suggested that these segments represent the stages of diffusion of the adsorbate species in the adsorbent particle and that they can be divided into three phases: (1) adsorption on the external surface, (2) intraparticle diffusion, and (3) adsorption at the sites inside the particle in equilibrium (El-Khaiary & Malash, 2011; Mohseni-Bandpi et al., 2016; Santos et al., 2022). It is asserted that the correct interpretation of the intraparticle mechanism can be achieved by adjusting this model to the experimental data (El-Khaiary & Malash, 2011; González-López et al., 2022; Tran et al., 2017).

Thus, the discussion of the kinetic models of intraparticle diffusion of Weber–Morris (1963) and Boyd et al. (1947) defined by Equation 5.15, based on multistage, is the most adopted since it is the most correct and accepted form in the current literature (El-Khaiary & Malash, 2011; González-López et al., 2022; Tran et al., 2017).

$$q_t = K_d t^{1/2} + C \tag{5.15}$$

The application and interpretation of the Weber–Morris models were correctly executed, with the multistage theory (Weber & Carrell Morris, 1963) serving as the basis for this process. As illustrated in Figure 5.5, the equilibrium time data for Pb^{2+} adsorption can be modeled using the Weber–Morris intraparticle diffusion model (Santos et al., 2022). The figure demonstrates the correct and incorrect ways of fitting the models.

As demonstrated in Figure 5.5, the plot of q_e versus $t^{1/2}$ is expressed in at least three linear stages. This finding indicates that the adsorption of the metal cations under investigation occurred through three distinct stages of diffusion. The initial linear region signifies the diffusion of metal cations to the adsorbent's exterior

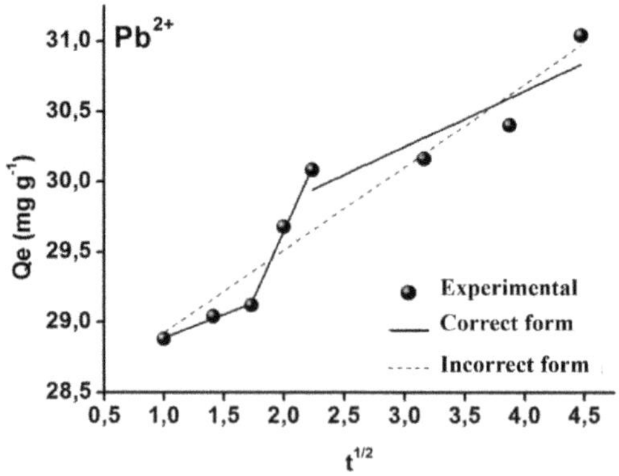

FIGURE 5.5 Adjustment of the Weber–Morris intraparticle diffusion model to the experimental adsorption data. Conditions: pH: 5.0; Temperature: 25°C ± 2°C, adsorbent mass: 50 mg, stirring speed: 250 rpm. (Reproduced with permission from Santos et al., 2022.)

TABLE 5.7
Parameters of the Weber–Morris Linear Kinetic Model of Intraparticle Diffusion for Pb^{2+} Cations

	Parameters		Statistical Functions	
Stages	**K_d (mg g^{-1} min$^{-1/2}$)**	**C (mg g^{-1})**	**R^2**	**SSE**
MultiStage (Correct Form)				
First	0.331	28.5	0.974	0.000388
Second	1.91	25.8	0.993	0.00164
Third	0.400	29.0	0.677	0.122
Single-Stage (Incorrect Form)				
Single	0.590	28.3	0.902	0.338

Source: Reproduced with permission from Santos et al. (2022).

surface, which occurs instantaneously. In the second region, the adsorption process is gradual, with a limiting velocity controlling diffusion within the particle. The third region is indicative of the final equilibrium phase, wherein intraparticle diffusion commences its decline due to the presence of a low concentration of species in solution (Kera et al., 2017; Ozcan et al., 2018; Santos et al., 2022; Viegas et al., 2014). It is also observed that the linear adjustment, disregarding the multistage, does not present a satisfactory value of R^2 and higher values of SSE (Table 5.7).

Several studies have addressed the interpretation of the Weber–Morris intraparticle diffusion model. For example, Abdel-Ghani et al., 2016; Chakraborty et al., 2019; Mallakpour & Tabesh, 2019 have correctly addressed this, while others have not evidenced the use of multistage to explain the process of intraparticle diffusion (Apiratikul & Pavasant, 2008; Visa, 2016). Some studies have misinterpreted this process (Hubicki & Wołowicz, 2009; Simonin & Bout, 2016; Ramachandran et al., 2011).

In the study by López-Luna et al. (2019), the adsorption of arsenic with manganese ferrite nanoparticles was investigated. The Weber–Morris intraparticle diffusion model, which considers multistage, was utilized in this analysis (Figure 5.6). The deviation from the straight lines of the origin can be attributed to the difference in mass transfer rate in the initial and final stages of adsorption. A rapid rate of surface diffusion of arsenic from solution to adsorbent surfaces characterized the initial adsorption stage. The rates of this process (k_{d1}) and the thickness of the solute boundary layers (C_1) increased with increasing nominal arsenic concentrations (Table 5.8). However, in the second stage, intraparticle or pore diffusion was the predominant course of the adsorption process, exhibiting slower rates (k_{d2}). The intraparticle diffusion stage was characterized by a significant increase in boundary layer thickness (C_2), whose values remained below those of the adsorption at equilibrium (q_e), resulting in the absence of a third equilibrium stage.

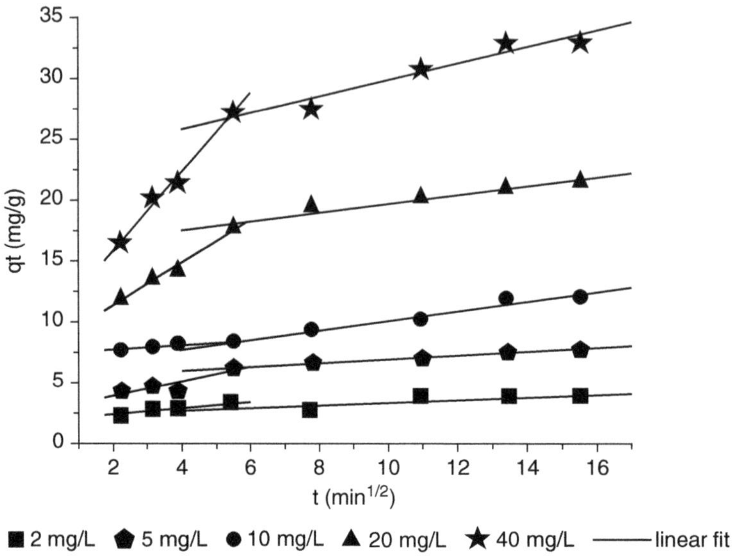

FIGURE 5.6 Intraparticle diffusion model of arsenic removal by manganese ferrite nanoparticles. (Adapted from López-Luna et al., 2019.)

TABLE 5.8
Intraparticle Diffusion Model of Arsenic Removal by Manganese Ferrite Nanoparticles

	Parameters					
[As (III)]	K_{d1} (mg g^{-1} min$^{-1/2}$)	C_1 (mg g^{-1})	R^2	K_{d2} (mg g^{-1} min$^{-1/2}$)	C_2 (mg g^{-1})	R^2
(mg L^{-1})		First Stage			Second Stage	
2	0.25	1.93	0.98	0.11	2.37	0.61
5	0.58	2.77	0.76	0.15	5.43	1.00
10	0.19	7.37	0.83	0.39	6.28	0.97
20	1.74	7.87	0.98	0.36	16.12	0.94
40	3.25	9.29	0.98	0.67	23.08	0.94

Source: Adapted from López-Luna et al. (2019).

ADJUSTMENTS TO FIXED-BED MODELS

Linearization of the Fixed-Bed Model

A significant challenge in adsorption research concerns the scale-up of fixed-bed batch operations in continuous flow. The study of the breakthrough curve is one of the primary experiments conducted to determine significant parameters associated

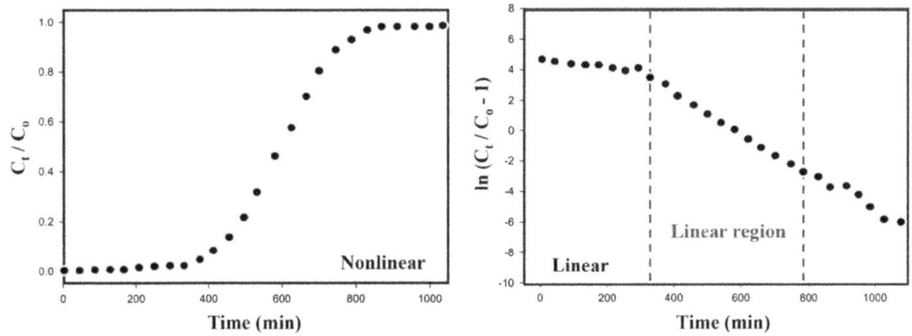

FIGURE 5.7 Nonlinear and linear curve fitting of a breakthrough curve of a fixed-bed model. (Adapted from González-López et al., 2012.)

with mass transport effects (González-López et al., 2012; Raulino et al., 2014; Wang et al., 2024). This curve delineates the temporal progression of the ratio between the adsorbate concentration at the column outlet (C) and its concentration at the inlet (C_0) (González-López et al., 2022). It is observed that this ratio is represented by the typical sigmoidal (S-shaped) curve, as demonstrated in Figure 5.7. Initially, the C/C_0 is zero, indicating total contaminant removal. As the bed becomes increasingly saturated, the concentration at the outlet begins to rise until it reaches the breaking point, which is typically defined as $C/C_0 = 0.05$, at which point the efficiency begins to decline. The region between the breaking point and the exhaustion point, near $C/C_0 = 1$, is designated the mass transfer zone, wherein the majority of the adsorption occurs (González-López et al., 2022; Wang et al., 2024). After exhaustion, the column will no longer retain any contaminants. Consequently, regeneration or replacement of the column is necessary (Raulino et al., 2014). A more thorough exposition of this system is provided in the Chapter 6. The curve's form is paramount in estimating system size and operational parameters.

The main models employed to describe the behavior of fixed-bed adsorption and derive the parameters of the breakthrough curve are the Thomas, Yoon–Nelson, and Bohart–Adams models (González-López et al., 2022). These models can be considered mathematically equivalent and linearized by applying the logarithmic function. The Thomas model is the most widely used of these and can be defined in its nonlinear and linear forms by Equations 5.16 and 5.17, respectively (González-López et al., 2022; Raulino et al., 2014). The model under consideration is predicated on the assumption that the flow under investigation is of a plug flow nature. Furthermore, it is postulated that a Langmuir isotherm describes the equilibrium and that the second-order reversible kinetics are applicable (González-López et al., 2012, 2022). The primary errors identified in linearization include the violation of the linear regression assumptions of normality and homoscedasticity, the distortion of the error distribution, particularly at the extremes of the curve, due to the over-weighting of extreme values by the logarithmic function, the production of satisfactory R^2 even when the model does not accurately represent the physics of the process, and the generation of false impressions in critical regions, such as the start/end of breakthrough curves (González-López et al., 2022; Wang et al., 2024). As demonstrated in Figure 5.5, the linear model cannot adjust extreme values by limiting itself to a range of the curve. This can result in spurious correlations and overestimated or underestimated values of the constants, with high experimental errors (Anako Opotu et al., 2022; Tran et al., 2017). Consequently, the choice of method directly impacts the interpretation of the fixed-bed capacity and adsorption rate.

Another common error that should be avoided is the use of an incorrect and simplified form of the linear Bohart–Adams model (Equation 5.18), in which the unit term is removed from the left hand side of the equation for use in the study of bed depth service time (BDST). This establishes a linear relationship between service time and bed depth (González-López et al., 2022). This spurious form significantly impacts the model's capacity to depict rupture curves accurately. The model can only capture the initial and exponential components of the curve, which limits its ability to provide a comprehensive representation.

$$\frac{C_t}{C_o} = \frac{1}{1 + \exp\left(\left(\frac{K_{Th}}{Q}\right)(q_{Th}m - C_oQt)\right)} \tag{5.16}$$

$$\ln\left(\frac{C_t}{C_o} - 1\right) = \frac{K_{Th}q_{Th}m}{Q} - K_{Th}C_ot \tag{5.17}$$

$$\ln\left(\frac{C_t}{C_o} - 1\right) = \frac{K_{BA}N_{BA}L}{V} - K_{BA}C_ot \tag{5.18}$$

Raulino et al. (2014) applied the Thomas model in its linear and nonlinear forms to evaluate a pilot-scale fixed-bed column with green coconut shells for Zn, Ni, and Cu adsorption at different flow rates (100, 200, and 300 mL min^{-1}). Figure 5.8 and Table 5.9 illustrate the results obtained by the researchers. It is evident that an increase in flow rate results in a greater discrepancy between the linear model's fit and the actual data. Conversely, the application of the nonlinear model leads to a reduction in error, as evidenced by the SSE. This is further substantiated by the observation that the adsorption capacity values (q_{Th}) of the nonlinear model are more closely aligned with the experimental values.

ADJUSTMENTS TO THERMODYNAMIC EQUATIONS

As demonstrated in Chapter 4, the field of thermodynamics provides a framework for understanding the energy changes associated with adsorption. This knowledge is crucial for elucidating the mechanism, irreversibility, and spontaneity of the process. The thermodynamic functions, ΔG^o, ΔS^o, and ΔH^o, are typically estimated through equilibrium experiments at varying temperatures. This approach

TABLE 5.9
Parameters by Linear (L) and NL Regression Analysis with Thomas Model for Metal Ions Cu (II), Ni (II), and Zn (II) for the Three Flow Rates

Metal	Flow Rate ($mL\ min^{-1}$)	K_t ($mL\ min^{-1}\ mg^{-1}$) L	K_t ($mL\ min^{-1}\ mg^{-1}$) NL	q_{Th} ($mmol\ g^{-1}$) Theoretical L	q_{Th} ($mmol\ g^{-1}$) Theoretical NL	q_{Th} ($mmol\ g^{-1}$) Experimental	SSE L	SSE NL
Cu^{2+}	100	0.102	0.113	0.242	0.242	0.233	0.061	1.36 E-29
	200	0.091	0.270	0.258	0.234	0.218	9.142	2.31 E-7
	300	0.292	0.299	0.191	0.180	0.172	0.428	1.69 E-29
Ni^{2+}	100	0.462	0.460	0.075	0.075	0.071	1.50 E-3	3.77 E-22
	200	1.008	1.141	0.096	0.091	0.093	0.357	4.44 E-31
	300	3.478	3.464	0.054	0.049	0.051	1.420	2.99 E-14
Zm^{2+}	100	0.412	0.410	0.075	0.075	0.071	3.35 E-4	1.17 E-28
	200	1.203	1.119	0.071	0.071	0.069	6.737	4.07 E-31
	300	2.716	1.991	0.057	0.050	0.050	1.059	6.84 E-13

Source: Reproduced with permission from Raulino et al. (2014).

necessitates a substantial experimental effort, which in turn restricts the availability of data (González-López et al., 2022; Ozcan et al., 2018; Shikuku et al., 2018a; Wang et al., 2024).

Various equations, including Gibbs, Van't Hoff, and Clausius–Clapeyron, can be used to estimate these parameters (Wang et al., 2024). The Van't Hoff equation (5.19) is the most widely used equation of its type, and its linear form (5.20) allows for the estimation of ΔH^o and ΔS^o from the slope and intercept of a ln K versus 1/T graph (González-López et al., 2022; Shikuku et al., 2018a).

However, it has been demonstrated that linearization generates larger deviations due to the non-normalized error distribution (El-Khaiary & Malash, 2011; González-López et al., 2022). However, the linear form is still favored due to its simplicity, particularly in cases with few experimental points, which hinders the convergence of the nonlinear form. Another recurring problem concerns the definition of the equilibrium constant K. It should be dimensionless, given its appearance within the logarithmic function. However, many works utilize constants with units (L/mol or L/mg), resulting in significant errors and incorrect ΔH^o and ΔS^o values (González-López et al., 2022).

$$K_e = e^{\frac{\Delta S^o}{R} - \left(\frac{\Delta H^o}{R}\right)\frac{1}{T}} \tag{5.19}$$

$$\ln K_e = \frac{\Delta S^o}{R} - \left(\frac{\Delta H^o}{R}\right)\frac{1}{T} \tag{5.20}$$

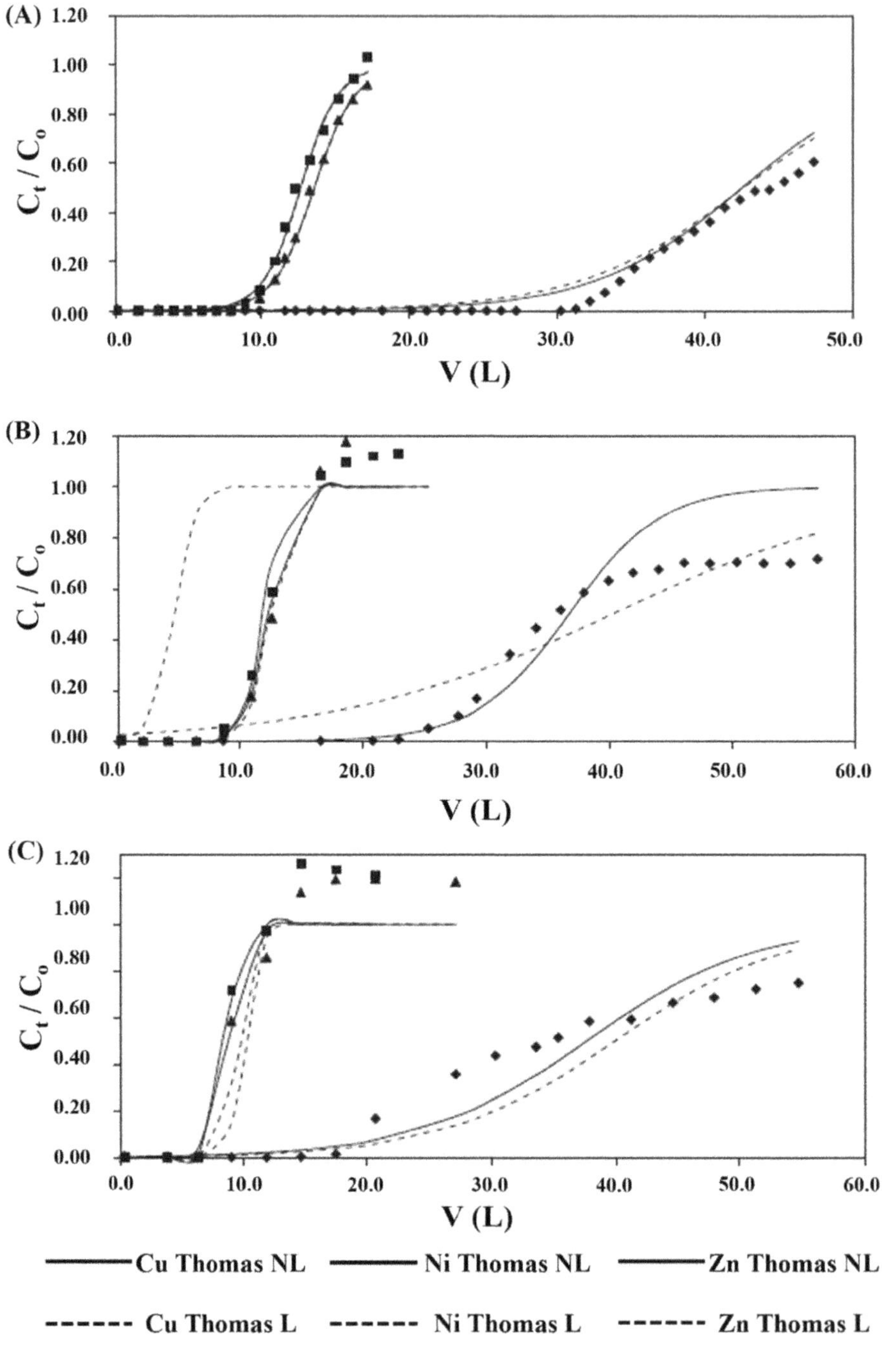

FIGURE 5.8 Comparison of experimental and theoretical breakthrough curves for metal ions at varied flow rates of (a) 100 mL min^{-1}, (b) 200 mL min^{-1}, and (c) 300 mL min^{-1} (NL, nonlinear; L, linear). (Reproduced with permission from Raulino et al., 2014.)

INCORRECT USE OF THE COEFFICIENT OF DETERMINATION (R^2)

A recurrent practice in adsorption studies entails the implementation of multiple theoretical models to the experimental data. The subsequent comparison of the parameters obtained from these fits facilitates the identification of the model that most accurately corresponds to the experimental data. The most frequently employed instrument to compare the fit and selection of the theoretical model is the coefficient of determination, R^2, as previously mentioned. This coefficient is the square of the correlation coefficient, R, between the dependent variable and the predicted value.

Tellinghuisen and Bolster (2011) state that, mathematically, R^2 is represented by Equation 5.21, in which SSE is the sum of the squared deviations of the points of the regression curve and SS_{tot} is the total sum of the squared deviations.

$$R^2 = 1 - \frac{SSE}{SS_{tot}} \tag{5.21}$$

Despite the apparent simplicity of interpretation of the R^2 value, its application is not consistently appropriate for evaluating the fit and comparison of models (Anako Opotu et al., 2022; González-López et al., 2022; Shikuku et al., 2018b; Tellinghuisen & Bolster, 2011). Consequently, El-Khaiary and Malash (2011) emphasized the most evident problems using R^2:

- The R^2 index is vulnerable to extreme data points, resulting in an illusory assessment of the quality of the fit. When data points are linearized, the effect becomes more intricate.
- The R^2 is affected by the interval of the independent variable (study range), increasing in value with an increase in the independent variable and decreasing with a decrease in the interval.
- The R^2 will be elevated when the model presents more parameters (lower degrees of freedom). In other words, an increase in R^2 is observed in cases where the degrees of freedom decrease due to error.

The initial two cases can be avoided by adapting the model to the experimental data without altering it (linearization of the model) and by conducting a thorough analysis of the extreme points. However, the third case is more intricate and will be addressed in greater detail in the following section.

The utilization of R^2 to evaluate the suitability of linear and nonlinear models is a prevalent practice. However, the optimal statistical criterion would be Marquardt's percent standard deviation (MPSD), which quantifies the percentage error between the experimental values and those predicted by the model, adjusted by the number of data points and defined by Equation 5.22, where n and p represent the number of experimental points and model parameters, respectively (Kumar et al., 2008). This statistical tool is regarded as more rigorous than other criteria, such as the root mean square error (RMSE), as it imposes a greater penalty on significant deviations in data exhibiting low variability. López-Luna et al. (2019)

utilized the MPSD to assess the suitability of the nonlinear PFO model for arsenic removal (Table 5.6). The analysis yielded favorable outcomes, as evidenced by the low and near-zero MPSD values, which indicated a robust fit between the model and the observed data.

$$\text{MPSD} = \sqrt{\frac{1}{n-p}\sum\left(\frac{q_{e,exp} - q_{e,cal}}{q_{e,exp}}\right)^2} \times 100 \tag{5.22}$$

ERROR FUNCTIONS USED IN ADSORPTION STUDIES

Comparison of Various Adsorption Isotherm Models

In adsorption equilibrium studies, the various isothermal models are typically adjusted to the experimental data, with the model exhibiting the optimal fit selected based on the highest R^2 value or its closest proximity to the unit value. From the selected model, assumptions concerning the phenomena and mechanisms involved in the studied adsorptive process are presented. These assumptions are related to the surface properties of the adsorbent, as well as adsorber–adsorbent and adsorber–adsorber interactions (Zhang et al., 2013).

However, when the adjusted isothermal models present values close to the R^2, statistical tools must be used to evaluate their fit. Otherwise, erroneous conclusions may be obtained (Núñez-Gómez et al., 2019). Among these tools, the chi-square test (χ^2), the F test, and the Student's t-test are particularly important (El-Khaiary & Malash, 2011; Hadi et al., 2012; Lima et al., 2012).

For illustrative purposes, the F test was applied to the experimental data of the linear type II and nonlinear isotherms (Table 5.3). Given the results, employing the R^2 values for the linear type II and nonlinear isotherms (0.9807 and 0.9906, respectively) and the F_{cal} value (3.70), considering a 95% confidence level and degree of freedom equal to 9, the nonlinear model ($F_{cal} > F_{tab}$) best fitted the experimental data.

Although the exclusive use of R^2 is not appropriate in regression analysis, as already reported, several studies still use this value as the only tool in comparing linear and nonlinear isothermal models (Emily et al., 2016; Salmani Nuri et al., 2019; Tanhaei et al., 2016; Zaheer et al., 2019).

In addition to the static tools highlighted above, error functions are highly important in determining the best theoretical model. Table 5.10 highlights the most commonly used error functions in adsorption studies (Lima et al., 2012; Ng et al., 2002; Subramanyam & Das, 2014).

Among the models mentioned, SSE is one of the most applied models in adsorption studies since it is suitable for high-concentration data. However, HYBRID presents itself as the best model for low concentrations, including the system's degrees of freedom (Lima et al., 2012; Ng et al., 2002).

It is important to note that the value of R^2 does not include the degree of freedom, as indicated in Equation 5.18. Consequently, the isolated utilization of R^2 to compare fits between isothermal models containing two, three, or even four parameters, for

TABLE 5.10
Error Functions Used in Adsorption Studies

Error Functions	Definition
Sum of squares of errors (SSE)	$\sum_{i=1}^{n}(q_{exp}-q_{cal})_i^2$
Hybrid fractional error function (HYBRID)	$\frac{100}{P-n}\sum_{i=1}^{n}\left[\frac{(q_{exp}-q_{cal})}{q_{exp}}\right]_i^2$
The sum of absolute errors (EABS)	$\sum_{i=1}^{n}\left\|q_{exp}-q_{cal}\right\|_i$
Nonlinear chi-square test (X^2)	$\sum_{i=1}^{n}\frac{(q_{exp}-q_{cal})_i^2}{q_{cal}}$
The average relative error (ARE)	$\frac{100}{n}\sum_{i=1}^{n}\left\|\frac{(q_{exp}-q_{cal})}{q_{exp}}\right\|_i$

Source: From Lima et al. (2012) and NG et al. (2002).

example, is employed incorrectly. The observed phenomenon can be attributed to the fact that an increase in the number of parameters results in a decrease in the number of degrees of freedom, frequently resulting in a reduction in error. This suggests that models with more parameters may offer a more accurate data representation. However, this is not always the case (Amiri et al., 2019; El-Khaiary & Malash, 2011; Milani et al., 2018; Nadeem et al., 2009).

Some authors have proposed an additional indicator, in addition to R^2 and error (SSE), to compare models with different numbers of parameters. Akaike's information criterion (AIC) is another indicator (Akaike, 1974; El-Khaiary & Malash, 2011; Zhang et al., 2013). The AIC is a well-determined statistical method, based on information theory and the theory of maximum likelihood. This method is employed to ascertain the most probable model and to quantify the most likely model (Akaike, 1974; El-Khaiary & Malash, 2011; Figueira et al., 2017). The following Equation 5.23 gives the general formula for calculating the AIC:

$$AIC = 2N_p - 2\ln L \qquad (5.23)$$

L is the maximum value of the likelihood function for the model, but in practice, when using nonlinear or linear regression with independent normal errors, it is not calculated directly. As an alternative approach, it is customary to substitute L with the SSE, on the assumption of normality, as outlined in Equation 5.24 (Akpa & Unuabonah, 2011; El-Khaiary & Malash, 2011).

$$AIC = N\ln\left(\frac{SSE}{N}\right) + 2N_P \qquad (5.24)$$

The AIC sign lacks inherent meaning and can assume a positive or negative value. This is due to the index's susceptibility to transformations of units of the variables involved in the process, with the proviso that consistency in units is always maintained. The primary concern pertains to the disparity in the AIC values of two models, wherein lower values of the index, ranging from $+\infty$ to $-\infty$, are indicative of a superior model fit. Mathematicians demonstrated that the AIC is inadequate for a limited sample size (N). Consequently, the corrected AIC (AICc) is more precise, as it is calculated using Equation 5.25 (Akpa & Unuabonah, 2011; El-Khaiary & Malash, 2011).

$$AIC_C = AIC + \frac{2N_p\left(N_p+1\right)}{N-N_p-1} \tag{5.25}$$

In general, AICc can only be calculated when the number of data points is at least two more points than the number of parameters. Since Np refers to the number of variable parameters in the isotherm model, N is the number of recorded data points.

This statistical method, AIC, was employed by Bouabidi et al. (2018) in the fitting of two isotherm models with only two parameters (Langmuir and Freundlich) and two models with three parameters (Redlich–Peterson and Sips). The study utilized the R^2, SSE, AIC, and AICc values calculated for the various isotherms to ascertain the model that most accurately describes the experimental data. The authors concluded that the Sips model best fits the results based on a comparison using only R^2. However, the Redlich–Peterson model is preferable when the AICc is employed, given its lower AICc value. Consequently, utilizing the AICc would constitute a more consistent method for comparing the adequacy of the four models.

Equation 5.26 can be used to calculate the probability that a given isotherm model fits accurately with the experimental data relative to an alternative model, using the AIC as the basis for calculation (Akpa & Unuabonah, 2011; El-Khaiary & Malash, 2011).

$$P_A = \frac{e^{0.5\Delta}}{1+e^{0.5\Delta}} \tag{5.26}$$

Where Δ is the absolute value of the difference in AICc between the two models, this probability is also known as Akaike's weight. Another way to compare the values of the AICc is by employing the rate or ratio of Evidence, defined by Equation 5.27 (Malash & El-Khaiary, 2010; Milani et al., 2018).

$$\text{Evidence ratio} = \frac{1}{e^{-0.5\Delta}} \tag{5.27}$$

Thus, the evidence ratio indicates that the model with the lowest AIC is $1/e^{-0.5\Delta}$ times more likely to be correct than the alternative model.

SOLVED PROBLEMS

– Errors in Adsorption Modeling

Exercise 5.1 Absolute Error and Percentage Error

Experimental and calculated values of qt ($mg\ g^{-1}$) for different times:

Time (Minutes)	$\lvert q_{texp} \rvert$	$\lvert q_{tcalc} \rvert$
10	10.5	9.8
30	23.2	22.5
60	30.0	29.0
90	32.5	31.0

Absolute Error: $EA = \lvert q_{texp} - q_{tcalc} \rvert$
Percentage Error: $EP = (\lvert q_{texp} - q_{tcalc} \rvert / q_{texp}) \times 100$
Results:

Time (minutes)	$\lvert$ EA ($mg\ g^{-1}$) $\rvert$	EP (%)
10	0.7	6.67
30	0.7	3.02
60	1.0	3.33
90	1.5	4.62

Conclusion: Low errors indicate a good model fit.

Exercise 5.2 RMSE and R^2 for Langmuir Isotherm

Data:

C_e ($mg\ L^{-1}$)	$\lvert q_{exp}$ ($mg\ g^{-1}$) $\rvert$	q_{calc} ($mg\ g^{-1}$)
10	8.5	8.2
20	15.0	14.8
30	18.5	18.0
40	20.0	19.5

$$RMSE = \sqrt{\left[\left(\Sigma\left(q_{exp} - q_{calc}\right)^2\right)/n\right]} = 0.384\ mg\ g^{-1}$$

$$R^2 = 1 - \left[\Sigma\left(q_{exp} - q_{calc}\right)^2 / \Sigma\left(q_{exp} - \bar{q}_{exp}\right)^2\right]\left(\text{with } \bar{q}_{exp} = 15.5\right)$$

Conclusion: Low RMSE and R^2 close to 1 indicate excellent fit.

Exercise 5.3 Chi-Square Error (x^2)

Data:

C_e	q_{exp}	q_{calc}
5	5.2	5.0
10	9.6	9.4
20	13.8	13.6

$$\chi^2 = \Sigma\left[\left(q_{exp} - q_{calc}\right)^2 / qc_{alc}\right] = 0.0152$$

Conclusion: A low χ^2 value indicates an excellent fit between the data and the model.

Exercise 5.4 Comparison of Models

Comparison between Thomas and Yoon–Nelson models (fixed-bed column):

Models	RMSE	R^2	χ^2
Thomas	0.65	0.980	0.021
Yoon–Nelson	0.58	0.984	0.018

Conclusion: The Yoon–Nelson model has the lowest error and the highest R^2, making it the best fit.

Exercise 5.5 Fitting the Freundlich Isotherm

Experimental Data

C_e (mg L^{-1})	q_e (mg g^{-1})
5	8.1
10	12.0
20	16.5
30	19.2
40	21.3

a. Calculation of ln (C_e) and ln (q_e):

C_e	q_e	ln (C_e)	ln (q_e)
5	8.1	1.609	2.091
10	12.0	2.303	2.485
20	16.5	2.996	2.803
30	19.2	3.401	2.955
40	21.3	3.689	3.058

b. Linear Adjustment:
Equation: $\ln(q_e) = \ln(K_f) + (1/n) \cdot \ln(C_e)$
Straight Line Fit: $\ln(K_f) \approx 1.69 \rightarrow K_f \approx 5.42\ \text{mg g}^{-1}$

$$1/n \approx 0.39 \rightarrow n \approx 2.56$$

c. Conclusion: $n > 1$ indicates favorable adsorption.

Exercise 5.6 Thomas Model (Fixed-Bed)

Experimental Data

Time (Minutes)	C/C_0	$\ln[(C_0/C) - 1]$
10	0.05	2.94
20	0.12	2.00
30	0.25	1.10
40	0.46	0.30
50	0.70	−0.51
60	0.88	−2.04
70	0.95	−3.00

Equation Adjustment:
Thomas Linear Equation:

$$\ln[(C_0/C) - 1] = k_{Th.}\, q_0 \,.\, m/Q - k_{Th} \text{-} C_0 t$$

Plotting the Data:

$$\text{Slope} = -\,k_{Th} \,.\, C_0$$

$$\text{Intercept} = k_{th} \,.\, q_0 \,.\, m/Q$$

Simulated Parameters:

$$C_0 = 100\ \text{mg L}^{-1}, \quad m = 2\ \text{g}, \quad Q = 10\ \text{mL min}^{-1}$$

$$k_{Th} \approx 0.055\ \text{mL mg-min}^{-1}$$

$$q_0 \approx 25.4\ \text{mg g}^{-1}$$

PROPOSED PROBLEMS

5.1 Utilize the nonlinear equation a of the Freundlich isotherm model below to adjust the data in Table 5.2 (q_e and C_e) nonlinearly and calculate the constants K_F and n. Demonstrate the R^2 and SSE of the regression.

$$q_e = K_F C_e^{1/n}$$

5.2 Use the nonlinear equation provided above to derive the linear equation of the Freundlich isotherm model. Provide the structure of the equation's plot and the methodology for calculating the constants.

5.3 Utilize the Freundlich linear model and the data presented in Table 5.2 (q_e and C_e) to construct a graph and calculate the model parameters K_F and n. The R^2 and SSE of the curve should be presented.

5.4 Based on the results obtained from the previous questions, which model most effectively adjusted the adsorption data for Pb^{2+}?

5.5 Calculate the Akaike information criterion (AIC) and the Akaike information criterion with a correction for small samples (AICc) of the two models and compare the values. According to these values, does the best fit model remain unchanged from the previous question?

5.6 Calculate the Akaike probability (PA) and the Evidence ratio between the linear and nonlinear models of the Freundlich isotherm and justify determining which model is more appropriate for modeling experimental data.

5.7 Employing the data provided below concerning Cd^{2+} adsorption, analyzing the impact of the inclusion and exclusion of the point (0,0) in the linear fit of the type I Langmuir model. The R^2, SSE, and parameter values obtained should be used to justify.

C_e (mg L^{-1})	q_t (mg g^{-1})
0.45	2.058
0.26	5.453
0.39	6.372
0.71	11.029
0.68	23.375
2.35	41.453
52.80	91.174

5.8 Analyze the data presented below using the Weber–Morris model interpretation. This analysis should include identifying, explaining, and calculating the K_d value for each stage.

t (Minutes)	q_t (mg g^{-1})
2.2	15.5
3.0	19.8
4.1	23.5
5.1	26.5
6.2	28.5
8.2	30.5
10.3	31.8
12.3	33.0

5.9 Why can the exclusive use of R^2 lead to the selection of incorrect models in adsorption studies?

5.10 A study compared four isotherm models: Langmuir, Freundlich, Sips, and Redlich–Peterson. The Sips model had a higher R^2 and a lower SSE, but the Redlich–Peterson model had a lower AICc. Which model should be considered superior? Justify your answer.

REFERENCES

Abdel-Ghani, N. T., Rawash, E. S. A., & El-Chaghaby, G. A. (2016). Equilibrium and kinetic study for the adsorption of p-nitrophenol from wastewater using olive cake based activated carbon. *Global Journal of Environmental Science and Management*, 2(1), 11–18.

Akaike, H. (1974). A new look at the statistical model identification. *IEEE Transactions on Automatic Control*, 19(6), 716–723.

Akpa, O. M., & Unuabonah, E. I. (2011). Small-sample corrected akaike information criterion: An appropriate statistical tool for ranking of adsorption isotherm models. *Desalination*, 272(1–3), 20–26.

Amiri, M. K., Ghaemi, A., & Arjomandi, H. (2019). Experimental, kinetics and isothermal modeling of carbon dioxide adsorption with 13X zeolite in a fixed bed column. *Iranian Journal of Chemical Engineering (IJChE)*, 16(1), 54–64.

AnakoOpotu, L. et al. (2022). Errors and inconsistencies in scientific reporting of aqueous phase adsorption of contaminants: A bibliometric study. *Cleaner Materials*, 5, 100100.

Apiratikul, R., & Pavasant, P. (2008). Sorption of Cu^{2+}, Cd^{2+}, and Pb^{2+} using modified zeolite from coal fly ash. *Chemical Engineering Journal*, 144(2), 245–258.

Ayawei, N., Ebelegi, A. N., & Wankasi, D. (2017). Modelling and interpretation of adsorption isotherms. *Journal of Chemistry*, 2017, 3039817.

Bergmann, C. P., & Machado Machado, F. (2015). *Carbon Nanomaterials as Adsorbents for Environmental and Biological Applications*. 1. ed. Springer International Publishing, Cham.

Bolster, C. H., & Hornberger, G. M. (2008). On the use of linearized langmuir equations. *Soil Science Society of America Journal*, 72(6), 1848–1848.

Bouabidi, Z. B. et al. (2018). Steel-making dust as a potential adsorbent for the removal of lead (II) from an aqueous solution. *Chemical Engineering Journal*, 334, 837–844.

Boyd, G. E., Adamson, A. W., & Myers, L. S. J. (1947). The exchange adsorption of ions from aqueous solutions by organic zeolites. *Journal of the American Chemical Society*, 69(11), 2836–2848.

Chakraborty, P. et al. (2019). Linearity and non-linearity analysis of isotherms and kinetics for ibuprofen remotion using superheated steam and acid modified biochar. *Process Safety and Environmental Protection*, 126, 193–204.

El-Khaiary, M. I., & Malash, G. F. (2011). Common data analysis errors in batch adsorption studies. *Hydrometallurgy*, 105(3–4), 314–320.

Emily, C. N. et al. (2016). Adsorption of caffeine and ciprofloxacin onto pyrolitically derived water hyacinth biochar: Isothermal, kinetic and thermodynamic studies. *Journal of Chemistry and Chemical Engineering*, 10(4), 185–194.

Fang, L. et al. (2018). A novel method for the sequential removal and separation of multiple heavy metals from wastewater. *Journal of Hazardous Materials*, 342, 617–624.

Figueira, P. et al. (2017). Periodic mesoporous organosilica with low thiol density: A safer material to trap Hg(II) from water. *Journal of Environmental Chemical Engineering*, 5(5), 5043–5053.

Gonen, F., & Biçer, G. (2017). Telon blue AGLF adsorption by NiO based nanomaterials: Equilibrium, kinetic and thermodynamic approach. *Journal of the Turkish Chemical Society, Section A: Chemistry*, 4(3), 675–675.

González-López, M. E. et al. (2012). A discussion on linear and non-linear forms of Thomas equation for fixed-bed adsorption column modeling. *Revista Mexicana de Ingeniería Química*, 8(3), 97–104.

González-López, M. E. et al. (2022). A critical overview of adsorption models linearization: Methodological and statistical inconsistencies. *Separation and Purification Reviews*, 51(3), 358–372.

Guo, X., & Wang, J. (2019). Comparison of linearization methods for modeling the Langmuir adsorption isotherm. *Journal of Molecular Liquids*, 296, 111850.

Hadi, M. et al. (2012). Prediction of optimum adsorption isotherm: Comparison of chi-square and log-likelihood statistics. *Desalination and Water Treatment*, 49(1–3), 81–94.

Hamzaoui, M., Bestani, B., & Benderdouche, N. (2018). The use of linear and nonlinear methods for adsorption isotherm optimization of basic green 4-dye onto sawdust-based activated carbon. *Journal of Material and Environmental Sciences*, 9(4), 1110–1118.

Hubicki, Z., & Wołowicz, A. (2009). Adsorption of palladium(II) from chloride solutions on Amberlyst A 29 and Amberlyst A 21 resins. *Hydrometallurgy*, 96(1–2), 159–165.

Hussain, A. et al. (2003). Application of the Langmuir and Freundlich equations for P adsorption phenomenon in saline-sodic soils. *International Journal of Agriculture and Biology*, 5(3), 349–356.

Igwe, J. C., & Abia, A. A. (2007). Adsorption isotherm studies of Cd(II), Pb(II) and Zn(II) ions bioremediation from aqueous solution using unmodified and EDTA-modified maize cob. *Ecletica Quimica*, 32(1), 33–42.

Karri, R. R., Sahu, J. N., & Jayakumar, N. S. (2017). Optimal isotherm parameters for phenol adsorption from aqueous solutions onto coconut shell based activated carbon: Error analysis of linear and non-linear methods. *Journal of the Taiwan Institute of Chemical Engineers*, 80, 472–487.

Kera, N. H. et al. (2017). Selective removal of toxic Cr(VI) from aqueous solution by adsorption combined with reduction at a magnetic nanocomposite surface. *Journal of Colloid and Interface Science*, 503, 214–228.

Kumar, K. V., Porkodi, K., & Rocha, F. (2008). Comparison of various error functions in predicting the optimum isotherm by linear and non-linear regression analysis for the sorption of basic red 9 by activated carbon. *Journal of Hazardous Materials*, 150(1), 158–165.

Langmuir, I. (1919). The adsorption of gases on plane surfaces of glass, mica and platinum. *Journal of the American Chemical Society*, 40(1914), 1361–1403.

Lima, A. C. A. De et al. (2012). Modified coconut shell fibers: A green and economical sorbent for the removal of anions from aqueous solutions. *Chemical Engineering Journal*, 185–186, 274–284.

López-Luna, J. et al. (2019). Linear and nonlinear kinetic and isotherm adsorption models for arsenic removal by manganese ferrite nanoparticles. *SN Applied Sciences*, 1(8), 1–19.

Malash, G. F., & El-Khaiary, M. I. (2010). Piecewise linear regression: A statistical method for the analysis of experimental adsorption data by the intraparticle-diffusion models. *Chemical Engineering Journal*, 163(3), 256–263.

Mallakpour, S., & Tabesh, F. (2019). Tragacanth gum based hydrogel nanocomposites for the adsorption of methylene blue: Comparison of linear and non-linear forms of different adsorption isotherm and kinetics models. *International Journal of Biological Macromolecules*, 133, 754–766.

Milani, P. A. et al. (2018). Agricultural solid waste for sorption of metal ions: Part I—characterization and use of lettuce roots and sugarcane bagasse for Cu(II), Fe(II), Zn(II), and Mn(II) sorption from aqueous medium. *Environmental Science and Pollution Research*, 25(36), 35895–35905.

Mohseni-Bandpi, A. et al. (2016). Improvement of zeolite adsorption capacity for cephalexin by coating with magnetic Fe_3O_4 nanoparticles. *Journal of Molecular Liquids*, 218, 615–624.

Nadeem, R., Nasir, M. H., & Hanif, M. S. (2009). Pb (II) sorption by acidically modified Cicer arientinum biomass. *Chemical Engineering Journal*, 150(1), 40–48.

Nagy, B. et al. (2017). Linear and nonlinear regression analysis for heavy metals removal using Agaricus bisporus macrofungus. *Arabian Journal of Chemistry*, 10, S3569–S3579.

Nebaghe, K. C. et al. (2016). Comparison of linear and non-linear method for determination of optimum equilibrium isotherm for adsorption of copper(II) onto treated Martil sand. *Fluid Phase Equilibria*, 430, 188–194.

Ng, J. C. Y., Cheung, W. H., & Mckay, G. (2002). Equilibrium studies of the sorption of Cu(II) ions onto chitosan. *Journal of Colloid and Interface Science*, 255(1), 64–74.

Núñez-Gómez, D. et al. (2019). Adsorption of heavy metals from coal acid mine drainage by shrimp shell waste: Isotherm and continuous-flow studies. *Journal of Environmental Chemical Engineering*, 7(1), 102787.

Ozcan, K. et al. (2018). Modelling of adsorption kinetic processes—errors, theory and application. In S. Edebali (Ed.), *Advanced Sorption Process Applications*. IntechOpen. https://doi.org/10.5772/intechopen.80495.

Qiu, Q. et al. (2018). Adsorption of heavy metal ions using zeolite materials of municipal solid waste incineration fly ash modified by microwave-assisted hydrothermal treatment. *Powder Technology*, 335, 156–163.

Ramachandran, P., Vairamuthu, R., & Ponnusamy, S. (2011). Adsorption isotherms, kinetics, thermodynamics and desorption studies of reactive orange16 on activated carbon derived from Ananas comosus (L.) carbon. *Journal of Engineering and Applied Sciences*, 6(11), 15–26.

Raulino, G. S. C. et al. (2014). Treatment influence on green coconut shells for removal of metal ions: Pilot-scale fixed-bed column. *Environmental Technology (United Kingdom)*, 35(14), 1711–1720.

SalmaniNuri, O., Irannajad, M., & Mehdilo, A. (2019). Reagent adsorption on modified mineral surfaces: Isotherm, kinetic and thermodynamic aspects. *Journal of Molecular Liquids*, 291, 111311, 2019.

Santos, B. A. et al. (2022). Magnetic zeolite from fly ash as a cost-effective adsorbent for Cu^{2+}, Pb^{2+}, Zn^{2+}, and Cd^{2+} removal from aqueous media: A comprehensive study. *International Journal of Environmental Analytical Chemistry*, 104(16), 3913–3935.

Shalaby, T. et al. (2018). Geochemistry of El-Salam Canal and the adjacent groundwater in north Sinai, Egypt: An application to a water treatment process using magnetic zeolite nanoparticles. *Applied Water Science*, 8(4), 1–17.

Shikuku, V. O. et al. (2018a). Single and binary adsorption of sulfonamide antibiotics onto iron-modified clay: Linear and nonlinear isotherms, kinetics, thermodynamics, and mechanistic studies. *Applied Water Science*, 8(6), 1–12.

Shikuku, V. O., Kowenje, C. O., & Kengara, F. O. (2018b). Errors in parameters estimation using linearized adsorption isotherms: Sulfadimethoxine adsorption onto kaolinite clay. *Chemical Science International Journal*, 23(4), 1–6.

Shyam, R. et al. (2013). Single and binary adsorption of heavy metals on fly ash samples from aqueous solution. *Journal of Molecular Liquids*, 178, 31–36.

Silva, L. T. V. Da et al. (2023). Comparative study on the performance of two synthetic zeolite 4A on the removal of heavy metals from aqueous solution—effect of coal fly ash as Al and Si source. *Desalination and Water Treatment*, 310, 157–166.

Simonin, J. P., & Bout, J. (2016). Intraparticle diffusion-adsorption model to describe liquid/solid adsorption kinetics. *Revista Mexicana de Ingeniería Química*, 15, 161–173.

Subramanyam, B., & Das, A. (2014). Linearised and non-linearised isotherm models optimization analysis by error functions and statistical means. *Journal of Environmental Health Science and Engineering*, 12(1), 1–6.

Tanhaei, B. et al. (2016). A magnetic mesoporous chitosan based core-shells biopolymer for anionic dye adsorption: Kinetic and isothermal study and application of ANN. *Journal of Applied Polymer Science*, 133(22), 1–11.

Tellinghuisen, J., & Bolster, C. H. (2011). Using R2 to compare least-squares fit models: When it must fail. *Chemometrics and Intelligent Laboratory Systems*, 105(2), 220–222.

Tran, H. N. et al. (2017). Mistakes and inconsistencies regarding adsorption of contaminants from aqueous solutions: A critical review. *Water Research*, 120, 88–116.

Viegas, R. M. C. et al. (2014). How do the HSDM and Boyd's model compare for estimating intraparticle diffusion coefficients in adsorption processes. *Adsorption*, 20(5–6), 737–746.

Visa, M. (2016). Synthesis and characterization of new zeolite materials obtained from fly ash for heavy metals removal in advanced wastewater treatment. *Powder Technology*, 294, 338–347.

Wang, Y. et al. (2024). Guideline for modeling solid-liquid adsorption: Kinetics, isotherm, fixed bed, and thermodynamics. *Chemosphere*, 349(September 2023), 140736.

WeberJr., W. J., & Carrell Morris, J. (1963). Kinetics of adsorption on carbon from solution. *Journal of the Sanitary Engineering Division*, 89(2), 31–59.

Zaheer, Z. et al. (2019). Adsorption, equilibrium isotherm, and thermodynamic studies to the removal of acid orange 7. *Materials Chemistry and Physics*, 232, 109–120.

Zhang, Z. et al. (2013). Comparative study on adsorption of two cationic dyes by milled sugarcane bagasse. *Industrial Crops and Products*, 42(1), 41–49.

6 Fixed-Bed Adsorption

Giselle Santiago Cabral Raulino,
Luiz Thiago Vasconcelos da Silva,
Ari Clecius Alves de Lima, Carla Bastos Vidal,
Diego de Quadros Melo, and
Ronaldo Ferreira do Nascimento

The choice of the adsorption system operating mode depends mainly on the volume and composition of the fluid to be treated, the time required or available for treatment of this fluid, as well as the adsorbent regeneration conditions (John et al., 2024). Adsorption isotherms have been used in the preliminary screening of an adsorbent, as they give an indication of the adsorption effectiveness for removing specific impurities, that is, its selectivity, as well as the maximum amount of adsorbate that can be adsorbed by a particular adsorbent unit (Yusoff et al., 2025). However, these experiments carried out in batch or transient mode are limited to the treatment of small volumes of effluent, in addition to not providing data for the exact dimension of continuous treatment systems (Yusoff et al., 2025; Gupta et al., 1998). Added to this is the inconvenience and relatively high costs of labor and handling of solid particles (filling, emptying, and cleaning the reactor) when using transient state operations.

The use of fixed beds to simulate adsorption processes is a widely used tool for process design, as it reduces the cost of tests performed in the laboratory and in the pilot plant before expansion (Ortiz et al., 2019). Thus, in the practical operation of large-scale adsorption processes, continuous flow fixed-bed columns are frequently used (Aksu et al., 2002; Yusoff et al., 2025; Lodeiro et al., 2016; Moreira et al., 2009; Raulino et al., 2014). This type of system is one of the most efficient configurations for treating a large volume of effluent and reusing the adsorbent (adsorption–desorption), allowing greater column efficiency.

A conventional fixed-bed system consists of a column where adsorbent particles with specific characteristics are placed in contact with the solution containing the adsorbate(s) to be treated, as can be seen in Figure 6.1. The idea is to use adsorbents with a well-developed surface area, with low mass transfer limitations, to improve the service time of the bed (Greenwald et al., 2015). In this configuration, the solute concentration in both phases (liquid and solid) varies (Aksu et al., 2002). Thus, the adsorption process by fixed-bed columns is generally monitored by measuring the analyte concentration in the effluent as a function of time, using different analytical techniques (Dichiara et al., 2015; Afroze et al., 2016).

During the adsorption process, adsorbate accumulates on the surface (or in the pores) of the adsorbent particles, provided that the equilibrium state is reached (Worch, 2012; Cooney, 1999). Initially, the adsorbate concentration in the effluent that percolates in the column is low and may be zero, since the adsorbent can

DOI: 10.1201/9781003654391-6

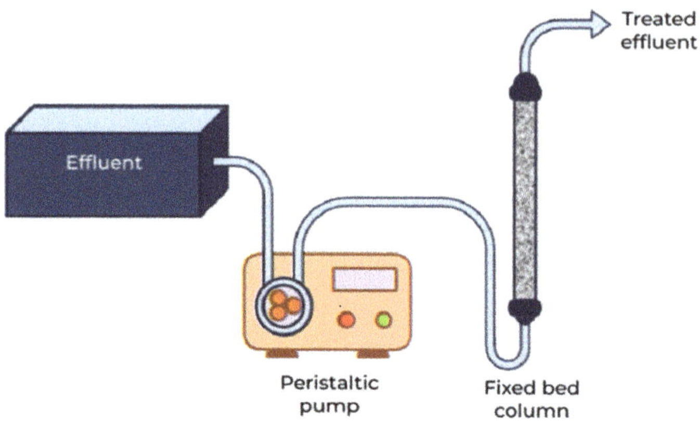

FIGURE 6.1 Schematic of a fixed-bed column adsorption system.

effectively retain the (adsorbed) compounds that enter the column. This equilibrium process continues successively, layer by layer, from the inlet to the outlet of the column. However, due to the slow adsorption kinetics, there is no defined boundary between adsorbate-saturated absorbent layers and adsorbate-free layers. Instead, equilibrium occurs in a more or less wide zone of the adsorbent bed, referred to as the mass transfer zone (MTZ) or adsorption zone (Worch, 2012; Cooney, 1999). After a certain time, the adsorbate concentration in the effluent solution increases as the adsorbent particles retain the adsorbate (Cooney, 1999).

BREAKTHROUGH CURVE

The graph of adsorbate concentration in the effluents as a function of time (or volume) is called the breakthrough curve and has a typical sigmoidal shape, since the effluent concentration rises throughout the process, going from zero to a concentration equal to the influent solution when the adsorbent is completely saturated (Reguer et al., 2011), as shown in Figure 6.2.

When designing and optimizing fixed-bed adsorption processes, kinetic data from breakthrough curves are required (Hussein & Ahmed, 2016). The breakthrough curve is demonstrated in the form of a graph in relation to the effluent/influent concentration (C/C_0) in relation to time (t) or volume (V) (Benstoem et al., 2017). This curve explains the dynamics and efficiency of a continuous adsorption system. The behavior of the breakthrough curve is related to the adsorption shape, and the other demand is affected by the diffusion stages in the fixed bed (Chu, 2010). The breakthrough time and the shape of the breakthrough curve are very important characteristics to determine the operation and dynamics of an adsorption column and are closely related to the MTZ where adsorption occurs (Sousa et al., 2010; Aksu et al., 2002; Tan & Hameed, 2017).

In an ideal breakthrough curve, it is assumed that adsorbate removal is complete in the initial stages of operation (Figure 6.2). After a certain time or percolated volume, unretained adsorbate is found in the effluent solution to the column until it reaches the

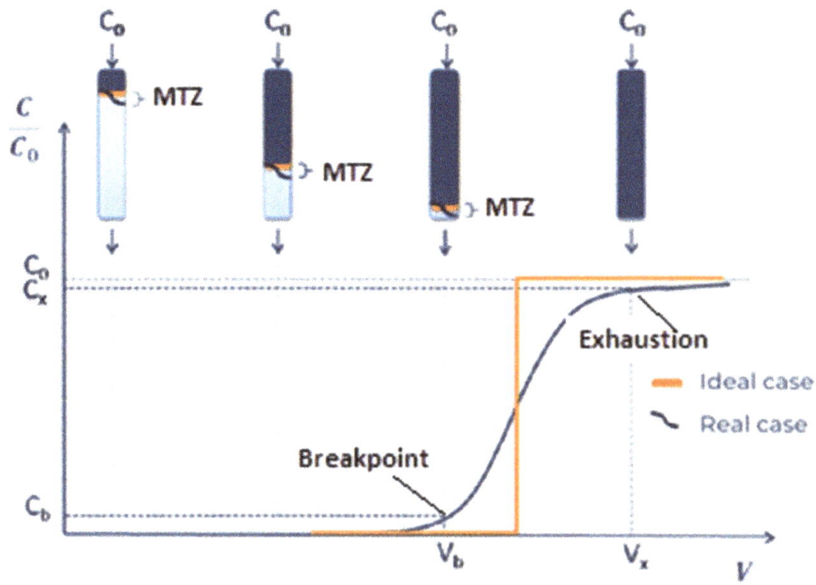

FIGURE 6.2 Schematic representation of the mass transfer zone in a fixed-bed column for an ideal breakthrough curve.

breakthrough point. Generally, the breakthrough point, designated by C_b, is chosen as the point at which the concentration of adsorbate effluent to the column is 5% of the initial concentration (C_0), that is, the concentration of adsorbate entering the column. This point can also be determined by the maximum concentration permitted by legislation permitted for a given pollutant in a given effluent. For example, CONAMA resolution 430/2011 (Brasil, 2011) establishes that the maximum concentration of copper in an effluent is 1.0 mg L^{-1}. Thus, an effluent whose copper concentration before treatment is 100 mg L^{-1} will require that the rupture point, if this effluent is treated by a fixed-bed column containing adsorbent material, be 1.0% of the initial concentration coinciding with the concentration of 1.0 mg L^{-1}. The concentration of adsorbate effluent to the column increases until the exhaustion point is reached. In the same way, as illustrated in Figure 6.2, a concentration of around 90% C_0 is chosen for the exhaustion point, C_x. The amount of total effluent mass passing through a unit of cross-section until the rupture point and the nature of the rupture curve between the rupture volume – V_b (corresponding to C_b) and the exhaustion volume – V_x (corresponding to C_x) are important for the design of a fixed-bed adsorber (Gupta et al., 1998).

Parameters such as actual adsorption rate, mechanism of the adsorption process, nature of adsorption equilibrium, fluid velocity, solute concentration in the feed, and height of the adsorbent bed contribute to the shape of the breakthrough curve (Wilkins et al., 2021). Usually, the breakthrough point decreases with decreasing bed height, increasing adsorbent particle size, increasing fluid velocity through the bed and increasing initial solute concentration in the feed (Aksu et al., 2002; Vijayaraghavan et al., 2005; Sousa et al., 2010; Muhamad et al., 2010; Srivastava et al., 2008).

MASS TRANSFER ZONE (MTZ)

The performance of a column and its breakthrough curve are closely related to the length and shape of the MTZ that develops in the column during the adsorption process. The assumption of an adsorption zone, that is, the MTZ, already mentioned above, provides the basis for a simple design that allows for the scale-up of small-scale experiments in the laboratory (Yusoff et al., 2025). The MTZ is defined based on Figure 6.2.

The MTZ can be considered as a region within the column where the adsorbate concentration varies from 90% to 5% of the feed value; that is, it develops between the section of the column that is saturated and the section that still contains unsaturated adsorbent. This is the region where most of the mass transfer occurs. At the beginning of the process, the adsorbent is hit with a high concentration of adsorbate. Theoretically, this would be the region of greatest mass transfer. In an ideal system, where mass transfer processes are considered to be infinitely fast and dispersion is absent, the length of the MTZ would be reduced to zero and would present a linear and vertical shape in the rupture curve, triggering an immediate increase in the effluent concentration to the value of the influent concentration to the column, as can be seen in Figure 6.2 (orange line). However, in a real system, it takes time for the MTZ to establish itself. This is due to the resistance to mass transfer existing due to the liquid film located in the vicinity of the particle, the liquid flow rate, temperature, among other factors (Naja & Volesky, 2006). Thus, the MTZ configuration assumes another profile (blue line), with a spreading of this zone occurring, as can be seen in Figure 6.2 for a real case, also varying according to the situations described above.

This partial saturation zone moves through the column in the flow direction at a certain velocity, which, although slower than the fluid velocity, is predominantly determined by the adsorbate loading, adsorbent capacity, and column feed rate. The stronger the adsorbate–adsorbate interaction, the greater the difference between the MTZ velocity and the fluid velocity. Before the MTZ reaches the column outlet, the effluent adsorbate concentration is generally equal to 0. When the MTZ reaches the end of the column, the adsorbate concentration in the effluent begins to increase gradually. On the breakthrough curve, this phenomenon is reflected in the breakthrough point, V_b, described above. After this point, the adsorbate concentration at the column outlet increases due to the advancement of the MTZ and the reduction of the adsorption capacity. The column is operational until the exhaustion point, when practically all the MTZ has reached the end of the column. After this point, the effluent is practically free adsorbate and the concentration is equal to the initial concentration, C_0.

The breakthrough curve reflects the shape of the MTZ. The more open the shape of the curve, the greater the resistance to mass transfer. The curve becomes narrower as the resistance to mass transfer decreases. The smaller this region, the closer to ideality the system is, the longer the column service time during the feed stage and the more complete the desorption of the solute (Yusoff et al., 2025; John et al., 2024; Gupta et al., 1998; Cooney, 1999; Naja & Volesky, 2006; Moreira, 2008; Worch, 2012).

In addition to the advantages presented at the beginning of the chapter, fixed-bed adsorption processes, compared to batch adsorption processes, are more advantageous because the adsorbent is always in contact with the inlet concentration C_0, which results in a high driving force throughout the fixed-bed adsorption process and adsorbate is completely removed until the breaking point is reached. In batch adsorption processes, on the other hand, the mass transfer driving force, and therefore also the adsorption rate, decreases during the process due to the decreasing concentration in the reactor and very low residual concentrations can only be achieved if very high doses of adsorbent are applied, therefore being less advantageous processes (John et al., 2024; Worch, 2012).

It should be noted that all the discussion carried out so far considers a solution containing only one adsorbate. In cases where there are several components that can be adsorbed, an individual MTZ occurs for each component traveling at different speeds through the adsorbent bed. In multicomponent systems, competition processes for adsorbent sites and, as a consequence, displacements of previously adsorbed components occur and the breakthrough curves are somewhat different (Worch, 2012), as can be seen in Figure 6.3.

The MTZ of the weaker adsorbed component (component 1 in Figure 6.3) travels at a faster speed through the bed, reaching the unsaturated adsorbent first and thus being readily adsorbed. When the more strongly adsorbed component (component 2 in Figure 6.3) reaches the layers already containing the weaker adsorbed component, a new equilibrium is established, with a partial displacement of the component

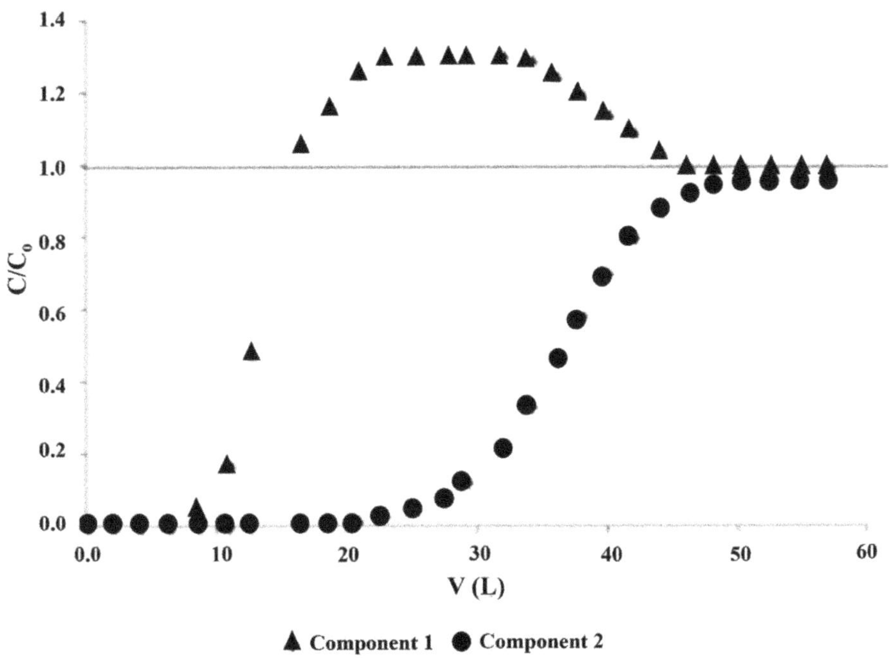

FIGURE 6.3 Breakthrough curves for a system bicomponent.

present. As a result of this phenomenon, the concentration of the weaker adsorbed component in the region between the two MTZs is greater than the initial concentration C_0, as can be seen in Figure 6.3.

The same processes occur for systems containing three or more components, with displacement processes being more prominent for the weaker adsorbed components, and therefore, the effluent concentration of these components will be higher among the MTZs of the components present in the solution.

Therefore, in multicomponent adsorption systems, care must be taken when choosing the column's operating point. In fact, the complete removal of compounds only occurs up to the point at which the most weakly adsorbed component breaks down. From that point onward, even if the most strongly adsorbed components can still be removed, the concentration of the weakly adsorbed component increases with the passage of time or the volume of percolated solution and therefore; for example, in the case of effluent treatment, it becomes an inconvenience in these types of processes.

After use, the column containing the adsorbent undergoes a process of desorption of the compounds, through washing with eluent solutions (competing agent) and can then be reused for a new adsorption cycle. The objective of the desorption study was to remove the retained adsorbate with the smallest volume of eluent solution until obtaining the maximum possible concentration, without destroying the capacity of the adsorbent, making it reusable in several retention/desorption cycles (Yusoff et al., 2025).

In the study of adsorption column, some operating parameters that describe the behavior of the breakthrough curve, and consequently of the MTZ, can be obtained from the experimental data. The operating parameters are useful in the development of a fixed-bed column for the treatment of adsorbates of known concentration. A detailed description of these parameters is given as follows.

APPARENT DENSITY (P_{AP}) AND PACKING DENSITY (P_E)

The apparent density is the ratio of mass per unit volume of the adsorbent in the bed and is determined by the ASTM D2854 method (American Society for Testing and Materials). From the apparent density, the apparent volume (V_{ap}) for a given mass can be calculated. The packing density of the bed (ρ_E) is then determined, represented by Equation 6.1:

$$\rho_E = \frac{M_L}{V_L} \tag{6.1}$$

where

M_L is the total mass of the particle bed inside the column (g), and
V_L is the total (internal) volume of the empty column (mL).

The porosity of the bed is determined from the value of ρE (Equation 6.2), which represents the fraction of empty spaces inside the column filled with the adsorbent, which can be expressed by the following equation:

$$\varepsilon = 1 - \frac{\rho_E}{\rho_{ap}} = \frac{V_L - V_{ap}}{V_L} \tag{6.2}$$

where

ρ_E is the packing density (g cm^{-3} or g mL^{-1});
ρ_{ap} is the apparent density (g cm^{-3} or g mL^{-1});
V_L is the total internal volume of the empty column (mL);
V_{ap} is the total volume of particles in the bed (apparent volume – mL).

TIME TO ESTABLISH THE MTZ (T_X)

The portion between C_x (exhaustion point) and C_b (breakpoint) is known as the MTZ. The total time to move the MTZ along the column until the bed exit is given by Equation 6.3 (Gupta et al., 1998; Kundu & Gupta, 2005):

$$t_x = \frac{V_x}{F_m} \tag{6.3}$$

where

t_x is the time to move the MTZ to the bed exit (min);
F_m is the volumetric flow (mL min^{-1}); and
V_x is the exhaust volume (mL).

TIME REQUIRED TO MOVE THE ZAP IN THE COLUMN (T_δ)

The time required to move the MTZ along the column is given by Equation 6.4 (Gupta et al., 1998; Kundu & Gupta, 2005; Gupta et al., 2000):

$$t_\delta = \frac{V_x - V_b}{F_m} \tag{6.4}$$

where

t_δ is the time required to move the MTZ in the column (min);
V_x is the exhaust volume (mL);
V_b is the rupture volume (mL); and
F_m is the volumetric flow (mL min^{-1}).

ADSORBENT BED DEPTH RATIO

The ratio of the adsorbent bed depth (length) to time (U) is given by Equation 6.5 (Gupta, 1997; Kundu & Gupta, 2005; Gupta et al., 2000):

$$U = \frac{\delta}{D} = \left(\frac{t_\delta}{t_x - t_f}\right) \tag{6.5}$$

where

δ is the length of the ZTM (cm);
D is the depth of the bed (cm); and
t_f is the time for the formation of the ZTM, from the rupture point to the exhaustion point.

TIME REQUIRED FOR INITIAL FORMATION OF THE MTZ (T_F)

The time required for the formation of the ZTM is given by Equation 6.6 (Gupta et al., 1998; Kundu & Gupta, 2005; Gupta et al., 2000):

$$t_f = (1-F)\, t_\delta \tag{6.6}$$

where

F is the fractional capacity of the adsorbent in the adsorption zone to continue removing solute after the breaking point. The fractional capacity of the adsorbent is given by Equation 6.7 (Gupta et al., 1998; Kundu & Gupta, 2005; Gupta et al., 2000):

$$F = \int_{V_b}^{V} \frac{(C_0 - C)\cdot d_v}{C_0 (V_x - V_b)} \tag{6.7}$$

COLUMN SATURATION PERCENTAGE

The column saturation percentage is obtained by Equation 6.8 (Gupta et al., 1998; Kundu & Gupta, 2005; Gupta et al., 2000):

$$\%S = \left[1 + \left(\frac{\delta\,(F-1)}{D}\right)\right] * 100 \tag{6.8}$$

COLUMN REMOVAL CAPACITY

The maximum removal capacity of metals in the column is given by Equation 6.9 (Cooney, 1999):

$$Q = \frac{C_0 * F_m}{m_s} \int_{t=0}^{t=x} \left(1 - \frac{C}{C_0}\right) dt \tag{6.9}$$

where

Q is the maximum adsorption capacity in the column (mg g^{-1});
C_o is the initial concentration of the solution (mg L^{-1} or mol L^{-1});
C is concentration of the metal in a given volume (mg L^{-1} or mol L^{-1});
m_s is the mass of adsorbent in grams (g);

F_m is the volumetric flow in (L min^{-1}); and
t is the time in minutes.

EFFECTS OF OPERATION PARAMETERS ON COLUMN ADSORPTION

To illustrate the effects of flow velocity, bed height and solute concentration on the breakthrough curve performance, we will apply the above equations to results reported in the literature involving experiments of laboratory and semi-pilot scale, on adsorption of inorganic (Oliveira et al., 2014; Raulino et al., 2014; Sousa et al., 2010) and organic pollutants (Yusoff et al., 2025).

In this context, in a semi-pilot-scale situation of a fixed-bed system, Oliveira et al. (2014) studied low-cost adsorbents to remove metal ions from synthetic effluents. The experiments were carried out in a column system for the conditions described in Figure 6.4. For this, sugarcane bagasse adsorbent was investigated with a multielement solution of Cu^{2+}(216 mg L^{-1}), Zn^{2+} (60 mg L^{-1}), and Ni^{2+}(7 mg L^{-1}), while the experiments with green coconut fiber were carried out with a solution of Cu^{2+} (126 mg L^{-1}), Zn^{2+} (42 mg L^{-1}), and Ni^{2+}(6 mg L^{-1}).

The hydraulic detention time (HDT) is a typical design and operating parameter for columns. Long residence times can lead to a decrease in contaminant removal, while shorter times do not allow for effective contact interaction between the sorbent and sorbate (Ayoob et al., 2007; Cooney, 1999). The HDT is calculated by the ratio between the column volume and the flow rate. The calculated hydraulic load (ratio between flow rate and cross-sectional area) was 6.6 mL min^{-1} cm^{-2}, and the hydraulic retention time was 229 min.

Another important parameter is the adsorbent utilization rate (AUR), which is the ratio between the weight of the biosorbent in the column and the amount of liquid that has passed into the column at the time of breakthrough (Cooney, 1999).

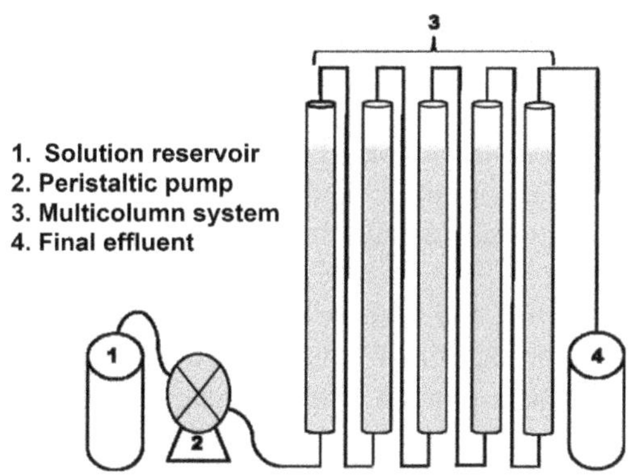

FIGURE 6.4 PVC columns (160 cm long and 5.4 cm internal diameter each) connected in series and a pump that provided continuous circulation of the solution through the column system. A volume of 50 L of synthetic multielement solution (pH = 5.0) was percolated and 50 mL aliquots were collected every 15 min at the column outlet. (Reproduced with permission from Oliveira et al., 2014).

TABLE 6.1
Breakthrough Times (t_b), Exhaustion Times (t_x), Adsorbent Usage Rate, and Adsorption Capacities (When $C/C_0=0.5$) Obtained from Breakthrough Curves of Sugarcane Bagasse and Green Coconut Fibers

Adsorbent	Metal Ion	t_b (min)	t_x (min)	AUR (mg L^{-1})	q_e (mg g^{-1})
Sugarcane bagasse	Cu^{2+}	22	202	0.57	1.417
	Ni^{2+}	22	127	0.57	0.029
	Zn^{2+}	22	142	0.57	0.286
Green coconut fibers.	Cu^{2+}	140	>250	0.10	2.772
	Ni^{2+}	20	185	0.62	0.032
	Zn^{2+}	20	170	0.62	0.248

Source: Modified with permission from Oliveira et al. (2014).

The AUR provides information on the amount of adsorbent needed to treat a fixed volume of solution. The AUR values calculated can be seen in Table 6.1

The breakthrough curves obtained for adsorption of Cu^{2+}, Ni^{2+}, and Zn^{2+} using sugarcane bagasse and green coconut fibers are shown in Figure 6.5a and b. It can be seen that the breakthrough occurred rapidly for the three metal ions studied. However, the copper was saturated on sugarcane bagasse, as shown in Figure 6.5a. In Figure 6.5b, in contrast the copper did not reach saturation, indicating a high affinity of green coconut fibers for these metal ions. The concentration of metal ions in synthetic solutions aimed to simulate an effluent from an electroplating industry.

The higher concentration of copper than nickel and zinc ions may have led to this higher absorption efficiency of copper, rather than nickel and zinc. This concentration gradient favors greater adsorption of the higher concentration adsorbates compared to the others present in the solution. Sousa et al. (2010), studying the absorption of metal ions using green coconut shell powder in a column system, found the same results.

For educational purposes, the values of rupture and exhaustion volume, volumetric flow rate, column length, initial concentration of the ions, etc., were added in Tables 6.1 and 6.2, which were created using an electronic spreadsheet (using Excel, Origin, etc. software).

According to the data in Table 6.1, the AUR is similar for the three metals studied, as well as the breakthrough time. This indicates that only 0.57 g of adsorbent is required to treat 1 L of contaminated solution using sugarcane bagasse, and 0.62 g L^{-1} for green coconut fibers, except for copper, with an AUR of 0.1 g L^{-1}. The low AUR values are due to the high height of the column bed, leading to a high HRT and high metal ion removal, which are parameter very important concepts for industrial applications.

For educational purposes, the values of rupture and exhaustion time, and column saturation were added to Table 6.2, which were created using an electronic spreadsheet (using Excel, Origin, etc. software).

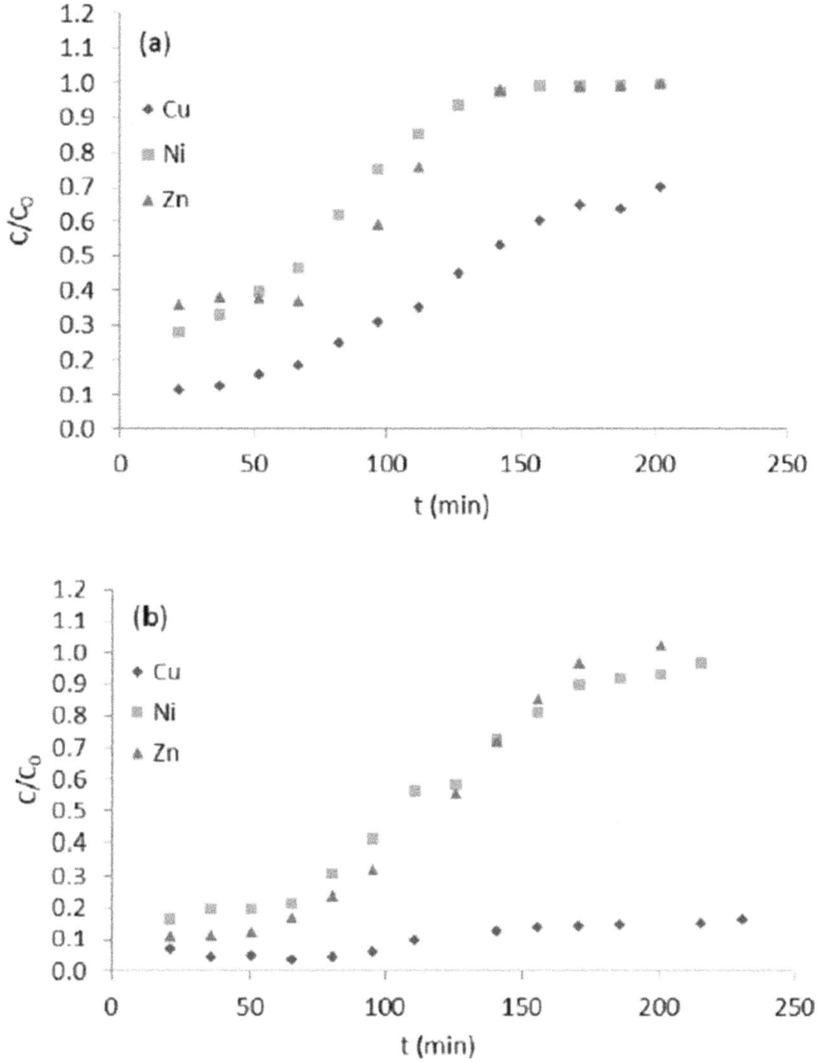

FIGURE 6.5 Breakthrough curves of multicomponent system (Cu^{2+}, Ni^{2+}, and Zn^{2+}) for (a) sugarcane bagasse and (b) green coconut fiber. Conditions: pH = 5, mass of adsorbent of 1 kg, volumetric flow rate of 80 mL min^{-1}, and room temperature (28 ± 2°C). (Reproduced with permission from Oliveira et al., 2014)

Table 6.2 shows the parameters t_x, t_δ, t_f, F, δ, and percentage saturation (%S) of the column obtained from the breakthrough curves in Figure 6.2a and b. The results show that the total time to establish the primary adsorption zone (t_x) was maximum for copper for both adsorbents used. The time needed to move the adsorption zone through the column (t_δ) was between 105 and 360 min. The time to form the primary adsorption zone (t_f) was between 18–42 min for sugarcane bagasse and 14–170 min for green coconut fibers. The length of the PAZ was between 782 and 993 cm for

TABLE 6.2
Parameters t_x, t_f, t_δ, F, δ, and Percentage Saturation of Column for Sugarcane Bagasse and Green Coconut Fibers

Adsorbent	Metal ion	t_x (min)	t_δ (min)	t_f (min)	f	δ (cm)	S (%)
Sugarcane bagasse	Cu^{2+}	202.50	180.40	18.00	0.90	782.24	90.24
	Ni^{2+}	127.10	105.00	5.00	42.53	993.27	49.71
	Zn^{2+}	142.10	120.00	4.99	41.61	955.29	58.60
Green coconut fibers	Cu^{2+}	500.00	360.00	170.46	0.53	873.93	48.28
	Ni^{2+}	215.50	195.00	30.00	0.85	840.98	83.83
	Zn^{2+}	173.75	153.25	14.41	0.91	769.45	90.95

Source: Modified with permission from Oliveira et al. (2014).

sugarcane bagasse and 769 and 873 cm for green coconut fiber. This parameter is related to the region in the rupture curve between the time of rupture and exhaustion. The values obtained for the three metal ions for both adsorbents studied were close to or greater than the length of the column (800 cm), probably due to the rupture occurring most of the time at the first point collected.

It can be observed from the results that the total time to move the ZTM (t_x) was maximum for copper and minimum for zinc. The time required to move the adsorption zone along the column (t_δ) is between 24 and 223 min. The time for formation of the ZTM (t_f) is between 9 and 80 min. Up to the point where the effluent concentration is approximately half that of the influent ($C/C_0=0.5$), the column is not yet saturated, presenting percentages of 52%–62% saturation.

The breakthrough curves obtained by Sousa et al. (2010) for the adsorption of multiple metal ions on green coconut shells are shown in Figure 6.6. The bed height, flow rate and initial concentration were set at 10 cm, 2.0 mL min^{-1}, and 100 mg L^{-1}, respectively. The results indicated that the breakthrough and exhaust volumes for lead and copper ions occurred at 150; 270 mL and 90; 250 mL, respectively. A similar trend was observed for cadmium, zinc, and nickel ions; however, the breakthrough and exhaustion volumes occurred earlier.

The effects of the hydraulic loading rate were investigated at different flow rates (2–15 mL min^{-1}), initial concentration 100.0 mg L^{-1}; pH 5.0; bed height 10.0 cm, and hydraulic loading rate varying from 2.11 to 15.8 mL min^{-1} cm^2. The results obtained by authors indicated that the HDT increased from 1.33 to 9.98 min, resulting in a reduction in the time needed to reach breakthrough. Han et al. (2006) argued that the higher the hydraulic load, the lower the adsorption, which is due to the short contact time between the adsorbent and the adsorbate.

The results of the aforementioned parameters are shown in Table 6.3. It can be observed that the total time to move the TMZ (t_x) is maximum for Pb^{2+} (135 min) and minimum for Ni^{2+} (40 min), while the times for Cu^{2+}, Zn^{2+}, and Cd^{2+} are between these values. The time needed to move the adsorption zone along the column (t_δ) is between 20 and 80 min. The time taken to form the primary adsorption zone (t_f) is between 12 and 40 min. The fractional capacity of the column (f), in the adsorption

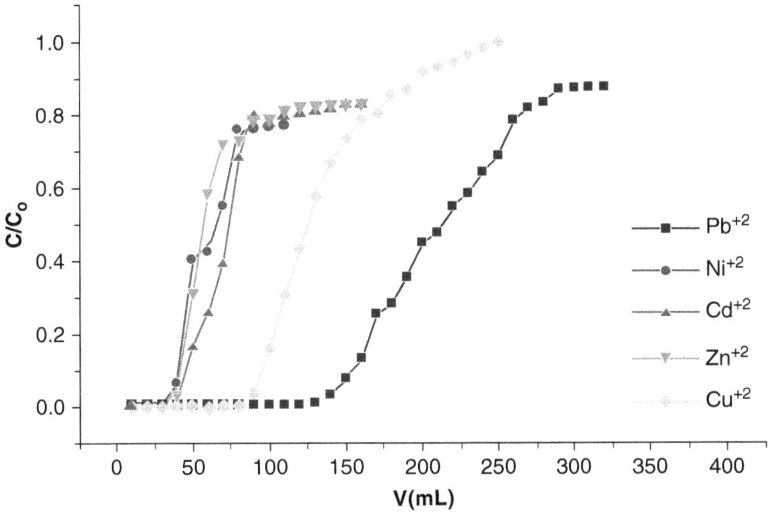

FIGURE 6.6 Multicomponent breakthrough curve on green coconut shell. Condition: multicomponent model solution 100 mg L^{-1}; pH 5.0; flow 2.0 mL min^{-1}; bed height 10 cm; room temperature (28 ± 2°C). (Reproduced with permission from Sousa et al., 2010).

TABLE 6.3
Parameters t_x, t_f, t_δ, F, δ, and the Percentage of Column Saturation

	Parameters					
Metal Ion	**t_x (min)**	**t_δ (min)**	**f**	**t_f (min)**	**δ (cm)**	**SAT (%)**
Cu^{2+}	125.0	80.0	0.500	40.0	9.4	53.0
Ni^{2+}	40.0	20.0	0.320	13.7	7.6	48.1
Zn^{2+}	45.0	25.0	0.390	15.2	8.4	40.2
Cd^{2+}	45.0	25.0	0.500	12.49	7.68	61.60
Pb^{2+}	135.0	60.0	0.500	30.00	5.37	52.95

zone at the point of rupture, to continue removing adsorbate from the solution is 0.500; 0.316; 0.500; 0.394; and 0.500 for Pb^{2+}, Ni^{2+}, Cd^{2+}, Zn^{2+}, and Cu^{2+}, respectively. The percentages of saturation observed were in the following order: $Pb^{2+} > Cd^{2+} > Cu^{2+} > Zn^{2+} > Ni^{2+}$.

To illustrate the effects of parameters such as flow rate, bed height, and concentration of an organic solution on the breakthrough curve performance, the results obtained by Yusoff et al. (2025) using functionalized amine in the adsorption of naproxen are explored. The breakthrough curves for naproxen were obtained for the different conditions of flow rate (3–10 mL min^{-1}), bed height (0.5–1.5 cm), and naproxen concentration (2–8 mg L^{-1}).

The parameters calculated by authors for the specific variables are given in Table 6.4. To this end, the breakthrough time (q_b, mg g^{-1}), saturation time (q_s, mg g^{-1}), MTZ, cm), volume treated at breakthrough time (V_b), saturation time (V_s, L),

TABLE 6.4
Parameters of the Breakthrough Curve Parameters for Naproxen Adsorption on EDA-Modified (PE/PP-g-GMA) Adsorbents in Packed Bed Columns

Q (mL min⁻¹)	Z (cm)	$[NPX]_o$ (mg·L⁻¹)	t_b (h)	V_b (L)	T_s (h)	V_s (L)	q_b (mg g⁻¹)	q_s (mg g⁻¹)	R_b (%)	R_s (%)	MTZ (cm)
3	1.0	5	27	4.86	102	18.36	116.88	279.00	96.19	60.78	0.735
5	1.0	5	12	3.60	81	24.30	86.98	326.10	96.64	53.68	0.852
7	1.0	5	9	3.78	47	19.74	91.54	300.33	96.87	60.86	0.809
10	1.0	5	0.50	0.30	7	4.20	7.31	58.42	97.40	55.64	0.929
7	0.5	5	0.25	0.11	19	7.98	5.15	185.88	98.00	46.59	0.500
7	0.5	5	0.25	0.11	19	7.98	5.15	185.88	98.00	46.59	0.500
7	0.5	5	0.25	0.11	19	7.98	5.15	185.88	98.00	46.59	0.500
7	0.5	5	0.25	0.11	19	7.98	5.15	185.88	98.00	46.59	0.500

Source: Adapted with permission from Yusoff et al. (2025).

percentage removal of naproxen at breakthrough (R_b, %), and saturation time (R_s, %) were calculated. According to theoretical considerations, the breakthrough time (T_B) is 5%, while the exhaustion time (T_S) is 95%, both calculated on the basis of the final solute concentration.

From the results presented in Table 6.4, it is concluded that the optimum conditions for maximum breakthrough, achieved for a C/C_0 value of 0.05, were the following: flow rate of 7 mL min^{-1}, bed height of 1.0 cm (0.2 g of adsorbent) for a solute concentration of 5 mg L^{-1}. This resulted in a shorter saturation time, a higher column capacity (300.33 mg g^{-1}) for the treatment of 3.78 L of naproxen solution (5 mg g^{-1}), resulting in a removal of 96.87% after 47 h of saturation, while the MTZ was 0.809 cm.

MODELING FOR THE DESIGN OF FIXED-BED COLUMNS

The data collected in laboratory-scale or pilot plant tests serve as the basis for the design of large-scale fixed-bed absorber columns. There are two different approaches for this purpose. In the first, an appropriate theoretical model is chosen to calculate the breakthrough curves, requiring parameters such as mass transfer coefficients and diffusivity. To do this, it is necessary to use the mass balance equations of the process being studied. However, in order to solve these equations, it is often useful to use computer programs, since it is not feasible to do this manually. Here we will only give a brief introduction to mass balance for continuous systems.

The second approach is almost entirely empirical, where experiments in columns of different lengths (pilot-scale columns) are carried out, setting parameters such as hydraulic load, particle size, and initial concentration. The use of kinetic parameters and isotherms can also be used, but will not always be accurate. And because it is empirical, as the name implies, it is based on experimental studies and is simpler to carry out, but more time-consuming.

MASS BALANCE EQUATION FOR FIXED-BED COLUMNS

In the study of mass balance in continuous systems, some considerations must be made. It is assumed that the system is isothermal, there is no competition effect between solutes and no variation in the axial velocity of the liquid in the radial direction. The variation in velocity after the adsorbed solute, the channeling effect on the column walls (provided that the particle size/column diameter ratio is above 20), and the variation in interstitial velocity are negligible (Cooney, 1999).

The mass balance equation for a solute in a fixed-bed column is obtained considering a small variation in length ΔZ that has a cross-sectional area equal to S, as in Figure 6.7:

Where ϵ is the void fraction of the bed or interstices and Z is the axial distance of the coordinate.

The solute mass balance along the cross-section can be expressed as:

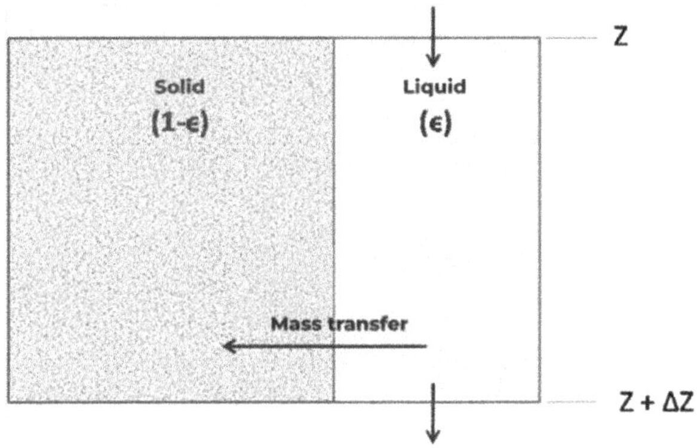

Substituting by terms of an equation, we have (Equation 6.10):

$$\epsilon vS\,(C)_Z - \epsilon vS\,(C)_{Z+\Delta Z} = \epsilon S\Delta Z\left(\frac{\partial C}{\partial t}\right)_Z + (1-\epsilon)S\Delta Z\left(\frac{\partial \bar{q}}{\partial t}\right)_Z \tag{6.10}$$

where

t is the time (min);
C is the concentration of the solute in the fluid phase (mg L^{-1} or mol L^{-1});
q is the average concentration of the solute in the solid phase (mg L^{-1} or mol L^{-1}); and
v is the average axial velocity of the fluid in the interstices ($m^2\ s^{-1}$).

If we consider Q as the volumetric flow rate and S is the cross-sectional area of the bed then, Q, S, and v can be related by the equation $Q = \epsilon vS$

Dividing equation 6.10 by ϵSΔZ and rearranging:

$$v\left[\frac{(C)_{Z+\Delta Z} - (C)_Z}{\Delta Z}\right] + \left(\frac{\partial C}{\partial t}\right)_Z + \frac{(1-\epsilon)}{\epsilon}\left(\frac{\partial \bar{q}}{\partial t}\right)_Z = 0 \tag{6.11}$$

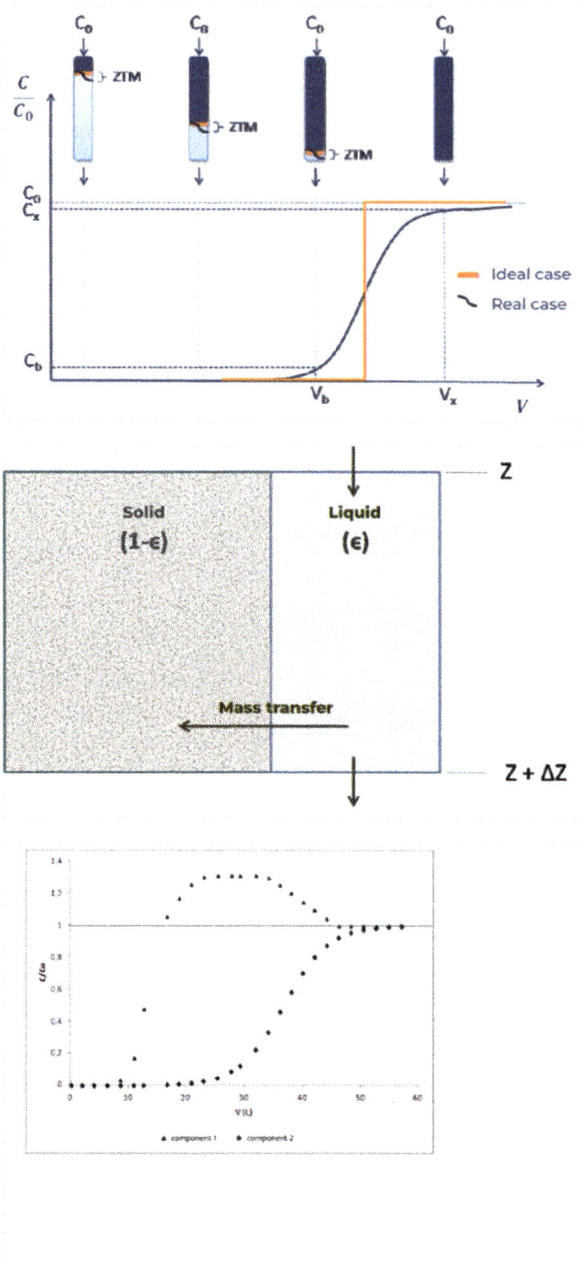

FIGURE 6.7 Schematic diagram of a section of an adsorber bed. (Adapted from Cooney, 1999).

If we take the limit of $\Delta Z \rightarrow 0$, in order to shorten the column to a differential point in space, the first term will simply be the derivative v ∂C∂Zt. Our Equation 6.12 will therefore be the following equation:

$$v\left(\frac{\partial C}{\partial Z}\right)_t + \left(\frac{\partial C}{\partial t}\right)_Z + \left(\frac{1-\in}{\in}\right)\left(\frac{\partial \bar{q}}{\partial t}\right)_Z = 0 \qquad (6.12)$$

This equation can be used to describe the behavior of a solute in an adsorbent, showing how the concentration of the solute in the fluid and solid phase varies with time and position (Cooney, 1999).

EMPIRICAL MODELS

Empty Bed Contact Time (EBCT)

In some full-scale systems, where a standard behavior is not observed, the breakthrough curve will depend on the depth of the adsorbent bed. This dependence can be determined by performing experiments in a pilot-scale column, at the same flow rate, varying the depths (or heights) of the adsorbent bed. Then, the breakthrough times of each curve are determined, obtained from the breakthrough concentration that must be the same for all breakthrough curves. The empty bed contact time (EBCT) is the measure of how long a portion of the fluid remains in the column, based on the fact that the column does not contain adsorbent material (Cooney, 1999; Srivastava et al., 2008). The EBCT is given by Equation 6.13:

$$\text{EBCT}\ (\text{min}) = \frac{V_L}{H \times A_{sec}} \qquad (6.13)$$

where

V_L is the total internal volume of the empty column (mL);
H is the hydraulic load (mL $\text{min}^{-1}\text{cm}^{-2}$); and
A_{sec} is the cross-sectional area (cm^2).

Another important parameter is the AUR, which is given by the ratio between the mass of the adsorbent in the column and the volume of solution treated at the rupture point (Cooney, 1999).

$$\text{UR} = \frac{M_L}{V_b} \qquad (6.14)$$

where

M_L is the mass of adsorbent material inside the column (g), and
V_b is the volume of solution treated at the breaking point (L).

TABLE 6.5
Comparison of the Performance of the Fixed Bed Filled with BFA for Phenol Adsorption under Different Conditions

C_0 (mg dm^{-3})	Q (dm^{3}min^{-1})	D (cm)	Z (cm)	EBCT (min)	T_b (min)	V_b (dm^3)	U_r (g dm^{-3})
100	0.01	2.54	40	19.64	115.39	1.15	46.05
100	0.01	2.54	60	29.46	216.96	2.17	36.73
100	0.01	2.54	75	36.83	496.41	4.96	20.07
100	0.01	2.54	90	44.19	652.11	6.52	18.33
100	0.01	2.54	60	29.46	216.96	2.17	36.73
100	0.02	2.54	60	15.21	166.35	3.33	24.73
100	0.03	2.54	60	9.82	71.05	2.13	37.39
100	0.04	2.54	60	7.37	13.55	0.54	147.04
100	0.02	2.00	60	6.29	59.91	1.2	28.38
100	0.02	2.54	60	15.21	166.35	3.33	24.73
100	0.02	4.00	60	47.14	691.82	13.84	18.43
50	0.02	2.54	60	15.21	248.18	4.96	16.57
100	0.02	2.54	60	15.21	166.35	3.33	24.73
200	0.02	2.54	60	15.21	80.55	1.61	51.07
500	0.02	2.54	60	15.21	19.69	0.39	208.94

Source: Adapted with permission from Srivastava et al. (2008).

This parameter is of great practical importance, as it shows how much adsorbent will be used to treat a certain volume of effluent.

Srivastava et al. (2008) carried out column studies for the removal of phenol using bagasse ash (BFA) as an adsorbent. The effects of adsorbent bed length (Z) (40–90 cm), volumetric flow rate (Q) (01–0.04 dm^{3}min^{-1}), bed diameter (D) (2–4 cm), and initial concentration (C0) (50–500 mg dm^{-3}) were investigated with an influent pH of 6.5. Several models were applied to the data obtained, including the EBCT. The results are shown in Table 6.5.

Figure 6.8 shows the graph of EBCT versus RH obtained from the results of Srivastava et al. (2008). It can be seen that the higher the EBCT, the smaller the amount of adsorbent is required per unit of volume treated, resulting in a constant; that is, no matter how much the EBCT is increased, the RH will remain constant. Generally, this minimum RH value is desired in real applications. Other parameters should also be evaluated. For example, the flow rate used in the study cited of 0.0–0.04 dm^{3}min^{-1} (approximately 10–40 mL min^{-1}) makes a real process very expensive and difficult to perform, considering that the flow rate of a real effluent is in the order of liters per minute.

THOMAS MODEL

This model assumes continuous flow-bed behavior and uses the Langmuir isotherm for equilibrium and reversible second-order reaction kinetics (Thomas, 1944; Thomas, 1948). It is applicable for favorable and unfavorable adsorption conditions.

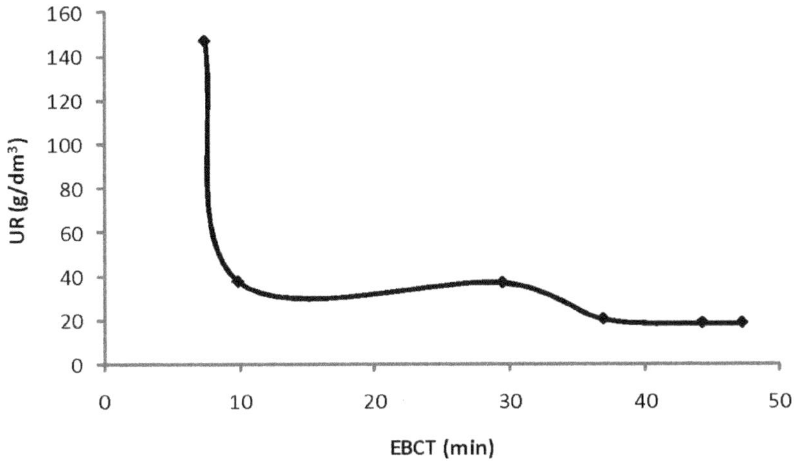

FIGURE 6.8 Dependence of the adsorbent usage rate on the EBCT. (Adapted with permission from Srivastava et al., 2008).

Traditionally, this model is used to determine the maximum adsorption capacity of an adsorbent in continuous systems. The Thomas model is expressed by Equation 6.15 (Vijayaraghavan et al., 2005; Srivastava et al., 2008; Yan & Viraraghavan, 2001; Aksu & Gönen, 2004; Sousa Neto et al., 2013; Yaghmaeian et al., 2014):

$$\frac{C}{C_0} = \frac{1}{1 + e^{\left[\frac{K_t(q_0 m_s - C_0 V_e)}{F_m}\right]}} \tag{6.15}$$

where

C is the concentration of the metal in a given volume (mg L^{-1} or mol L^{-1}); C_0 is the initial concentration of the solution (mg L^{-1} or mol L^{-1}); K_t is the Thomas constant; q_0 is the maximum concentration of the solute in the solid phase, that is, the maximum adsorption capacity (mg g^{-1}); m_s is the mass of adsorbent in grams (g); F_m is the volumetric flow in (L min^{-1}); and V_e is the effluent volume to the column (mL).

The linearized form of the Thomas equation takes the following form:

$$\ln\left(\frac{C_0}{C} - 1\right) = \frac{K_t q_0 m_s}{F_m} - K_t q_0 t \tag{6.16}$$

where

$$\frac{V_e}{F_m} = t \tag{6.17}$$

The bed adsorption capacity q_0 and the coefficient K_t can be obtained through the intercept and slope, respectively, of a curve obtained by plotting

$\ln\left(\frac{C_0}{C} - 1\right)$ versus t or V_e from the breakpoint C_b to the exhaustion point C_x.

YOON–NELSON MODEL

Yoon and Nelson developed a relatively simple model relating the adsorption of gases to activated carbon. Simpler does not require characteristics of the adsorbent or bed, such as mass or adsorption capacity. The application of the Thomas and Yoon–Nelson models in the removal of organic contaminants (such as dyes, pesticides, drugs, and phenols) in fixed-bed adsorption systems is widely studied in the technical-scientific and practical literature in water and effluent treatment (Yusoff et al., 2025; John et al., 2024; İpek et al., 2013).

The nonlinear Equation 6.18 for this model is expressed as:

$$C/C_0 = \frac{1}{1+\exp\left[K_{KY}\left(\tau - t\right)\right]} \tag{6.18}$$

The linearized Yoon–Nelson equation is described as follows:

$$t = \tau + \left(\frac{1}{K_{KY}}\right)\ln\left(\frac{C_e}{C_0 - C_e}\right) \tag{6.19}$$

where

C_e and C_0 are the effluent and affluent concentrations (mg L^{-1}) at a given time t and t=0, K_{YN} (L min^{-1}) is the Yoon–Nelson constant, τ (min) is the time required for 50% breakthrough of the adsorbate. From a linear graph of ln $(C_0/C_e–C_0)$ as a function of time t, the values of K_{YN} and τ were determined from the intercept and slope of the graph.

The application of the Thomas and Yoon–Nelson models is particularly useful in environmental, chemical, and process engineering to predict the efficiency of removal of contaminants from effluents on a laboratory or industrial scale; special attention will be given to their applications in modeling fixed-bed adsorption (packed bed) to describe the removal behavior of inorganic and organic contaminants in aqueous solutions. Thus, both models are valuable tools for the design and optimization of full-scale effluent treatment systems.

In this context, the experimental data obtained by Oliveira et al. (2014) were fitted by the Thomas and Yan–Nelson models. Nonlinear regression analysis was carried out. The sum of squares of the errors (SSE) was examined for each set of experimental data, and the parameters of K_{TH} and q_{TH}, as well as K_{YN} and T were determined for the smallest error values in each case by fitting and optimizing the functions themselves using the solver add-on for Microsoft Excel®.

Figures 6.9 and 6.10 show a comparison of the curves obtained from experimental data and theoretical models predicted by Thomas and Yoon–Nelson. These figures show that the Thomas model can be considered a more suitable kinetic model to describe the adsorption of Cu^{2+}, Zn^{2+}, and Ni^{2+} in a fixed-bed column of sugarcane bagasse and green coconut fibers. However, at the initial stage of adsorption, a small deviation between the model prediction and the experimental data was observed, mainly for zinc by sugarcane bagasse.

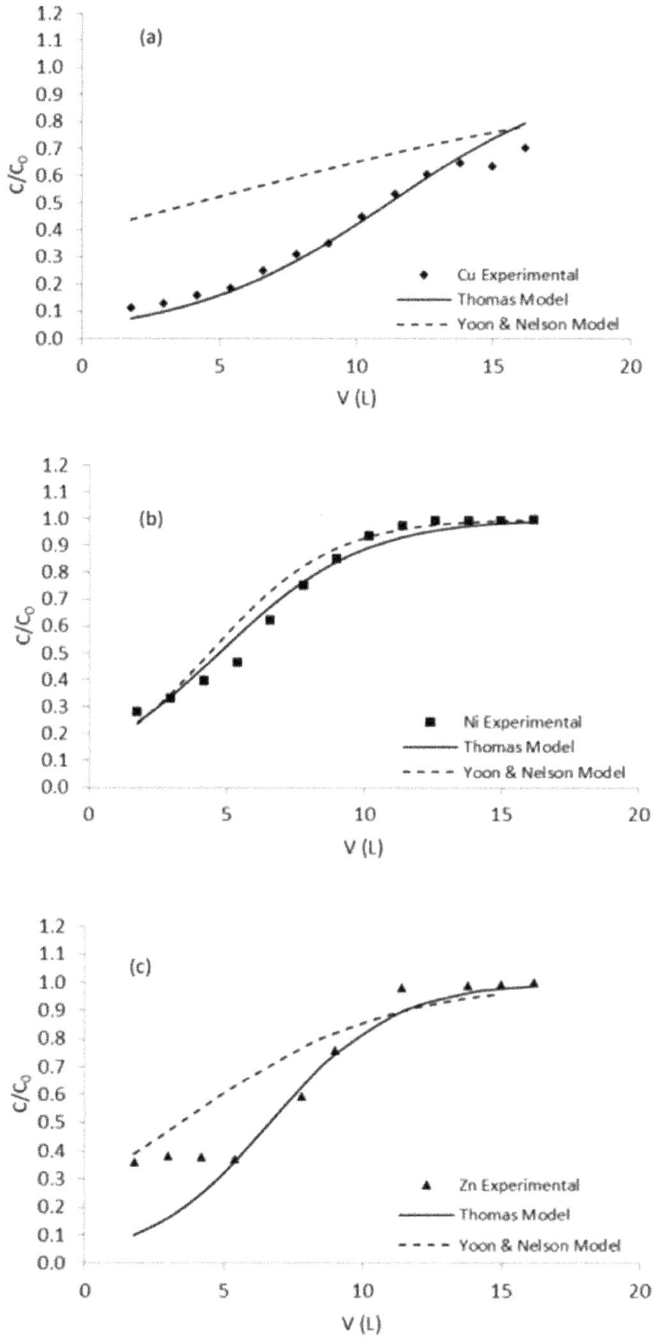

FIGURE 6.9 Comparison of the experimental and theoretical breakthrough curves obtained for (a) Cu^{2+}, (b) Ni^{2+}, and (c) Zn^{2+} by sugarcane bagasse. (Reproduced with permission from Oliveira et al., 2014).

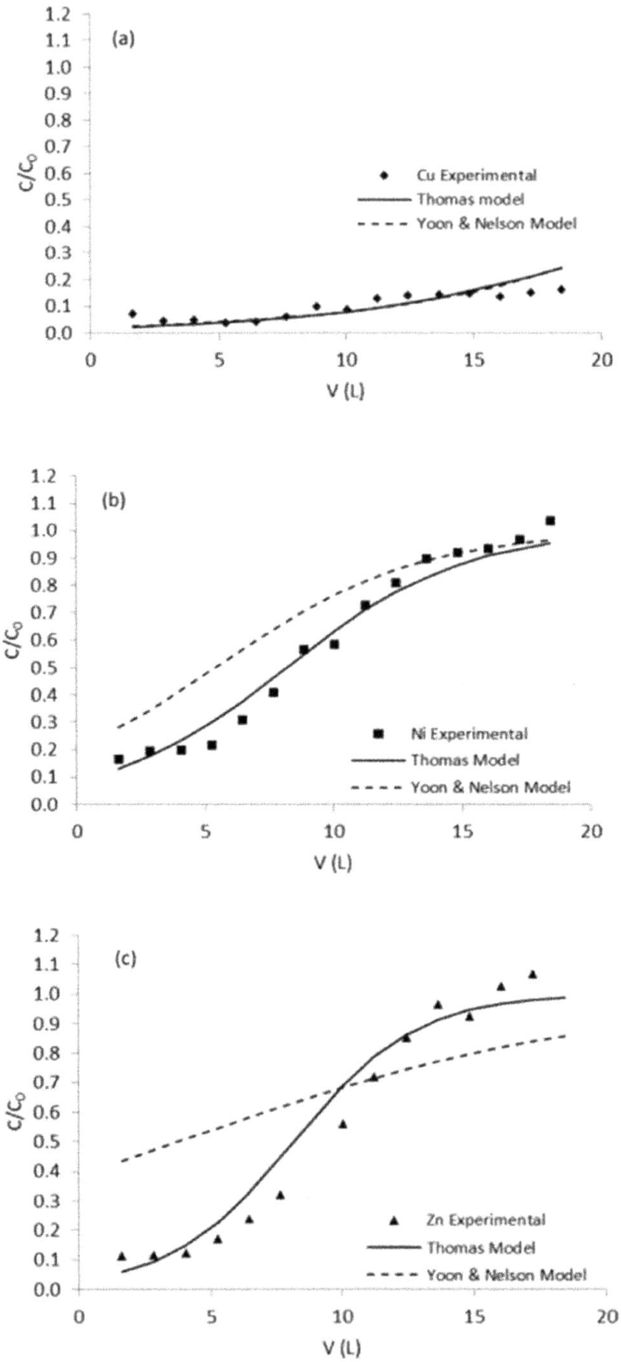

FIGURE 6.10 Comparison of the experimental and theoretical breakthrough curves obtained for (a) Cu^{2+}, (b) Ni^{2+}, and (c) Zn^{2+} by green coconut fiber. (Reproduced with permission from Oliveira et al., 2014).

TABLE 6.6
Calculated Parameters for Thomas and Yoon–Nelson Model from Sugarcane Bagasse

	Thomas Model			Yoon–Nelson Model			
Metal Ions	K_{TM}	q_{TM}	SSE	K_{YN}	τ	τ_{exp}	SSE
Cu^{2+}	9.9×10^{-2}	2.42	2.2×10^{-20}	8.4×10^{-3}	51.2	142.1	8.87
Ni^{2+}	2.0	0.07	4.0×10^{-18}	3.6×10^{-2}	54.6	72.1	1.3×10^{-1}
Zn^{2+}	2.0	0.07	5.7×10^{-1}	2.2×10^{-2}	42.9	97.1	3.36×10^{-1}

Source: Reproduced with permission from Oliveira et al. (2014).

TABLE 6.7
Calculated Parameters for Thomas and Yoon–Nelson Model from Green Coconut Fiber

	Thomas Model			Yoon–Nelson Model			
Metal Ions	K_{TM}	Q_{TM}	SSE	K_{YN}	τ	τ_{exp}	SSE
Cu^{2+}	1.0×10^{-1}	3.23	3.0×10^{-33}	1.0×10^{-2}	426.0	320.5	2.2×10^{-29}
Ni^{2+}	3.78	0.05	7.29×10^{-15}	2.0×10^{-2}	67.3	110.5	2.58
Zn^{2+}	8.0×10^{-1}	0.35	3.6×10^{-20}	9.8×10^{-3}	47.0	125.5	9.1×10^{-1}

Source: Reproduced with permission from Oliveira et al. (2014).

The Thomas model is widely used to evaluate column performance. It can be used to predict the breakthrough curve and the maximum absorption of the solution by the adsorbent. These parameters are essential for the successful design of an adsorption column (İpek et al., 2013). The Yoon–Nelson model fitted perfectly to the data obtained for nickel from sugarcane bagasse and for copper from green coconut shell fibers (Table 6.6).

Tables 6.7 and 6.8 show the parameter values of each model for each metal ion studied, where it can be seen that the experimental data fitted the Thomas model well, with low SSE values, while the Yoon–Nelson model showed high SSE values, reaching 8.87. The Thomas constant, k_{TH}, which characterizes the rate of transfer of metal ions from the solution to the adsorbent, is related to the slope of the curve. The steeper the slope, the lower the constant. The steeper the curve, the higher the constant. In the case of sugarcane bagasse, the value of the constant was lower for the copper ion, as it was for green coconut fibers. Comparing the adsorption capacities obtained by the Thomas model and the experimental results, it can be seen that the values were close, especially for green coconut fibers (Tables 6.7 and 6.8).

TABLE 6.8
Model Fitting Results for Thomas and Yoon–Nelson Dynamic Adsorption

Q (mL min^{-1})	Z (cm)	$[NPX]_o$ (mg L^{-1})	Thomas KT_H (L mg^{-1}h^{-1})	q_0, cal (mg g^{-1})	R^2	Yoon–Nelson K_{YN} (L min^{-1})	τ_{cal} (h)	R^2
3	1	5	0.014	279.37	0.991	0.071	62.08	0.991
5	1	5	0.014	315.05	0.982	0.070	35.97	0.982
7	1	5	0.026	304.19	0.998	0.131	28.97	0.998
10	1	5	0.172	58.64	0.998	0.859	3.91	0.998
7	0.5	5	0.052	172.58	0.990	0.262	8.22	0.990
7	1.5	5	0.014	415.25	0.996	0.070	59.32	0.996
7	1	2	0.045	176.74	0.989	0.091	42.08	0.989
7	1	8	0.040	281.63	0.997	0.316	16.76	0.997

Source: Reproduced with permission from Yusoff et al. (2025).

The q_{TH} for copper ions is about 34 times higher than that of zinc and nickel ions in sugarcane bagasse, while for green coconut fibers it is 64 times higher than that of nickel and 9 times higher than that of zinc. In fact, the higher concentration of copper ions in solution favors adsorption. Obviously, other factors such as electronegativity and ionic radius may be involved. Therefore, the Thomas model can be considered an adequate kinetic model to describe adsorption in a fixed bed. On the other hand, as the Yoon–Nelson model did not fit the data well, it was observed that the 50% breakthrough time (τ) obtained by the model was different from the experimental one.

Yusoff et al. (2025) obtained breakthrough curves using amine-functionalized ethylenediamine for naproxen removal and applied the Thomas and Yoon–Nelson models as shown in Figures 6.12, for data located between $0.05 < C/C_0 < 0.90$. Analysis of the coefficient of determination (R^2) was examined for the experimental data. The parameters of each model were determined for the lowest error values by fitting using the Microsoft Excel® solver.

Figure 6.11 shows the result of the nonlinear fit to the experimental data for various flow rates, bed heights, and initial concentrations of the analyte (naproxen). The regression analysis represented the experimental data for both the Thomas and Yoon–Nelson models with R^2 (>0.98). The data presented in Table 6.1 indicate that increasing the bed height leads to a decrease in K_{TH} and an increase in capacity (q_0). In addition, an increase in flow velocity increases K_{TH}. In relation to the Yoon–Nelson model, K_{YN} increased with increasing flow velocity and initial solute concentration, but the contrast occurred with increasing bed height. The authors concluded that there was no significant difference between the accuracy of the models for the various parameters and different experimental values studied to describe the breakthrough curves of naproxen in functionalized EDA.

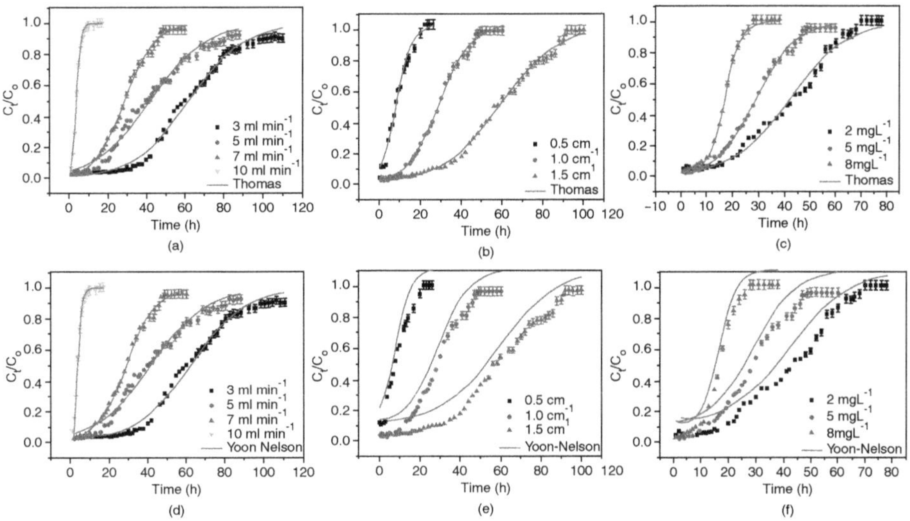

FIGURE 6.11 (a–c) Thomas model and (d–f) Yoon–Nelson model fitting curves of naproxen adsorption onto EDA-((PE/PP)-g-GMA) fixed-bed column under various influencing factors. (Reproduced with permission from Yusoff et al., 2025).

BREAKTHROUGH CURVES FOR A REAL EFFLUENT

Breakthrough curves are essential tools for evaluating the performance of bed-in-column systems for removing organic and inorganic contaminants from real aqueous effluents. It allows the evaluation of the performance of the column in real conditions, taking into account the associated complexities, such as the effect of contaminants, pH variation, influence of organic, and inorganic matter. Thus, the curves show the efficiency of the process over time and allow prediction of when the column adsorbent should be regenerated or replaced, and the optimization of full-scale effluent treatment systems (ref).

In this context, Sousa et al. (2010) investigated a real effluent to validate the models with experimental data and evaluate the effect of pH on the direct efficiency in the process. The procedure was applied to a real sample from the electroplating industry, containing Ni^{2+} (2.787 mg L^{-1}); Zn^{2+} (98.02 mg L^{-1}); and Cu^{2+} (320.4 mg L^{-1}). Figures 6.12 and 6.13 show the breakthrough curves obtained at pH 1.14 and pH 5.0, respectively. The apparent column was saturated with a volume of approximately 30 mL for all the metal ions at pH 1.14. At pH 5.0, saturation occurs at volumes of 70 mL for Ni^{2+}, 90 mL for Zn^{2+}, and 200 mL for Cu^{2+} (Table 6.9). The adsorption properties are highly sensitive to pH, and treatment of a larger quantity of solution is feasible when the pH of the sample is adjusted to 5.0.

The values obtained from the breakthrough curves were used to calculate the parameters t_x, t_f, t_δ, f, and the column saturation percentage. The total time to establish the primary adsorption zone (t_x) is 15 min at pH 1.14 and between 35 and 100 min at pH 5.0. The time required to move the adsorption zone down the column (t_δ) and the time required for the initial formation of the primary adsorption zone (t_f) are

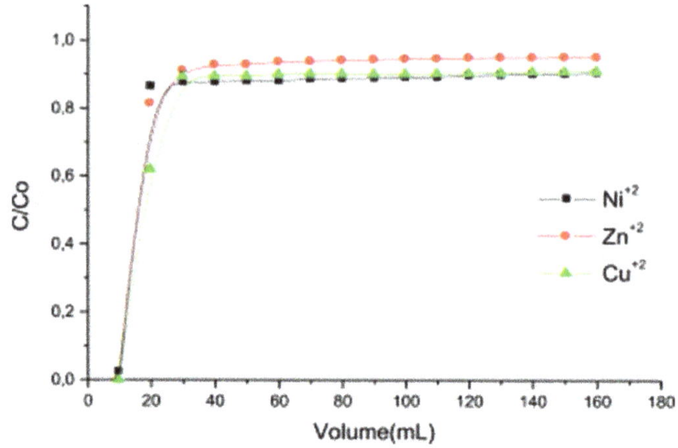

FIGURE 6.12 Breakthrough curve of real sample in the green coconut shell. Condition: pH 1.14; flow 2.0 mL min^{-1}; bed height 10 cm; room temperature (28 ± 2°C). (Reproduced with permission from Sousa et al., 2010.)

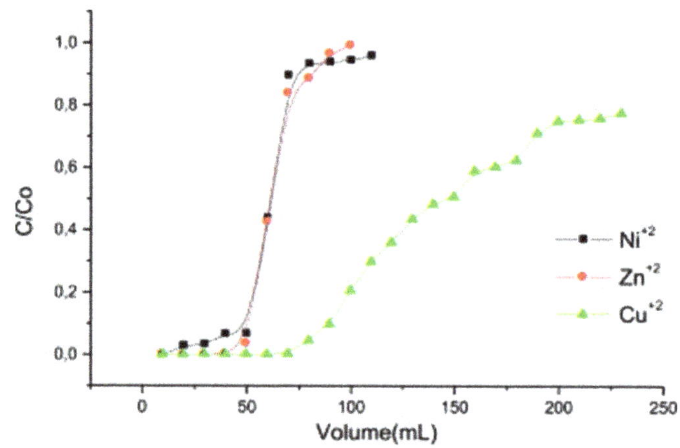

FIGURE 6.13 Breakthrough curve of real sample in the green coconut shell. Condition: pH 5.0; flow 2.0 mL min^{-1}; bed height 10 cm; room temperature (28 ± 2°C). (Reproduced with permission from Sousa et al., 2010.)

5–10 min at pH 1.14 and 10–60 min at pH 5.0. The adsorption capacities increased to 0.132; 5.96 and 42.50 mg g^{-1} for Ni^{2+}, Zn^{2+}, and Cu^{2+}, respectively, at pH 5.0. The saturation percentage at pH 1.14 was around 50% for all the metal ions, while at pH 5.0, saturation was 83.42% for Ni^{2+}; 71.45% for Zn^{2+}; and 56.40% for Cu^{2+} (Table 6.8). It is possible that, at pH 5.0, the competitive effects of protons for the adsorbent's active sites are reduced (Han et al., 2006).

With the K_t and q_0 values for each adsorbate, it is possible to obtain the total mass of adsorbent required to treat an effluent in a column, under the same conditions, requiring only the feed concentration of the adsorbate, the daily flow rate, and the volume of solution to be treated.

TABLE 6.9
Parameters t_x, t_f, t_δ, f, δ, Q, and Saturation (%S) of the Column for the Real Sample in the Effluent at pH 1.14 and 5.0

Metal Ion	t_x (min)	t_δ (min)	t_f (min)	f	δ (cm)	Q (mg g^{-1})	%S
pH 1.14							
Ni^{2+}	15	10	5.00	0.500	10.0	0.056	50.00
Zn^{2+}	15	10	5.00	0.500	10.0	1.98	50.15
Cu^{2+}	15	10	4.99	0.500	9.99	4.92	50.05
pH 5.0							
Ni^{2+}	35	10	4.98	0.502	3.33	0.132	83.42
Zn^{2+}	45	20	10.0	0.500	5.71	5.96	71.45
Cu^{2+}	100	60	30.36	0.494	8.62	42.50	56.40

Source: Reproduced with permission from Sousa et al. (2010).

BED DEPTH SERVICE TIME (BDST)

The data collected from laboratory and pilot plant tests will serve as a basis for the design of a large-scale fixed-bed adsorption column. Among the various parameters, the bed depth required for a specific adsorption time (service time) before adsorbent regeneration is an important parameter. The model (BDST) proposed by Bohart and Adams (1920), which correlates service life with bed depth, can be used to estimate the bed depth required for a given service life. This approach uses breakthrough time data obtained from various bed height (or depth) studies. This model is based on the surface reaction rate theory and can be represented by (Ayoob et al., 2007):

$$\ln\left(\frac{C_0}{C_b}-1\right)=\ln\left(e^{kN_0(x/V)}-1\right)-K\ C_0\ t \tag{6.20}$$

where

C_0 is the initial solute concentration (mg L^{-1}),
C_b is the concentration at the breaking point (mg L^{-1}),
K is the adsorption rate constant (L $mg^{-1}h^{-1}$),
N_0 is the adsorption capacity (mg L^{-1}),
x is the column bed height (cm),
V is the linear flow velocity (cm h^{-1}), and
t is the column service time under the above conditions (h).

At least nine individual column tests must be performed to collect sufficient data necessary to use the Bohart–Adams approach. However, Hutchins (1973) proposed a linear relationship ($t=ax+b$) between service time and bed depth, where it was considered the following equation:

$$e^{kN_0(x/V)} \gg 1 \tag{6.21}$$

Rearranging, we have the following equation:

$$t = \frac{N_0}{C_0V}x - \frac{1}{C_0K}\ln\left(\frac{C_0}{C_b} - 1\right) \tag{6.22}$$

where

a: angular coefficient $\left(\frac{N_0}{C_0V}\right)$

b: linear coefficient $\frac{1}{C_0K}\ln\left(\frac{C_0}{C_b} - 1\right)$

One of the problems with this model is that the adsorbent capacity of the bed changes when the bed height is changed (Ko et al., 2002). It also does not predict the behavior of the bed for multicomponent systems, where the competition effect is present.

Applying the BDST model to the same system as in Figure 6.4, we obtain the graphs in Figure 6.14, where the breakthrough and exhaustion times were plotted for each bed height for each metal ion. Here, only two bed heights were performed and the model was only applied for educational purposes. As stated in the literature (Muhamad et al., 2010; Srivastava et al., 2008), at least three bed heights must be performed to apply the model and obtain reliable data.

The equations relating to the graphs and the values of N_0 and K obtained from them are listed in Table 6.10 below:

- Column parameters:
- Diameter—6.2 cm
- Height—100 or 160 cm
- Cross-sectional area—30.2 cm^2
- Linear flow velocity—397.2 cm h

Since the lines of the graphs of 5% and 90% of bed saturation for the zinc ion were practically parallel, the calculation of the horizontal distance between them was performed as follows: obtain the values of x when y = 0, and subtract the values of x90% and x5%. This distance represents the height of the exchange zone or the MTZ, and the value of 43 cm was found for this ion, corroborating the theoretical value found in Table 6.10. For the Cu^{2+} and Ni^{2+}, the values found were 211 and 87 cm, different from the results in Table 6.1. This may be because the lines are not parallel for these two metals, causing the theoretical and experimental results to diverge. In this case, more studies of bed height should have been performed, as previously mentioned.

It is important to mention that this chapter briefly describes the main topics and models best known and used in the literature for adsorption in fixed-bed columns. The literature also shows many other more specific and complex models and studies of effluent treatment in columns. What was presented in this chapter reflects the basic considerations necessary to be applied in the development of an academic work.

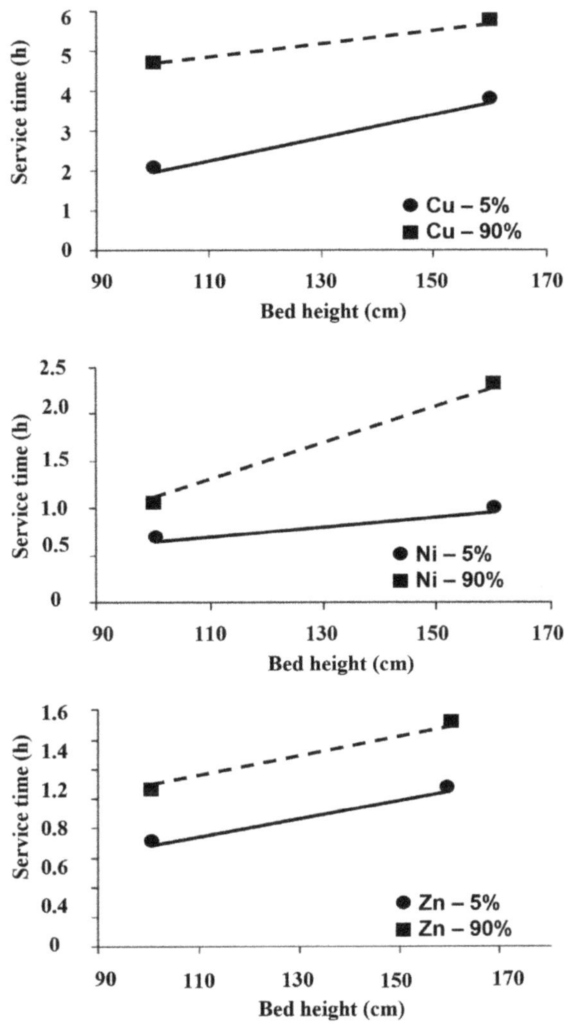

FIGURE 6.14 BDST graph for the metal ions Cu^{2+}, Ni^{2+}, and Zn^{2+} at breakthrough concentrations of 5% and 90%. Conditions: volumetric flow rate: 200 mL min^{-1}, multielement solution $C_0 \approx 200$ mg L^{-1}, pH = 4.8, bed height = 100 and 160 cm and room temperature (28 ± 2°C).

TABLE 6.10
Equations Related to the Graphs in Figure 6.6 and the Values of N_0 and K Obtained from Them

Metal Ion	5%	90%	N_0 (mg L^{-1})	K (L mg^{-1}h^{-1})
Cu^{2+}	y = 0.0283x – 0.833	y = 0.0167 x + 3.033	2198.7	–0.018
Ni^{2+}	y = 0.005x – 0.20	y = 0.0203 x + 0.9553	431.9	+0.0674
Zn^{2+}	y = 0.0063x – 0.0667	y = 0.007 x + 0.38	454.7	–0.28

PROBLEMS SOLVED

Solved Exercises about breakthrough curves on fixed-bed columns, with a focus on adsorption—including concepts such as breakthrough time, treated volume, and application of the Thomas and Yoon–Nelson models in areas, such as water treatment, carbon capture, air emissions control, and environmental remediation

Example 6.1: Breakthrough time

The breakthrough curve represents the output of an adsorbate over time in a continuous fixed-bed system. It is used to describe the performance of the adsorption column.

In a continuous adsorption experiment, the output concentration (C) as a function of time (t) was measured as shown in the table:

Tempo (min)	C_e (mg L^{-1})	C_e/C_0
0	0	0
20	2	0.040
60	5	0.100
40	10	0.200
80	20	0.400
100	45	0.900
120	48	0.960

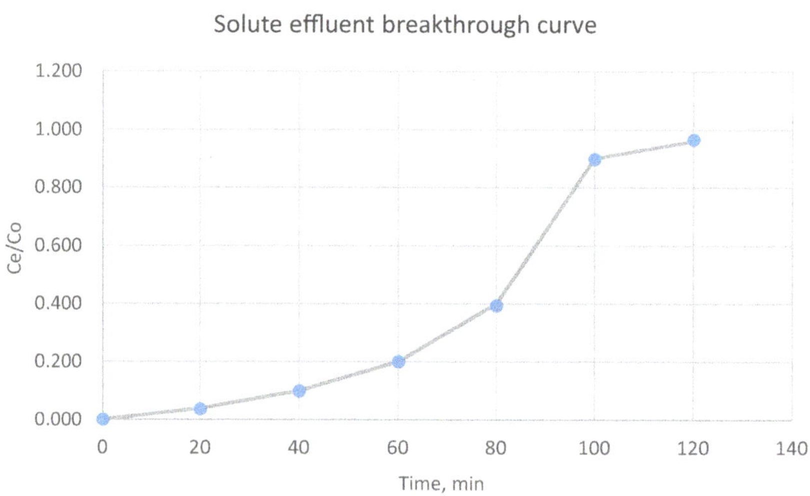

The inlet concentration (C_0) is 50 mg L^{-1}. Calculate the breakthrough time which is defined as the time at which $C = 0.05 - C_0 = 2.5$ mg L^{-1}.

Resolution:

- We are looking for the time when C reaches 2.5 mg L^{-1}.
- According to the table, C = 2 mg L^{-1} at 20 min and C = 5 mg L^{-1} at 40 min.
- By interpolation:

  $t_r = 20 + ((2.5 - 2)/(5 - 2)) * (40 - 20)$

  $t_r = 20 + (0.5/3) * 20 \approx 23.33$ min

  The estimated breakthrough time is approximately 23.3 min.

Example 6.2: Adsorption capacity until breakthrough

Knowing that the flow rate of the system is 10 mL min^{-1} and that the inlet concentration is 50 mg L^{-1}, calculate the mass of adsorbate retained until the breakthrough time (23.33 min).

Resolution:

- Volume treated until breakthrough = flow rate × time = 10 mL min^{-1} × 23.3 3 min = 233.3 mL = 0.2333 L
- Mass adsorbed = concentration × volume = 50 mg L^{-1} × 0.2333 L = 11.665 mg

The amount of adsorbate retained until breakthrough is approximately 11.67 mg.

Example 6.3: Efficiency of the bed

Knowing that total saturation occurs at 120 min (C ≈ C0), calculate the efficiency of the adsorption bed until rupture.

Resolution:

- Total volume treated until saturation = 10 mL min^{-1} × 120 min = 1,200 m L = 1.2 L
- Total possible mass = 50 mg L^{-1} × 1.2 L = 60 mg
- Efficiency = (mass to break/total mass) × 100

  = (11.67/60) × 100 ≈ 19.45%

The efficiency of the bed up to breakthrough time is approximately 19.5%.

Example 6.4: Calculation of the volume treated until the rupture time

Rupture occurs when the contaminant concentration at the column outlet (C_t) begins to increase significantly, indicating that the bed is becoming saturated. Rupture is normally defined when

$C_t/C_0 \geq 0.05$ to 0.10,

In this context:

a. A fixed-bed column operates at a constant flow rate of 20 mL min^{-1} to remove dyes from an industrial effluent. The rupture occurs after 100 min. What is the volume of effluent treated until the rupture?

$V = Q \times t = 20\ mL\ min^{-1} \times 100\ min = 2{,}000\ mL$ ou 2.0 L

b. A fixed-bed column operates at 25 mL min^{-1} treating an effluent containing zinc (Zn^{2+}). Rupture occurs after 80 min.

Volume treated = 25 mL min^{-1} × 80 min = 2,000 mL = 2.0 L

Example 6.5: Column saturation time

The same column as in the previous exercise reaches full saturation at 220 min. What is the efficiency of using the bed until it breaks?

Efficiency = 100 × (rupture time (t_x)/saturation time (t_s))

Efficiency = 100 × (100/220) = 45.45%

Example 6.6: Application of the Thomas model

A 10 g activated carbon adsorption column is used to treat effluent with $C_0 = 50\ mg\ L^{-1}$. The flow rate is 15 mL min^{-1}. The fitted Thomas curve generated the following data:

$k_{Th} = 0.005\ mL\ mg^{-1}\ min^{-1}$

$q_0 = 70\ mg\ g^{-1}$

CALCULATE THE THEORETICAL BREAKTHROUGH TIME (T) WHEN $C_T/C_0 = 0.5$

$\ln((C_0/C_t) - 1) = (k_{Th} * q_0 * m)/Q - k_{Th} * C_0 * t$

$\ln(1) = (0.005 \times 70 \times 10)/15 - 0.005 \times 50 \times t$

$0 = 3.5/15 - 0.25t$

$0 = 0.2333 - 0.25t$

$t = 0.2333/0.25 = 0.933$ min

(Note: too short a time suggests recalculating with another more realistic C_t/C_0, or reviewing the parameters).

Example 6.7: Application of the Yoon–Nelson model

Fitting the experimental data on caffeine adsorption in effluent with the Yoon–Nelson model, we obtained the following equations:

- $k_{YN} = 0.02\ min^{-1}$, $\tau = 120$ min
- Calculate C_t/C_0 at time t = 140 min
- Yoon–Nelson Model:

  $\ln(C_t/(C_0 - C_t)) = k_{YN}(t - \tau)$

  $\ln(C_t/(C_0 - C_t)) = 0.02(140 - 120) = 0.4$

  $(C_t/(C_0 - C_t) = e^{0.4} \approx 1.4918 \Rightarrow C_t = 1.4918(C_0 - C_t)$

  $C_t + 1.4918C_t = 1.4918C_0$

  $C_t = (1.4918/2.4918)C_0 \approx 59.8\%$ de C_0

Example 6.8: CO_2 capture by adsorption

An industrial plant is using a CO_2 capture system by adsorption using a solid adsorbent with a specific surface area of 1,000 m^2/g. The gas stream entering the capture unit contains 10% CO_2 and the gas flow is 1,000 $m^3\ h^{-1}$. Knowing that the adsorption capacity of the material is 2.5 mmol of CO_2 per gram of adsorbent, calculate

1. The amount of CO_2 present in the incoming flow.
2. The mass of adsorbent required to capture 90% of the CO_2 in the gas stream.

Data:

- Gas flow = 1,000 $m^3\ h^{-1}$
- Concentration of CO_2 in the gas stream = 10% (vol/vol)
- Adsorption capacity of the material = 2.5 mmol CO_2/g
- Specific surface area of the adsorbent = 1,000 $m^2\ g^{-1}$

Solution:

1. Amount of CO_2 present in the incoming stream:

   First, we need to calculate the amount of CO_2 present in the gas stream entering the system.

   The gas stream has 10% CO_2; that is, 10% of the total volume is CO_2.
   The amount of CO_2 in the incoming stream (by volume) is
   Volume of CO_2 = Gas flow × (Concentration of CO_2/100)
   Substituting the values:
   Volume of CO_2 = (1,000 $m^3\ h^{-1}$) × (10/100) = 100 $m^3\ h^{-1}$
   Therefore, the flow of CO_2 is 100 $m^3\ h^{-1}$.

   Now, to convert this volume value into moles of CO_2, we use the ideal gas equation. For simplicity, we'll use the molar volume of CO_2 at 25°C and 1 atm, which is approximately 24.45 L mol^{-1} (or 0.02445 $m^3\ mol^{-1}$).

   The amount of CO_2 in moles is
   Mols of CO_2 = Volume of CO_2/Molar volume
   Substituting the values:
   Mols of CO_2 = (100 $m^3\ h^{-1}$)/(0.02445 $m^3\ mol^{-1}$) = 4085.4 mol h^{-1}
2. Mass of adsorbent needed to capture 90% of the CO_2:

   Now, we want to capture 90% of the CO_2 present in the gas stream. The total amount of CO_2 that needs to be captured is 90% of 4085.4 mol h^{-1}:

   CO_2 to be captured = 0.9 × 4085.4 mol h^{-1} = 3676.86 mol h^{-1}

   The adsorption capacity of the material is given as 2.5 mmol of CO_2 per gram of adsorbent. Let's convert this to moles:

   Adsorption capacity = 2.5 mmol g^{-1} = 2.5×10^{-3} mol g^{-1}

   Now, to determine the mass of adsorbent needed, we use the relationship between the amount of CO_2 to be captured and the adsorption capacity:

Mass of adsorbent = CO_2 to be captured/Adsorption capacity
Substituting the values:
Mass of adsorbent = $(3676.86\ mol\ h^{-1})/(2.5 \times 10^{-3}\ mol\ g^{-1}) = 1.470{,}744\ g =$ 1470.74 kg

Final result:

- The amount of CO_2 present in the incoming flow is 100 $m^3\ h^{-1}$.
- The mass of adsorbent needed to capture 90% of the CO_2 in the gas stream is 1470.74 kg.

Example 6.9: Fixed-bed removal efficiency for CO_2 capture

Fixed-bed system with 20 g of adsorbent, CO_2 at 10% v/v, flow rate of 1.5 L min^{-1}. Breakthrough begins at 80 min, full saturation at 150 min. Calculate the total capacity in mg g^{-1}.

Resolution:

Total volume = $1.5 \times 150 = 225$ L
Useful volume = $(80 + 150)/2 \times 1.5 = 172.5$ L
CO_2 volume = $0.10 \times 172.5 = 17.25$ L
CO_2 moles = $17.25/22.4 = 0.77$ mol
CO_2 mass = $0.77 \times 44 = 33.\ 88$ g = 33,880 mg
Capacity = 33,880/20 = 1,694 mg g^{-1}

Example 6.10: Adsorbent sizing for VOC

An activated carbon filter removes 95% of toluene (molar mass = 92 g mol^{-1}) from an air stream containing 300 ppmv of toluene, at a flow rate of 500 $m^3\ h^{-1}$. The carbon has an adsorption capacity of 2.5 mmol g^{-1}. Calculate the mass of coal needed to operate for 10 h.

Resolution:

a. Convert ppmv to mole fraction of toluene:
   $300\ ppmv \times 10^{-6} = 0.0003$
b. Volume of toluene/h:
   $V = 0.0003 \times 500 = 0.15\ m^3\ h^{-1}$
   Mols of toluene/h:
   $n = 0.0240.15 = 6.25\ mol\ h^{-1}$
c. 95% removal for 10 h:
   $n_{total} = 0.95 \times 6.25 \times 10 = 59.38\ mol$
d. Adsorption capacity: 2.5 mmol g^{-1} = 0.0025 mol g^{-1}
   $m_{coal} = 59.38/0.0025$
   $m_{coal} = 23.752\ g = 23.75\ kg$
   Answer: 23.75 kg of activated carbon.

PROPOSED PROBLEMS

6.1 An effluent containing a dye is treated in a fixed-bed column filled with a natural adsorbent in environmental remediation. The experimental output data were obtained according to the table following:

Time (min)	Concentration (mg L^{-1})	C/C0 -
0	0	0.00
10	5	0.05
20	12	0.12
30	25	0.25
40	42	0.42
50	63	0.63
60	81	0.81
70	91	0.91
80	98	0.98

Operating Conditions:

- Initial concentration, $C_0 = 100\,mg\,L^{-1}$
- Flow rate $Q = 10\,mL\,min^{-1}$
- Adsorbent mass $m = 5\,g$
- Volume of solution treated until saturation $\approx 800\,mL$

Breakthrough Curve – According to the data in the table, solve

6.1a Obtain a Breakthrough Curve (C/C_0 vs Time)

6.1b Interpret as the curve at breaking point and saturation ($C/C_0 \approx 1$).

Application of Thomas model

According to the data in the table, solve to Linear Thomas Model

$\ln((C_0/C_t) - 1) = (k_{Th} * q_0 * m)/Q - k_{Th} * C_0 * t$

6.1c Obtain a fitted line of $\ln(C_0/C - 1) \times$ time. Is there an excellent linear correlation with the experimental data?

6.1d Calculate the Thomas constant (k_{Th})

6.1e Calculate the column capacity (q_0, mg g^{-1})

Application of Yoon–Nelson Model

According to the data in the table, solve to Yoon–Nelson model linear:

$\ln(C_t/(C_0 - C_t)) = k_{YN}\,(t - \tau)$

6.1f Obtain a fitted line of the plot of ln ($C/(C_0 - C)$) vs. time, also shows whether there is a very good linear fit, validating the Yoon–Nelson model

6.1g Calculate the Yoon–Nelson constant (k_{YN})

6.1h What is the time (τ) at which 50% of the adsorbent is saturated?

6.1i Calculate the column capacity (q_0, mg g^{-1})

6.2 A fixed-bed column removes CO_2 from a stream with an inlet concentration of 180 mg L^{-1} and a flow rate of 1,000 L min^{-1}.

Data: $k_{Th} = 0.002$ L (mg min)$^{-1}$, $q_0 = 40$ mg g^{-1}, m = 20,000 g, Q = 1,000 L min^{-1}, $C_0 = 180$ mg L^{-1}.

Use the Thomas model to calculate the time t for $C_t/C_0 = 0.5$.

6.3 With the data **for toluene removal:** $k_{YN} = 0.05$ min^{-1}, $\tau = 60$ min, t = 80 min. Use the Yoon–Nelson model to calculate the input toluene concentration

6.4 **Removal of CO_2 and VOCs from contaminated air.** A system treats air containing 0.1% CO_2 and 500 ppmv of toluene, with a flow rate of 800 m^3 h^{-1}. Efficiencies: 85% for CO_2 and 95% for toluene. Calculate the amount removed of each compound (mol h^{-1}).

6.5 **Simultaneous treatment of dyes and metals in effluent.** An effluent contains 150 mg L^{-1} of methylene blue and 80 mg L^{-1} of Cu^{2+}. The adsorbent removes 85% of the dye and 90% of the Cu^{2+}. Flow rate treated: 2 m^3 h^{-1}. Capacity of the adsorbent for Cu^{2+}: 1.8 mmol g^{-1}. Calculate

a. the mass removed per hour of each contaminant
b. The mass of adsorbent for Cu^{2+}.

6.6 **Fixed-Bed Removal Efficiency.** Fixed-bed system with 20 g of adsorbent, CO_2 at 10% v/v, flow rate of 1.5 L min^{-1}. Breakthrough begins at 80 min, full saturation at 150 min. Calculate the total capacity in mg g^{-1}.

6.7 **Mixture of pollutants NO and benzene.** Air with 200 ppmv NO and 150 ppmv benzene, flow rate 2,000 m^3 h^{-1}. Efficiencies: 85% (NO), 90% (benzene). Calculate

a. the mass of each pollutant removed (g h^{-1})
b. The total load removed (g h^{-1}).

6.8 **Regeneration and Adsorbent Sizing.** A fixed-bed adsorbs 500 g of pollutant per cycle before saturation. The adsorbent is regenerated and reused, maintaining 90% of the original capacity each cycle. How many cycles are needed before the capacity is less than 100 g?

6.9 **Saturation time of** fixed-bed column. Use the experimental data on caffeine adsorption: 5 kg adsorbent, $q_0 = 50$ mg g^{-1}, Q = 500 L min^{-1}, $C_0 = 300$ mg L^{-1} for calculate the s**aturation time of** the fixed-bed column.

6.10 An industrial plant uses a CO_2 capture system by adsorption, using a solid adsorbent with a specific surface area of 1,500 m^2 g^{-1}. The incoming gas stream contains 12% CO_2 and the volumetric gas flow is 2,000 m^3 h^{-1}. Knowing that the adsorption capacity of the material is 3.0 mmol of CO_2 per gram of adsorbent, calculate

a. The amount of CO_2 present in the incoming gas stream.
b. The amount of CO_2 to be captured if the system is designed to capture 80% of the CO_2 present.
c. The mass of adsorbent required to capture the amount of CO_2 calculated in item 2.

REFERENCES

Afroze, S., Sen, T. K., & Ang, H. M. (2016). Adsorption performance of continuous fixed bed column for the removal of methylene blue (MB) dye using Eucalyptussheathiana bark biomass. *Research on Chemical Intermediates*, 42(3), 2343–2364.

Aksu, Z., & Gönen, F. (2004). Biosorption of phenol by immobilized activated sludge in a continuous packed bed: Prediction of breakthrough curves. *Process Biochemistry*, 39(5), 599–613.

Aksu, Z., Açikel, U., Kabasakal, E., & Tezer, S. (2002). Equilibrium modelling of individual and simultaneous biosorption of chromium (VI) and nickel (II) onto dried activated sludge. *Water Research*, 36(12), 3063–3073.

Ayoob, S., Gupta, A. K., & Bhakat, P. B. (2007). Analysis of breakthrough developments and modeling of fixed bed adsorption system for As (V) removal from water by modified calcined bauxite (MCB). *Separation and Purification Technology*, 52(3), 430–438.

Benstoem, F., Nahrstedt, A., Boehler, M., Knopp, G., Montag, D., Siegrist, H., & Pinnekamp, J. (2017). Performance of granular activated carbon to remove micropollutants from municipal wastewater—a meta-analysis of pilot-and large-scale studies. *Chemosphere*, 185, 105–118.

Bohart, G. S., & Adams, E. Q. (1920). Some aspects of the behavior of charcoal with respect to chlorine. *Journal of the American Chemical Society*, 42(3), 523–544.

Brasil. (2011). Resolução n. ° 430, de 13 de maio de 2011. Dispõe sobre as condições e padrões de lançamento de efluentes, complementa e altera a Resolução n. ° 357, de 17 de março de 2005, do Conselho Nacional do Meio Ambiente-CONAMA. Diário Oficial da União.

Cooney, D. O. (1999). *Adsorption Design for Wastewater Treatment*. 1ª ed. CRC Press, Florida.

Chu, K. H. (2010). Fixed bed sorption: Setting the record straight on the Bohart–Adams and Thomas models. *Journal of Hazardous Materials,* 177(1–3), 1006–1012.

Dichiara, A. B., Weinstein, S. J., & Rogers, R. E. (2015). On the choice of batch or fixed bed adsorption processes for wastewater treatment. *Industrial & Engineering Chemistry Research,* 54(34), 8579–8586.

Greenwald, M. J., Redding, A. M., & Cannon, F. S. (2015). A rapid kinetic dye test to predict the adsorption of 2-methylisoborneol onto granular activated carbons and to identify the influence of pore volume distributions. *Water Research*, 68,784–792.

Gupta, V., Srivastava, S. K., & Tyagi, R. (2000). Design parameters for the treatment of phenolic wastes by carbon columns (obtained from fertilizer waste material). *Water Research*, 34(5), 1543–1550.

Gupta, V., Srivastava, S. K., & Mohan, S. (1998). Design parameters for fixed bed reactors of activated carbon developed from fertilizer waste for the removal of some heavy metal ions. *Waste Management*, 17(8), 517–522.

Han, R., Zou, W., Li, H., Li, Y., & Shi, J. (2006). Copper(II) and lead(II) removal from aqueous solution in fixed-bed columns by manganese oxide coated zeolite. *Journal of Hazardous Materials*, 137(2), 934–942.

Hussein, M. S., & Ahmed, M. J. (2016). Fixed bed and batch adsorption of benzene and toluene from aromatic hydrocarbons on 5A molecular sieve zeolite. *Materials Chemistry and Physics*, 181, 512–517.

Hutchins, R. A. (1973). New method simplifies design of activated carbon systems. *Chemical Engineering*, 80, 133–138.

İpek, İ. Y., Kabay, N., & Yüksel, M. (2013). Modeling of fixed bed column studies for removal of boron from geothermal water by selective chelating ion exchange resins. *Desalination*, 310, 151–157.

John, M. M., Benettayeb, A., Belkacem, M., Mitchel, C. R., Brahim, M. H., Benettayeb, I., Haddou, B., Al-Farraj, S., Alkahtane, A. A., Ghosh, S., Chia, C. H., Sillanpaa, M., Baigenzhenov, O., & Hosseini-Bandegharaei, A. (2024). An overview on the key advantages and limitations of batch and dynamic modes of biosorption of metal ions. *Chemosphere*, 357, 142051.

Ko, D. C. K., Lee, V. K. C., Porter, J. F., & Mckay, G. (2002). Improved design and optimization models for the fixed bed adsorption of acid dye and zinc ions from effluents. *Journal of Chemical Technology & Biotechnology: International Research in Process, Environmental & Clean Technology*, 77(12), 1289–1295.

Kundu, S., & Gupta, A. K. (2005). Analysis and modeling of fixed bed column operations on As (V) removal by adsorption onto iron oxide-coated cement (IOCC). *Journal of Colloid and Interface Science*, 290(1), 52–60.

Lodeiro, P., Herrero, R., & De Vicente, M. E. S. (2016). Batch desorption studies and multiple sorption–regeneration cycles in a fixed-bed column for Cd (II) elimination by protonated Sargassum muticum. *Journal of Hazardous Materials*, 137(3), 1649–1655.

Moreira, S. A., Sousa, F. W., & Oliveira, A. G. (2009). Remoção de metais de solução aquosa usando bagaço de caju. *Quimica Nova*, 32, 1717–1722.

Muhamad, H., Doan, H., & Lohi, A. (2010). Batch and continuous fixed-bed column biosorption of Cd^{2+} and Cu^{2+}. *Chemical Engineering Journal*, 158(3), 369–377.

Naja, G., & Volesky, B. (2006). Behavior of the mass transfer zone in a biosorption column. *Environmental Science & Technology*, 40(12), 3996–4003.

Oliveira, A. G., Ribeiro, J. P., Melo, D. Q., Sousa, F. W., Neto, V. O. S., Raulino, G. S. C., Cavalcante, R. M., & Nascimento, R. F. (2014). Evaluation of two biosorbents in the removal of metal ions in aqueous using a pilot scale fixed-bed system. *Orbital: The Electronic Journal of Chemistry*, 6(1), 47–55.

Ortiz, F. J. G., Rodríguez, M. B., & Yang, R. T. (2019). Modeling of fixed-bed columns for gas physical adsorption. *Chemical Engineering Journal*, 378, 121985.

Raulino, G. S. C., Vidal, C. A. B., Lima, A. C. A., Melo, D. Q., Oliveira, J. T., Nascimento, R. F. (2014). Treatment influence on green coconut shells for removal of metal ions: Pilot-scale fixed-bed column. *Environmental Technology*, 35, 1711–1720.

Reguer, A., Sochard, S., Hort, C., & Platel, V. (2011). Measurement and modelling of adsorption equilibrium, adsorption kinetics and breakthrough curve of toluene at very low concentrations on to activated carbon. *Environmental Technology,* 32(1), 757–766.

SousaNeto, V. D. O., Raulino, G. S. C., Paulo De Tarso, C. F., Araújo-Silva, M. A., & Nascimento, R. F. (2013). Equilibrium and kinetic studies in adsorption of toxic metal ions for wastewater treatment. In: Z. A. A.-O. Mu Naushad (Ed.), *A Book on Ion Exchange, Adsorption and Solvent Extraction* (Vol. 1, Cap. 1, pp. 1–8). Nova Science Publishers, New York.

Sousa, F. W., Oliveira, A. G., Ribeiro, J. P., Rosa, M. F., Keukeleire, D., & Nascimento, R. F. (2010). Green coconut shells applied as adsorbent for removal of toxic metal ions using fixed-bed column technology. *Journal of Environmental Management*, 91(8), 1634–1640.

Srivastava, V. C., Prasad, B., Mishra, M. I., Mall, I. D., & Swamy, M. M. (2008). Prediction of breakthrough curves for sorptive removal of phenol by bagasse fly ash packed bed. *Industrial & Engineering Chemistry Research*, 47(5), 1603–1613.

Tan, K. L., & Hameed, B. H. (2017). Insight into the adsorption kinetics models for the removal of contaminants from aqueous solutions. *Journal of the Taiwan Institute of Chemical Engineers*, 74, 25–48.

Thomas, H. C. (1948). Chromatography: A problem in kinetics. *Annals of the New York Academy of Sciences*, 49(2), 161–182.

Thomas, H. C. (1944). Heterogeneous ion exchange in a flowing system. *Journal of the American Chemical Society*, 66(10), 1664–1666.

Vijayaraghavan, K., Thilakavathi, M., Palanivelu, K., & Velan, M. (2005). Continuous sorption of copper and cobalt by crab shell particles in a packed column. *Environmental Technology*, 26(3), 267–276.

Wilkins, N. S., Rajendran, A., & Farooq, S. (2021). Dynamic column breakthrough experiments for measurement of adsorption equilibrium and kinetics. *Adsorption*, 27, 397–422.

Worch, E. (2012). *Adsorption Technology in Water Treatment: Fundamentals, Processes, and Modeling*. Walter de Gruyter, Berlin, Germany.

Yaghmaeian, K., Moussavi, G., & Alahabadi, A. (2014). Removal of amoxicillin from contaminated water using NH4Cl-activated carbon: Continuous flow fixed-bed adsorption and catalytic ozonation regeneration. *Chemical Engineering Journal*, 236, 538–544.

Yan, G., & Viraraghavan, T. (2001). Heavy metal removal in a biosorption column by immobilized *M. rouxii* biomass. *Bioresource Technology*, 78(3), 243–249.

Yusoff, N. A., Abdullah, I. C., Jamil, S. N. A. M., Ting, T. M., & Zainuddin, H. S. (2025). High-performance fixed-bed adsorption using active-site-controlled amine-functionalized adsorbent for naproxen removal: insights into isotherms, kinetics, and regeneration. *Journal of Water Process Engineering*, 74, 107802.

7 Experimental Methodology

*Giselle Santiago Cabral Raulino,
Ari Clecius Alves de Lima, Carla Bastos Vidal,
Diego de Quadros Melo,
Vicente de Oliveira Sousa Neto, and
Ronaldo Ferreira do Nascimento*

The application of adsorbent materials in adsorption processes follows a series of stages, of which we can mention three in particular:

Stage 1—Laboratory Tests: planning stage, synthesis, and operating conditions that require a smaller quantity of adsorbent materials and a higher adsorption capacity;
Stage 2—Pilot-Scale Tests: this stage is mainly aimed at modeling the operating conditions for a "scale up";
Stage 3—Full-Scale Tests: stage where the material is tested under real conditions. Most of the time, the materials developed do not reach this point, due to operational costs or low adsorption capacity;

These steps are just a guide for those wishing to begin their studies in adsorption. The steps are not necessarily prerequisites for each other and can be carried out independently. The question of following these steps is mainly related to the costs involved in starting work on steps 2 or 3. Working on a reduced laboratory scale reduces costs, and the necessary adjustments are easier and quicker to make. In addition, stage 1 provides important information for the development of stages 2 and 3, as well as serving as a basis for large-scale modeling, as seen in previous chapters. The description of the contents of this chapter focuses mainly on stages 1 and 2.

Adsorption experiments carried out on a laboratory scale can basically be classified into batch tests (or processes) (finite bath) and continuous flow fixed or expanded bed column tests. In batch tests, the material is placed in the feed, that is, at the start of the process, and after some time it is removed and analyzed. The most widely studied process uses a fixed volume of solution containing the adsorbate of varying concentration, with the mass of adsorbent used remaining fixed. From these experiments, relevant equilibrium, kinetic, and thermodynamic information is obtained, which is fundamental for describing the adsorption process. In column tests, the volume of solution to be treated varies with time. The adsorbate feed concentration remains constant, resulting in a high driving force for a fixed mass of adsorbent.

 DOI: 10.1201/9781003654391-7

It is a dynamic system in which the adsorbent material is confined in a column with known dimensions, and the solution to be treated is pumped through it.

In this chapter, we will see how the experimental procedures for batch and fixed-bed adsorption are developed, as well as the possible problems and solutions encountered during the adsorption experiments.

BATCH ADSORPTION TESTS

Batch experiments are simpler to carry out, and important information can be obtained for a column or pilot-scale experiment. This type of process (batch) is more commonly used for small quantities and the decision to maintain a batch or column adsorption mode on an industrial scale is associated with the volume of solution to be treated and the contact time required or available for the process. For large volumes and short contact times (hydraulics), a column system is more suitable. A batch agitation system for large volumes causes too high an increase in implementation costs.

Figure 7.1 shows examples of experimental setups for batch tests:

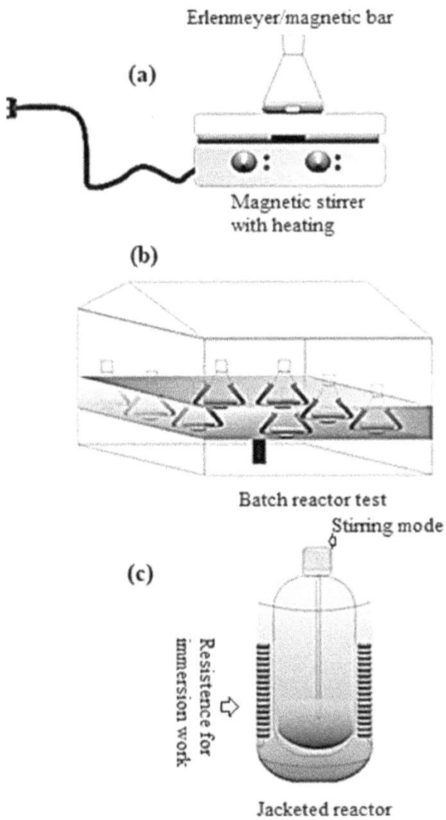

FIGURE 7.1 Types of batch test: magnetic stirrer test (a), stirred thermostatic bath test (b), jacketed batch reactor test (c).

Simple System: a few accessories are used, such as an Erlenmeyer flask and a magnetic stirrer (a),

Complete System: use a stirred thermostatic bath (b), or a jacketed batch reactor (c) (Vidal et al., 2011; Moura et al., 2011; Neto et al., 2011; Vidal et al., 2012; Lima et al., 2012; Melo et al., 2013).

The main variables studied in batch processes are temperature, adsorbent mass, adsorbate concentration, and agitation speed. For adsorption processes in which the adsorbate is in the liquid phase, the pH of the solution is also an important parameter in determining the optimum experimental conditions for kinetics and equilibrium studies. Another variable that can be included in this stage of the adsorption tests is the granulometry of the adsorbent, that is, the size of the particles being used in the study. This variable is closely related to the surface area. As we saw in Chapter 1, the greater the surface area of the adsorbent, the greater its adsorption capacity. Therefore, this variable can also be an important process parameter in adsorption tests.

Figure 7.2 shows an adsorption test with 16 experiments, studying four variables at two levels, that is, two values. The variables and levels chosen are temperature (T1 and T2), adsorbate concentration (C1 and C2), adsorbent mass (M1 and M2), and agitation in revolutions per minute (150 and 300 RPM). In experiment 1 (EXP1), a mass of adsorbent M1 is weighed and a fixed volume of solution containing adsorbate at concentration C1 is added to it in a flask, which can be an Erlenmeyer flask, for example. This flask is then placed under agitation at 150 RPM at temperature T1 for a long enough time to ensure that equilibrium between the phases is reached. This equilibrium time can be estimated and extrapolated from literature data. For example, the equilibrium time for the adsorption of metal ions on lignocellulosic

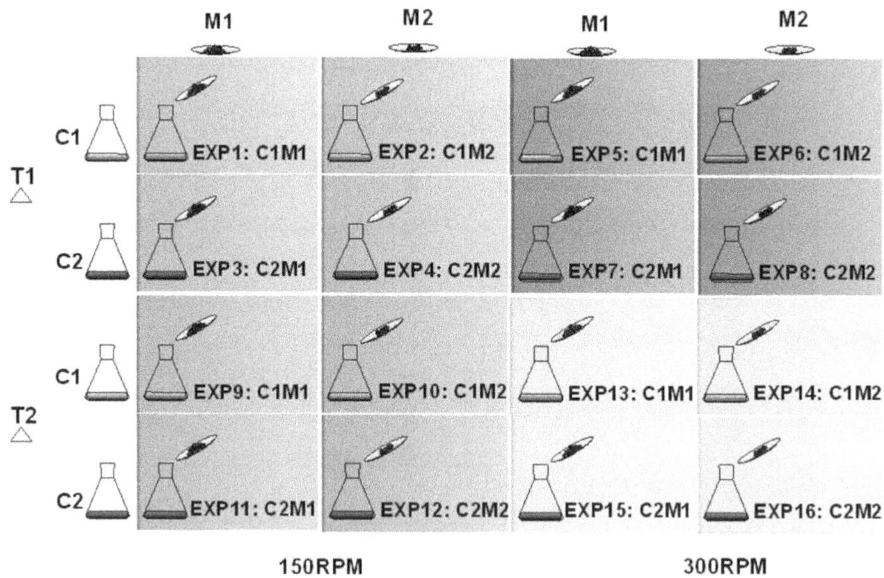

FIGURE 7.2 Schematic of batch adsorption tests.

materials is relatively short, ranging from minutes to a few hours (generally less than 24 h) (Sousa et al., 2010; Neto et al., 2011; Melo et al., 2014; Raulino, 2016). After this time, the final concentration of adsorbate in the fluid phase is measured, and consequently, the adsorption capacity is determined. In experiment 2 (EXP2), a mass M2 of adsorbent is weighed and a fixed volume of solution containing adsorbate at concentration C1 is added to it in another flask. This flask is also placed under agitation at 150 RPM and temperature T1 for a long enough time to ensure that equilibrium between the phases is reached. In other words, in this second experiment, the only variable changed compared to experiment 1 was the mass of the adsorbent. Each experiment is carried out successively, changing the conditions predetermined by the scheme, obtaining the adsorption capacities for each one.

We can see that for each value of each variable studied, there is a combination with each value of all the other variables studied. For example, for a given mass M1, tests are carried out at the two concentrations studied, C1 and C2, at the two temperatures studied, T1 and T2, and at the two agitation speeds studied, 150 and 300 RPM. Thus, within the range of values studied, an overview of the behavior of the adsorption process and the optimum conditions can be obtained, that is, the values of the variables at which the highest adsorption capacities are obtained, for the kinetics and equilibrium tests.

It is important to clarify that the variables to be studied depend on the operating conditions available, the characteristics of the adsorbent and adsorbate, and the objectives to be achieved in the study. For metal ions in aqueous media, for example, high pHs lead to the precipitation of these ions in the form of hydroxides, leading the study range to pH values between acidic and neutral.

Figure 7.3 shows a sequence of variables and possible information that can be obtained from adsorption experiments. In (a), we can obtain the equilibrium time and the percentage removal. In (b), as seen earlier, we can find the reaction rates by varying the concentration (c) and keeping the mass (M) fixed. Keeping C fixed and varying M, through the experiment shown in (c) and using the Arrhenius equation, it is possible to estimate the activation energy and thermodynamic parameters such as Gibbs free energy, enthalpy, and entropy of adsorption. With the test carried out in D, it is possible to verify the influence of diffusion through the film.

ADSORPTION KINETICS EXPERIMENTS

To carry out kinetics experiments, flasks are used, preferably closed ones containing a fixed mass of adsorbent. To this flask is added a fixed volume of solution containing the adsorbate at a given concentration. The pH and the speed of stirring must also be optimized and kept constant. Once the experiment is ready, you can carry out the next step in two ways, depending on the analysis technique to be used to determine the concentration of the adsorbate.

In the first mode, the experiment is carried out in just one container (in the case of duplicates, two containers) where aliquots of the supernatant are collected at predetermined time intervals. Care must be taken here because if the aliquot volume is too large, the volume at the end of the kinetics experiment will be different from the initial volume. For example, if 50 mL of solution and 1 g of adsorbent are mixed in

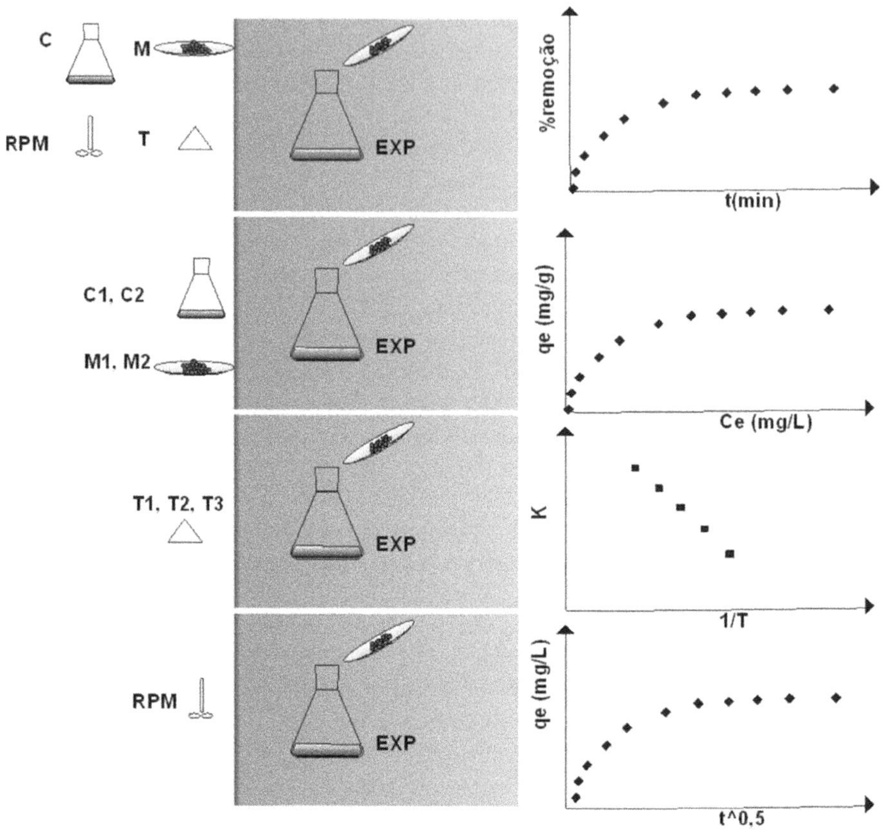

FIGURE 7.3 Scheme 2 of adsorption kinetics experiments.

a metal ion adsorption kinetics experiment, a minimum of 4 mL (undiluted) will be needed at each collection point to analyze the metal ion by flame atomic absorption spectrometry. If 10-time intervals are chosen, the final volume withdrawn will be 30 mL, and in the end, the experiment will be different from the initial one, making it unfeasible to follow this mode of operation. In the case of an adsorbate whose quantification is done by direct injection into a chromatographic system (e.g., gas chromatography) and whose volume order for analysis is μL, this possibility becomes possible.

In the second mode, one flask will be placed for each collection time. Thus, if ten times are chosen for the kinetics study, ten flasks will be needed. For example, for the times chosen of 10, 30, 60, 120, 180, 240, 300, 360, 420, 1,140 min, the procedure adopted is as follows. A flask containing the same mass of adsorbent and volume of solution, with a certain concentration of adsorbate, is placed in the flask. In 10 min, an aliquot is taken from the first flask. In 30 min, the second flask, and so on. The concentration of adsorbate is analyzed for each aliquot collected, and the adsorption capacity is calculated for each time studied. A graph of q (mg or mmol L^{-1}) is plotted against time to determine the time to reach adsorption equilibrium.

Some precautions must be observed in adsorption kinetics experiments. After the predetermined time of the experiment has elapsed, the aliquot of the adsorbate solution to be analyzed must be immediately separated from the adsorbent material to prevent the adsorption process from continuing and the result from being misleading. Another important detail is that if the kinetics are too fast, you should choose shorter time intervals in order to check the kinetic behavior between adsorbent and adsorbate in the early stages.

The initial concentration of the adsorbate is a parameter that interferes with adsorption kinetics. This is due to the increase in the concentration gradient of the adsorbate, which forces it to migrate from a more concentrated environment to a more dilute one. In terms of adsorption, this means an increase in the diffusion of the adsorbate in the pores of the adsorbent, leading to a longer time for equilibrium to be reached. It is therefore suggested that adsorption kinetics studies be carried out at different initial adsorbate concentrations in order to check the equilibrium time at high concentrations.

Models can be applied to the data obtained experimentally in the adsorption kinetics tests, as seen in Chapter 3, in order to explain the kinetic behavior and mechanism of the process studied.

ADSORPTION ISOTHERM EXPERIMENT

To carry out the adsorption isotherm study, the procedure is similar to the kinetics study. A fixed mass of adsorbent is added to flasks with a fixed volume of solution containing the adsorbate. While in the kinetics study, the variation in adsorption capacity as a function of time is studied, in the equilibrium isotherm study, the variation in adsorption capacity as a function of the initial concentration of the adsorbate is studied. Thus, a different concentration of adsorbate is added to each flask. The agitation, contact time between the adsorbate and adsorbent, and temperature, as determined in the previous sections, must be kept constant. After the determined contact time, the concentration of the adsorbate in each flask is determined. Based on the results, a graph is plotted relating the concentration of the adsorbate in the liquid phase (concentration of adsorbate in the solution) and in the solid phase (adsorption capacity), as in Figure 7.4. Through adsorption isotherm experiments, it is possible to obtain the maximum adsorption capacity. If it is observed in the first tests that temperature is a parameter that can influence the maximum adsorption capacity of

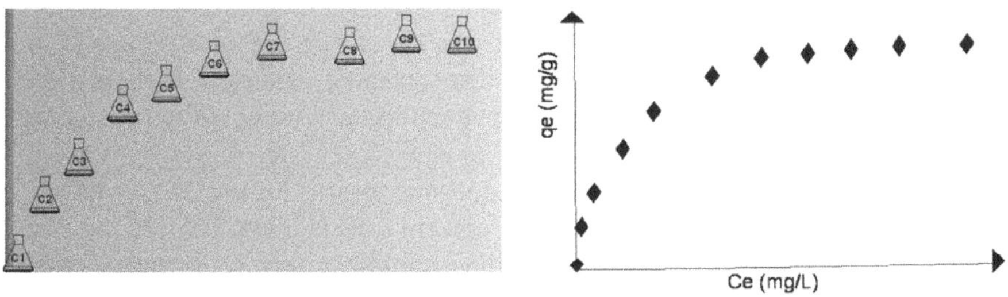

FIGURE 7.4 Adsorption isotherm test scheme.

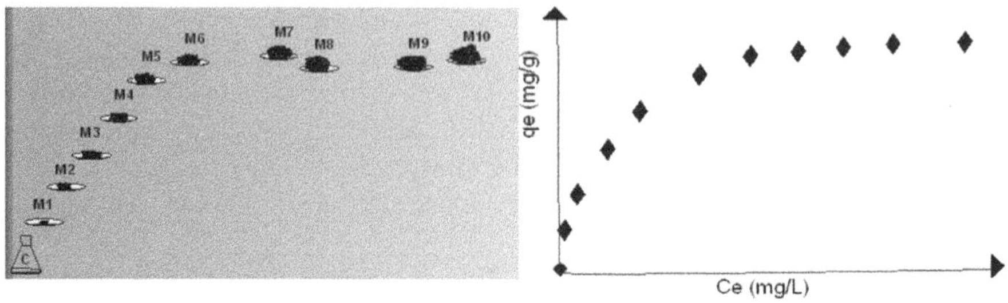

FIGURE 7.5 Adsorbent dosage test scheme.

the material under study, obtaining equilibrium isotherms at different temperatures can be interesting from the perspective of verifying how the adsorbent–adsorbate interaction behaves with temperature.

The researcher can also carry out an isotherm study by varying the mass of the adsorbent instead of the concentration of the adsorbate, and the experimental procedure is carried out in the same way as described earlier. Adsorbent dosage experiments (Figure 7.5) aim to verify whether, by increasing the dosage of the adsorbent, it is possible to remove the adsorbate (pollutant) to acceptable concentration levels, that is, to the values required by current legislation.

FIXED-BED ADSORPTION

Although batch processes are an important stage in adsorption studies, they have some technological disadvantages and high operating costs (for the continuous transportation of solid particles) compared to continuous processes. Continuous processes, on the other hand, involve a continuous flow of mass, in this case adsorbate, through the system at a constant rate at any point in the system over time, and are often used in large-scale processes. This technique is often more economical and basically consists of passing the solution containing the adsorbate through a bed containing the adsorbent solids (Treybal, 1955).

In general, fixed-bed adsorption studies are carried out in a system consisting basically of a container containing the solution to be treated, a peristaltic pump to control the flow rate, and a column filled with the adsorbent material and meshes at the ends of the column to prevent fluctuations in the material, as shown in Figure 6.1.

If you don't already have such a system in your laboratory, it can be assembled in a simple way from materials that can be found easily. The containers and columns can be made of glass or plastic. Care should only be taken with the connections to avoid leaks. Peristaltic pumps are generally the most commonly used due to their stability throughout the experiments. With the system set up, let's move on to the operating variables and parameters.

Adsorption in a fixed bed can be upward or downward (Figure 7.6a), while in an expanded/fluidized bed, it is carried out in upward flow (Figure 7.6b). In the case of fixed-bed column adsorption studies, the most suitable operation is upflow, mainly to avoid bed compaction or preferential paths. This mode of operation is also widely used in regeneration cycles.

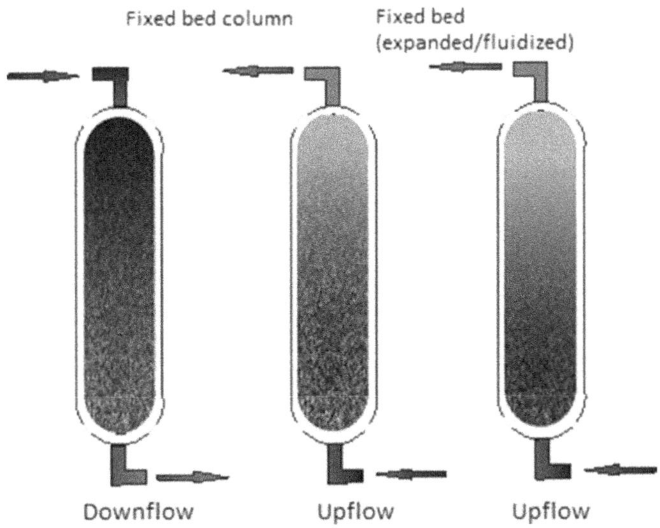

FIGURE 7.6 Schematic of a downflow and upflow adsorption column.

Expanded/fluidized bed column tests are most commonly used in systems where decreasing particle size can cause operational problems. The decrease in particle radius produces a contraction of the fluidized bed, unsaturated particles remain at the top of the bed, and a density gradient appears along the column, which leads to a stabilization of the fluidized bed (Menoud et al., 1998). In this type of bed, there must be a balance between the weight of the particle and the thrust, so that the particle does not leave the bed. Most of the works found in the literature are tests on fixed bed columns (Mccoy & Liapis, 1991; Mowla et al., 2013; Auta & Hameed, 2013; Acheampong et al., 2013; Sousa et al., 2010), but there are also applications in fluidized beds (Tsibranska & Hristova, 2010).

This chapter will only deal with fixed-bed adsorption.

Breakthrough Curve

The breakthrough curve is obtained when a solution initially containing an adsorbate (to be recovered) is brought into contact (continuously) with a bed of adsorbent previously free of adsorbate. Subsequently, the concentration of the adsorbate is determined at the outlet of the bed as a function of time or volume. The graph of concentration (C/C_0) versus time or volume generates a curve of the shape shown in Figure 7.7, known as the breakthrough curve. The technical reading of the breakthrough curve (Figure 7.7) indicates that initially, the upper adsorbent mass in the bed (for downward flow) adsorbs the adsorbate from the solution quickly, reducing its concentration at the column outlet. The distribution of adsorbate in the bed is indicated by the displacement of the vertical lines (1, 2, and 3). Initially, the effluent at the bottom of the bed is practically free of adsorbate (first line 1). In this situation, the top layer of the bed is practically saturated and adsorption takes place in a relatively narrow adsorption zone, where the concentration of the adsorbate changes rapidly.

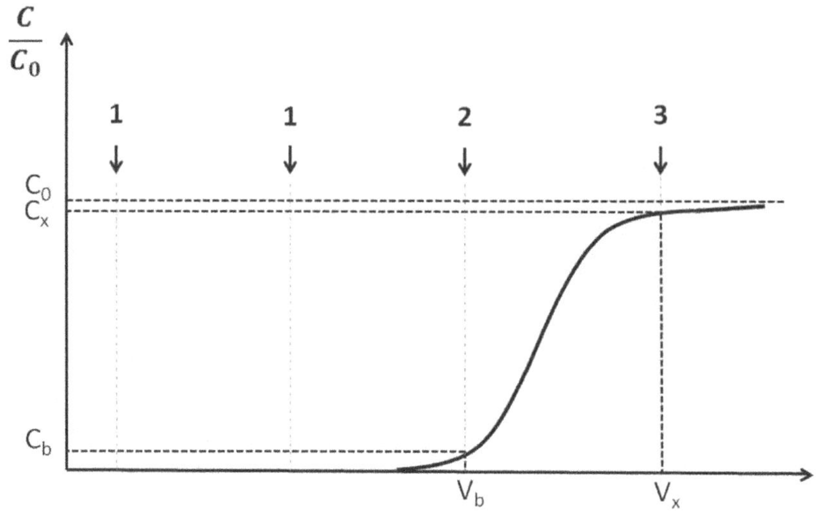

FIGURE 7.7 Diagram of a breakthrough curve.

In a second stage, characterized by the second vertical line 1, part of the bed is already saturated with the adsorbate, but the concentration in the effluent is still substantially zero. In the situation characterized by line 2, the adsorption zone reaches the bottom of the bed, and the concentration of the adsorbate in the effluent increases significantly. The system is said to be starting to break (breakthrough), while in the final situation (line 3), the concentration of the adsorbate in the effluent increases rapidly when the adsorption zone passes through the bottom of the bed, and the final concentration of the adsorbate (C_x) equals the value of the concentration in the initial solution (C_0). Continuing with the flow, for practical purposes, the bed is in equilibrium with the feed solution.

The removal rate and mechanism of the adsorption process, the nature of the adsorption equilibrium, the fluid velocity, the adsorbate concentration in the feed, and the height of the adsorbent bed all contribute to the shape of the breakthrough curve. Generally, the breakthrough time tends to decrease with decreasing bed height, increasing adsorbent particle size, increasing fluid velocity through the bed, and increasing initial adsorbate concentration in the feed.

How to Obtain a Breakthrough Curve

To understand how to obtain a breakthrough curve, we will describe an experimental procedure in which a PVC column 100 cm long and 6.2 cm in diameter was filled with a mass of 402 g of adsorbent material, in this case coconut shell powder, to remove copper ions from solution. The physical parameters of the adsorbent and some properties of the column are described in Table 7.1.

Chapter 6, which describes fixed-bed adsorption, shows how to obtain some of these parameters. The assembled system is similar to that in Figure 6.1. The solution is pumped from the container containing the solution to be treated to the column by a peristaltic

TABLE 7.1
Physical Parameters of the Adsorbent Material and the Column System

Property	Green Coconut Shell Powder
Column diameter (d_L) (cm)	6.2
Bed length (cm)	100
Total column area (cm^2)	2,007.15
Empty column volume (V_L) (cm^3)	3,017.54
Adsorbent mass in column (g)	402
Bulk density (ρap) (g cm^{-3})	0.188
Packing density (ρE) (g cm^{-3})	0.133
Particle volume (V_{ap}) (cm^3)	2,138.29
Bed porosity (ε)	0.292

Source: Raulino (2011).

pump in ascending flow, that is, it enters the lower part of the column and is collected at the upper part. For this column, we will use a flow rate of 200 mL min^{-1}, a synthetic solution containing copper ions (adsorbate) at an initial concentration of 100 mg L^{-1} and an initial pH of the solution equal to 5.0. This last parameter must be kept constant during the experiment, but depending on the nature of the adsorbate and the adsorbent, the pH of the solutions that will be collected at the column outlet may vary.

After the system is assembled and the solution has started to be pumped to the column, with the parameters established earlier, the treated solution will be collected at the column outlet. In the case of using peristaltic pumps, if you do not know the relationship between the flow rate and revolutions per minute (RPM), you can perform a preliminary test using distilled water in order to stabilize the flow rate and obtain this correlation.

Flow rate can be measured using a small measuring cylinder and a stopwatch, which measures the time it takes for the cylinder to fill up to a certain volume mark. Remember that flow rate is the relationship between volume and time. If the volume to be treated is large, you can collect fixed volumes of aliquots at predetermined time intervals. For example, in the system under study, there was approximately 100 L of solution to be treated and it was decided to collect 50 mL aliquots every 10 min. However, if the volume to be treated is small, as is the column, you should work with a much lower flow rate and collect smaller volumes. Sousa et al. (2010) used flow rates of 2–15 mL min^{-1}, bed height of 10 cm, and initial concentration of adsorbates (metal ions) of 100 mg L^{-1} to treat volumes of up to 1 L of solution.

In this study, the authors collected 10 mL of aliquots to be analyzed at the column outlet. It should also be borne in mind that the volume to be collected at the column outlet will depend on the technique used to analyze the adsorbates in the aliquots collected. For example, for the analysis of metal ions by atomic absorption spectrophotometry, depending on the equipment, a minimum volume of 3 mL is required, not counting dilutions and losses.

After the collections have been made and the aliquots analyzed, a graph of C/C_0 (final concentration divided by initial concentration) is plotted at each point collected versus time or volume, obtaining something similar to Figure 7.7.

It is important that when performing an adsorption experiment in a fixed-bed column, the exhaustion point is reached, or at least results close to it are obtained. Only then will it be possible to calculate the parameters related to the mass transfer zone and apply models, as well as predict the behavior of the column on larger scales.

There are many variables that can be evaluated in column adsorption studies, such as flow rate, bed height, column diameter, initial adsorbate concentration, presence of other substances, bed porosity, adsorbent mass, etc. However, the first four are the most studied in the literature. Let us now examine the importance of each of them.

FLOW RATE, BED HEIGHT, AND COLUMN DIAMETER

Figure 7.8a shows a column study scheme where the bed height varies and in Figure 7.8b the flow rate varies.

The variation in flow rate in an adsorbent fixed-bed system is related to the contact time, or hydraulic Detention time (HDT), between the adsorbent and the adsorbate. The HDT is the ratio between the volume of the column and the flow rate being operated. The higher the flow rate, the lower the HDT and vice versa. In a real application, the flow rate cannot be too low, to the point that the treatment becomes unfeasible due to the prolonged time; nor too high, to the point of not allowing sufficient contact between the adsorbent and adsorbate, thus reducing the column's service time. HDT values between 15 and 30 min are usual in real operations (Cooney, 1999). If the peristaltic pump being used only reports RPM, the flow rate can be measured with a measuring cylinder and a stopwatch. When you want to base your study on the HDT, it will depend on the volume of the column. For example, for a column with a diameter of 6.2 cm (radius $r = 3.1$ cm) and a height of 100 cm (h) (assuming that it is completely filled and the connections do not enter this height measurement), the volume of the column will be:

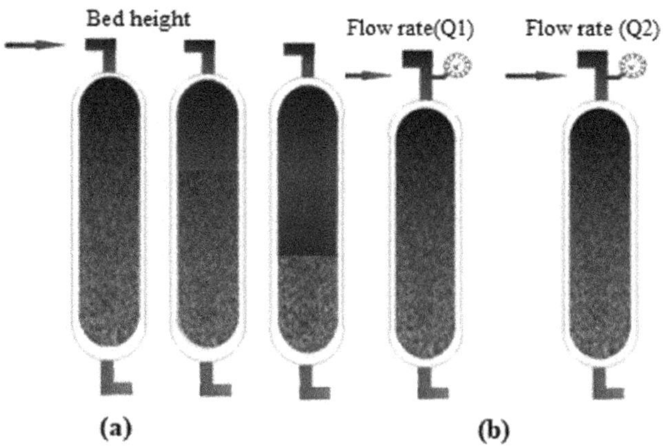

FIGURE 7.8 Column study scheme: (a) bed height and (b) flow rate.

$$V = \pi r^2 h \therefore V = 3.14 * 3.1^2 * 100 = 3{,}017.54 \text{ cm}^3$$

For a 15 min HDT, the required flow rate will be:

$$\text{HDT} = \frac{V\ (\text{mL})}{Q\left(\text{mL min}^{-1}\right)} \rightarrow 15 = \frac{3{,}017.54}{Q} \rightarrow Q = 201 \text{ mL min}^{-1}$$

Thus, a flow rate of approximately 200 mL min^{-1} would be necessary to achieve a HDT of 15 min in a column with the specifications earlier.

The same considerations made for flow rate can be applied to bed height and column diameter, since these three operational parameters are intrinsically related to DHT. These last two parameters mentioned are of utmost importance in the study of fixed-bed adsorption, since they provide predictions about the real-scale behavior of an adsorber bed through the models mentioned in Chapter 6 and other models cited in the literature. We also suggest that at least three values of each variable be studied, enabling a better prediction of the operation of the adsorbent bed.

Porosity

When varying the flow rate, bed height, and column diameter, it is important to keep the bed porosity constant for comparison purposes. Equation 6.16 shows how bed porosity is determined. By specifying a porosity value, it is possible to determine which adsorbent mass values should be studied for each specified flow rate, bed height, and diameter. For example, to the column mentioned above (I.D. = 6.2 cm and h = 100 cm), a mass of 402 g of a certain adsorbent is added whose apparent density is 0.188 g cm^{-3} and packing density is 0.133 g cm^{-3} (Chapter 6 also shows how to obtain these parameters using Equation 6.1). The bed porosity was found to be 0.292. If we double the length of the column, what mass should be added so that the porosity remains the same?

From Equation 6.2, we have:

$$\varepsilon = 1 - \frac{\rho_E}{\rho_{ap}} = \frac{V_L - V_{ap}}{V_L}$$

Since the apparent density is constant, the value to be varied will be the packing density. Since we want this value to also be constant, we will use its formula to calculate the mass required for this new bed height. The new volume will be:

$$V = \pi r^2 h \therefore V = 3.14 * 3.1^2 * 200 = 6{,}035.08 \text{ cm}^3$$

From the packing density, we then calculate the new mass for the new bed height.

$$\rho_E = \frac{M_L}{V_L} \therefore 0.133 \text{ g cm}^{-3} = \frac{M_L}{6{,}035.08 \text{ cm}^3} \rightarrow M_L = 802.6 \text{ g}$$

It is important to note that the greater the porosity of the bed, the greater the void volume and the lower the number of adsorbent sites. Therefore, the lower the porosity of the bed, the more adsorbent sites will be available and possibly the better the adsorbate removal will be. Be careful not to confuse bed porosity with particle porosity. In adsorption studies, the more porous the particle, that is, the more porous the adsorbent, in general, the better the adsorption capacity. In the case of bed porosity, the situation is reversed, as there is an increase in voids within the column.

Initial Adsorbate Concentration and Interfering Substances

The initial adsorbate concentration is a parameter that will allow us to determine the column's service time. The higher the initial adsorbate concentration, the faster the column will be saturated and the shorter its service time will be. Determining the initial adsorbate concentration will also depend on the actual effluent to be treated, since this is how the working parameters will be determined. Thus, let us take as an example an effluent from a fictitious industry that contains copper ions (Cu^{2+}) at a concentration of 20 mg L^{-1} and a daily flow rate (Q) of 100 L day^{-1}. Suppose that this industry wants to use the column mentioned above filled with 1,000 g of adsorbent material, whose adsorption capacity under given conditions is 80 mg g^{-1}. What will then be the column's service time?

To answer this question, we must first calculate the concentration of copper ions discharged daily:

$$C_{Cu} \times Q = 20 \text{ mg L}^{-1} \times 100 \text{ L dia}^{-1} = 2000 \text{ mg dia}^{-1}$$

The total capacity of the column for copper removal is then calculated:

$$q_{total} = q \times m = 80 \text{ mg g}^{-1} \times 1000 \text{ g} = 80{,}000 \text{ mg de Cu}$$

The total removal capacity of the column is then divided by the daily copper concentration:

$$\text{Service time} = 80{,}000 \text{ mg de Cu} \div 2000 \text{ mg dia}^{-1} = 40 \text{ day}$$

It would therefore take 40 days before the column would need to be replaced with a new one.

The last parameter we will address here is the presence of interfering substances. Many ions, such as cyanide and excess hydroxyl, can lead to the complexation or precipitation of metal ions in solution. In the case of organic compounds, the presence of inorganic salts can increase the adsorption of organics on coal, for example. This parameter depends on both the nature of the adsorbate and the adsorbent, as well as on each system being studied.

Of course, the considerations and calculations performed here are simplistic and serve only to estimate some parameters of the column. On a real scale, several processes must be taken into account. For example, is the hourly flow rate constant? Is the pollutant concentration constant? Are there other ions present in solution that can compete with the complexing sites or agents? What is the pH of the solution? All of this must be taken into account, and, therefore, many studies only deal with adsorption on a laboratory scale, or at most a pilot scale.

PROBLEMS SOLVED

Example 7.1: Calculating the Amount Adsorbed (q_e)

During a batch adsorption experiment, 100 mL of a dye solution with an initial concentration of 50 mg L^{-1} was placed in contact with 1.0 g of activated carbon. After equilibrium, the final concentration was 10 mg L^{-1}. How much dye is adsorbed per gram of adsorbent (q_e, in mg g^{-1})?

Resolution

The following data were used during the experiment:

- Initial concentration (C_i) = 50 mg L^{-1}
- Final concentration (C_f) = 10 mg L^{-1}
- Volume (V) = 100 mL = 0.1 L

Adsorbent mass (m) = 1.0 g
Substituting the values:

$$q_e = ((C_i - C_f) * V)/m$$

$$q_e = ((50 - 10) * 0.1)/1.0$$

$$q_e = (40 * 0.1)/1.0$$

$$q_e = 4.0 \text{ mg g}^{-1}$$

Example 7.2: Langmuir Isotherm

Knowing that q_{max} = 20 mg g^{-1} and K_L = 0.2 L mg^{-1}, calculate q_e for an equilibrium concentration of 10 mg L^{-1}.

Data:

- q_{max} = 20 mg g^{-1}
- K_L = 0.2 L mg^{-1}
- C_e = 10 mg L^{-1}

Substituting the values:

$$q_e = (q_{max} * K_L * C_e)/(1 + K_L * C_e)$$

Substituting the values:

$$q_e = (20 * 0.2 * 10)/(1 + 0.2 * 10)$$

$q_e = (40)/(1 + 2) = 40/3$; $q_e \approx 13.33$ mg g^{-1}, Therefore, the amount adsorbed at equilibrium is approximately 13.33 mg g^{-1}.

Example 7.3: Equilibrium Time

An adsorption experiment was monitored over 120 min. The following data were obtained for the amount adsorbed (q_e) as a function of time:

Time (min)	C_f (mg L^{-1})	q_e (mg g^{-1})
10	30	2.0
30	18	3.2
60	12	3.8
90	10	4.0
120	9.5	4.05

What is the approximate equilibrium time? Justify your answer based on the table.

Based on the table provided, we can see that:

- From 90 to 120 min, the value of qe varies from 9.1 to 9.2 mg g^{-1}, a very small variation.
- This indicates that the system reached equilibrium around 90 min.

The equilibrium time is approximately 90 min.

Example 7.4: Removal Efficiency

In an adsorption system, the initial concentration of a pollutant was 100 mg L^{-1}, and after treatment with an adsorbent, it dropped to 25 mg L^{-1}.

Data:

- C_i = 100 mg L^{-1}
- C_f = 25 mg L^{-1}
- % Removal = ((C_i – C_f)/C_i) * 100

Substituting the values:

- % Removal = ((100 – 25)/100) * 100
- % Removal = (75/100) * 100 = 75%

The removal efficiency was 75%.

Example 7.5: Graphical Interpretation

A graph of 1/q_e versus 1/C_e shows a straight line.

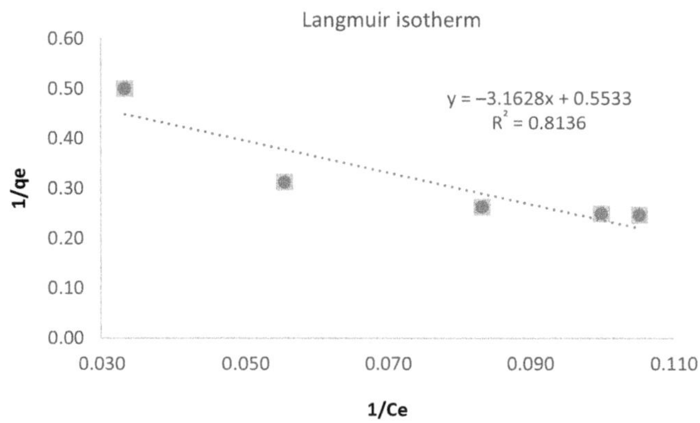

Question:

What kind of adsorption isotherm is this?

What parameters can be extracted from the slope and intercept of the straight line?

- The linearized Langmuir isotherm equation is:

$$1/q_e = (1/q_{max}) + (1/(q_{max} * K_L)) * (1/C_e)$$

- From the intercept, we can obtain ($1/q_{max}$):

$$1/q_{max} = 0.553\ (g\ mg^{-1}) \rightarrow q_{max} = 1.8073\ mg\ g^{-1}$$

- From the slope of the line, we can obtain ($1/q_{max} * K_L$):

$$1/(q_{max}.K_L) = 3.1628\ (g\ L^{-1}) \rightarrow 1/(1.8073\ mg\ g^{-1}.K_L)$$

$$= 3.1628\ g\ L^{-1} \rightarrow K_L = 1/(3.1628 \times 1.8077\ mg\ L^{-1})$$

$$K_L = 1/(5.7174\ mg\ L^{-1}) \rightarrow K_L = 0.1749\ L\ mg^{-1}$$

Thus, the parameters q_{max} and K_L can then be determined from the line.

Example 7.6: Mixture of Pollutants NO and Benzene

Air with 200 ppmv NO and 150 ppmv Benzene, flow rate 2,000 $m^3 h^{-1}$. Efficiencies: 85% (NO), 90% (Benzene). Calculate:

a. The mass of each pollutant removed ($g\ h^{-1}$)
b. The total load removed ($g\ h^{-1}$).

Resolution:

a. Convert ppmv to mole fraction of NO:

$$200\ ppmv \times 10^{-6} = 0.0002$$

Volume of NO

$$V_{NO} = 0.0002 \times 2000 = 0.4\ m^3\ h^{-1}$$

Quantity of moles of NO

$$n = 0.4/0.024 = 16.67\ mol\ h^{-1}$$

Mass of NO (30.0 g mol^{-1})

$$m = 16.67 \times 30 = 500 \text{ g h}^{-1}$$

NO removal

$$0.85 \times 500 = 425 \text{ g h}^{-1}$$

Idem for Benzene (78.11 g mol^{-1}):

$$\text{Volume} = 0.00015 \times 2000 = 0.3 \text{ m}^3 \text{ h}^{-1} \rightarrow \text{mols}(n)$$

$$= 12.5 \text{ mol h}^{-1} \rightarrow \text{mass}(m)$$

$$= 976.38 \text{ g h}^{-1} \rightarrow \text{removal} = 878.74 \text{ g h}^{-1}$$

b. Total load removed

$$\text{Total} : 425 + 878.74 = 1303.74 \text{ g h}^{-1}$$

Answer:

$$425 \text{ g h}^{-1}(\text{NO}), 878.74 \text{ g h}^{-1}(\text{benzene}), \text{total} = 1303.74 \text{ g h}^{-1}.$$

Example 7.7: Batch Adsorption Experiment—Toluene

1. Objective
   To evaluate the adsorption of the atmospheric pollutant toluene in an aqueous medium onto activated carbon using a batch system. Determine the amount adsorbed over time and adjust kinetic, isothermal, and thermodynamic models.
2. Materials and Equipment
   - Toluene solution (100 mg L^{-1})
   - Activated carbon (0.5 g)
   - Erlenmeyer flasks (250 mL)
   - Magnetic or orbital stirrer
   - Filter or centrifuge
   - UV-Vis spectrophotometer or chromatograph (GC)
   - Pipettes, beakers, and thermometer
   - Chronometer
3. Experimental Procedure
   1. Prepare 200 mL of toluene solution with a concentration of 100 mg L^{-1}.
   2. Add 0.5 g of activated carbon.
   3. Stir the mixture at 25°C.
   4. Take samples at times: 0, 10, 30, 60, 90, 120, and 180 min.
   5. Separate the adsorbent and analyze the toluene concentration (Ct).

4. Simulated Results

Tempo (min)	Ct (mg L^{-1})
0	100
10	65
30	42
60	25
90	15
120	12
180	10

5. Calculating the Amount Adsorbed (qt)
   Formula: $q_t = ((C_0 - C_t) \times V)/m$

$$C_0 = 100 \text{ mg } L^{-1}, \quad V = 0.2 \text{ L}, \quad m = 0.5 \text{ g}$$

   Example (t = 30 min): $q_{30} = ((100 - 42) \times 0.2)/0.5 = 11.6/0.5 = 23.2 \text{ mg g}^{-1}$
6. Kinetic Model—Pseudo-Second Order

$$t/q_t = 1/(k_2 - q_e^2) + t/q_e$$

   Graph t/q_t vs t → determine q_e and k_2.
7. Partial Conclusion
   The adsorption of toluene in batches allows the adsorption capacity and kinetic behavior of the system to be determined.
8. Study of Adsorption Isotherms
   Carry out the experiment by varying the initial concentration of toluene (e.g. 20–120 mg L^{-1}) with a constant mass of adsorbent.
   8.1. Isotherm of Langmuir

$$C_e/q_e = 1/(q_m \cdot b) + C_e/q_m$$

   Plot C_e/q_e vs C_e → determine q_m e b.
   8.2. Isotherm of Freundlich

$$\ln(q_e) = \ln(Kf) + (1/n) \cdot \ln(C_e)$$

   Plot $\ln(q_e)$ vs $\ln(C_e)$ → determine Kf e n.
9. Thermodynamic Study of Adsorption
   Perform the experiment at different temperatures (e.g., 298, 308, 318 K).
   Calculate $K_e = q_e/C_e$.
   Use Van't Hoff:

$$\ln(K_e) = -\Delta H°/(R \cdot T) + \Delta S°/R$$

$$\Delta G° = \Delta H° - T\Delta S°$$

$$R = 8314\ J\ (mol \cdot K)^{-1}$$

Plot ln(K_e) versus 1/T to obtain ΔH° and ΔS°.
ΔG° < 0 indicates spontaneous adsorption.

PROPOSED PROBLEMS

7.1 Remediation of a multisystem with adsorption in a soil–water system contaminated with a heavy metal. A soil contaminated with Cd^{2+} is partially saturated with water containing 100 mg L^{-1} of cadmium. After applying biochar, the concentration in the water dropped to 25 mg L^{-1}. 5 g of adsorbent per liter were applied. Calculate:

a. the amount of Cd^{2+} removed per liter.

b. The adsorptive capacity q_e in mg g^{-1}.

7.2 Removal of CO_2 and VOCs from contaminated air. A system treats air containing 0.1% CO_2 and 500 ppmv of toluene, with a flow rate of 800 $m^3 h^{-1}$. Efficiencies: 85% for CO_2 and 95% for toluene. Calculate the amount removed of each compound (mol h^{-1}).

7.3 **Adsorption in slurry with multiple contaminants.** A slurry contains 120 mg L^{-1} of NH_4^+, 60 mg L^{-1} of Pb^{2+}, and 30 mg L^{-1} of dye. Hybrid adsorbent (10 g L^{-1}) removes 70%, 90%, and 80%, respectively. Calculate the total load removed (mg g^{-1}).

7.4 **Calculating the Adsorption Capacity.** An adsorbent material was used to capture CO_2 at 25°C. At equilibrium, with a partial pressure of 0.8 atm, 1.2 mmol of CO_2 was adsorbed per gram of adsorbent. Calculate the adsorptive capacity of the material for CO_2 in mg g^{-1}.

7.5 **Fitting the Langmuir Isotherm**

Experimental data:

Pressure (atm): [0.2, 0.4, 0.6, 0.8]

q (mg g^{-1}): [20, 35, 45, 50]

Use the linearized form of the Langmuir isotherm to estimate q_{max} and K.

7.6 **Modeling with the Freundlich Isotherm**

An adsorbent is tested for the removal of a pollutant in an aqueous solution. The Freundlich isotherm is given by:

$$q_e = K_F . C_e^{(1/n)}, \text{ with } K_F = 5, \quad n = 2, \quad C_e = 10$$

Use the data and calculate the adsorption capacity.

7.7 **Fixed-Bed Removal Efficiency**

Fixed-bed system with 20 g of adsorbent, CO_2 at 10% v/v, flow rate of 1.5 L min^{-1}. Breakthrough begins at 80 min, full saturation at 150 min. Calculate the total capacity in mg g^{-1}.

7.8 Lead adsorption in effluent. A silica-based adsorbent is used to remove Pb^{2+} from an effluent containing 80 mg L^{-1} of the metal. After treatment with 1 g of adsorbent in 100 mL, the final concentration was 20 mg L^{-1}. Determine:

a. The mass of Pb^{2+} removed
b. The adsorptive capacity q_e

7.9 Removing contaminants from soil. A soil contaminated with a pesticide has a concentration of 150 mg kg^{-1}. After adding 10 g of adsorbent per kg of soil and 24 h of contact, the concentration dropped to 30 mg kg^{-1}. Determine the remediation efficiency.

7.10 H_2S removal in a treatment plant. Air with 200 ppmv of H_2S, flow rate 500 m^{3}h^{-1}. 90% removal with a zeolite filter. Calculate the amount removed (mol h^{-1} and g h^{-1}).

7.11 Adsorption Experiments: Batch and Fixed Bed—Atmospheric Pollutant

1. BATCH ADSORPTION

Pollutant: Toluene (simulated)
Solution Volume: 200 mL
Initial Concentration (C_0): 100 mg L^{-1}
Adsorbent Mass: 0.5 g
Temperature: 25°C

Experimental Results

Time (min)	C (mg L^{-1})
0	100
10	65
30	42
60	25
90	15
120	12
180	10

Calculation of the Amount Adsorbed (Q_T)

Equation:

$$q_t = ((C_0 - C_t) - V)/m$$

Data: V = 0.2 L; m = 0.5 g, for t = 30 min:

$$q_t = ((100 - 42) \times 0.2)/0.5 = (58 \times 0.2)/0.5 = 11.6/0.5 = 23.2 \text{ mg g}^{-1}$$

Calculation can be repeated for each time and used in the q_t versus t graph.

Kinetic model—Pseudo-second order
Equation:

$$t/q_t = 1/(k_2 - q_e^2) + t/q_e$$

Plot graph t/q_t versus t for linear adjustment and determination of q_e and k_2.

2. FIXED-BED ADSORPTION

Parameters:
Column Diameter: 2 cm
Bed Height: 10 cm
Flow Rate (Q): 50 mL min^{-1} = 0.05 L min^{-1}

$$C_0 = 100 \text{ mg L}^{-1}$$

Adsorbent mass: 10 g (simulated)
Outlet concentration results (Ct):

Time (min)	Ct (mg L^{-1})
0	0
10	5
20	15
30	40
40	70
50	90
60	100

Breakthrough Curve

Plot Ct/C_0 versus time to identify breakthrough and depletion time.
Total adsorption capacity (q_{total}):

$$q_{total} = \left(Q/m\right) \times \int_0^t (C_0 - C_t)\,dt$$

Using the trapezoidal rule:

$$\text{Area} \approx [(\Delta t/2) \times \Sigma(C_0 - C_t \text{ initial} + C_0 - C_t \text{ final})]$$

Calculation with the Values Provided

Pollutant: Toluene (simulated)
Solution volume: 200 mL
Initial concentration (C_0): 100 mg L^{-1}
Adsorbent mass: 0.5 g
Temperature: 25°C

Modeling with Thomas

Thomas equation:

$$\ln\left((C_0/C_t) - 1\right) = \left(k_{Th} \cdot q_0 \cdot m\right) / Q - k_{Th} - C_0 \cdot t$$

Adjust graph $\ln((C_0/C_t) - 1)$ versus t → straight line: angular coefficient = $-k_{Th} \cdot C_0$

Modeling with Yoon–Nelson

Yoon–Nelson equation:

$$\ln(C_t / (C_0 - C_t)) = k_{YN} - (t - \tau)$$

Adjust graph $\ln(C_t/(C_0 - C_t))$ versus $t \rightarrow \tau$ is the time at which $C_t = 0.5\ C_0$.

REFERENCES

Acheampong, M. A. et al. (2013). Removal of Cu(II) by biosorption onto coconut shell in fixed-bed column systems. *Journal of Industrial and Engineering Chemistry*, 19, 841–848.

Auta, M., & Hameed, B. H. (2013). Coalesced chitosan activated carbon composite for batch and fixed-bed adsorption of cationic and anionic dyes. *Colloids and Surfaces B: Biointerfaces*, 105, 199–206.

Cooney, D. O. (1999). *Adsorption design for wastewater treatment*. CRC Press, Florida.

Lima, A. C. A. et al. (2012). Modified coconut shell fibers: A green and economical sorbent for the removal of anions from aqueous solutions. *Chemical Engineering Journal*, 185–186, 274–284.

Mccoy, M. A., & Liapis, A. I. (1991). Evaluation of kinetic models for biospecific adsorption and its implications for finite bath and column performance. *Journal of Chromatography*, 548, 25–60.

Melo, D. Q. et al. (2013). Adsorption equilibria of Cu^{2+}, Zn^{2+}, and Cd^{2+} on EDTA-functionalized silica spheres. *Journal of Chemical Engineering Data*, 58, 798–806.

Melo, D. Q. et al. (2014). Removal of Cd^{2+}, Cu^{2+}, Ni^{2+}, and Pb^{2+} ions from aqueous solutions using tururi fibers as an adsorbent. *Journal of Applied Polymer Science*, 131(20), 1–12.

Menoud, P., Calvin, L., & Renken, A. (1998). Modelling of heavy metals adsorption to a chelating resin in a fluidized bed reactor. *Chemical Engineering and Processing: Process Intensification*, 37, 89–101.

Moura, C. P. et al. (2011). Adsorption of BTX (benzene, toluene, o-xylene, and p-xylene) from aqueous solutions by modified periodic mesoporous organosilica. *Journal of Colloid and Interface Science*, 363, 626–634.

Mowla, D., Karimi, G., & Salehi, K. (2013). Modeling of the adsorption breakthrough behaviors of oil from salty waters in a fixed bed of commercial. *Chemical Engineering Journal*, 218, 116–125.

Neto, V. O. S. et al. (2011). Use of coconut bagasse as alternative adsorbent for separation of copper (II) ions from aqueous solutions: Isotherms, kinetics, and thermodynamic studies. *Bioresources and Technology*, 6, 3376–3395.

Raulino, G. S. C. (2011). Sistema piloto de adsorção de íons de metais em coluna (leito fixo) utilizando como adsorvente o pó da casca de coco verde. 2011. 110 f. Dissertação (Mestrado em Saneamento Ambiental) – Engenharia Civil, Universidade Federal do Ceará, Fortaleza, 2011.

Raulino, G. S. C. (2016). Biossorção em sistema multielementar dos íons Pb^{2+}, Cu^{2+}, Ni^{2+}, Cd^{2+} e Zn^{2+} em solução aquosa usando a vagem seca do feijão (Phaseolus vulgaris L.) modificada: otimização usando planejamento fatorial. 2016. 164 f. Tese (Doutorado em Saneamento Ambiental) – Engenharia Civil, Universidade Federal do Ceará, Fortaleza, 2016.

Sousa, F. W. et al. (2010). Green coconut shells applied as adsorbent for removal of toxic metal ions using fixed-bed column technology. *Journal of Environmental Management*, 91, 1634–1640.

Treybal, E. R. (1955). *Mass-Transfer Operations*. McGraw-Hill, New York.

Tsibranska, I., & Hristova, E. (2010). Modelling of heavy metal adsorption into activated carbon from apricot stones in fluidized bed. *Chemical Engineering and Processing: Process Intensification*, 49, 1122–1127.

Vidal, C. B. et al. (2011). Adsorption of polycyclic aromatic hydrocarbons from aqueous solutions by modified periodic mesoporous organosilica. *Journal of Colloid and Interface Science*, 357, 466–473.

Vidal, C. B. et al. (2012). BTEX removal from aqueous solutions by HDTMA-modified Y zeolite. *Journal of Environmental Management*, 112, 178–185.

8 Statistical Tools for Adsorption Experiments

Ari Clecius Alves de Lima,
Breno Aragão dos Santos,
Diego de Quadros Melo, Carla Bastos Vidal,
Giselle Santiago Cabral Raulino, and
Ronaldo Ferreira do Nascimento

Mathematics serves as a starting point in the use of statistical tools. Without basic knowledge, it is difficult and sometimes impossible to interpret the data of a survey. Knowledge of statistics allows the user to choose and use appropriate statistical software according to need. This chapter will give some notions of regression analysis and the design of experiments. For more in-depth studies, it is recommended to read the specialized literature (Bates & Watts, 1988; Draper & Smith, 1998; Neto et al., 2003; Seber & Wild, 2003; Montgomery et al., 2006; Rodrigues & Iemma, 2009; Cornell, 2011).

The use of statistical tools is very important in the stages of data collection (sampling), data processing, and inference. Choosing good statistical software also helps in the development of the work. On the market, there are many commercial and free software. Most of the time, the researcher does not have the availability of statistical software (the license) and must choose between buying or using free software. There is excellent commercial statistical software such as SAS, SPSS, Statistica, Unscranbler, Stata, Origin, Minitab, Pirouett, MODDE, and the price varies according to the packages offered. Despite not being statistical software, Microsoft Excel spreadsheets are also used for many applications. However, there is excellent free software that is just as good as commercial ones, such as Octave, Scilab, R Project, and Portal Action.

Of these, the R Project (an open-source software) stands out for the availability of material available for consultation, for which a specialized magazine "The R Journal" was created. In addition, it is compatible with several platforms, has a link with the programming languages C, C++ and Fortran, and it is possible to obtain databases in R, including IBGE data. A guide for installing R Project can be obtained from the http://cran.r-project.org/doc/contrib/Itano-installation.pdf website. Changes to the graphics of R Project can be performed in R itself, in free image editing software such as Inkscape or in commercial software such as Corel Draw, for this it is necessary to save the file with the "metafile" format.

DOI: 10.1201/9781003654391-8

REGRESSION ANALYSIS (LINEAR AND NONLINEAR) APPLIED TO ADSORPTION

Regression analysis is a statistical technique that investigates the relationship between variables through a mathematical model. Simple linear regression analysis and simple nonlinear regression analysis for adsorption isotherms will be addressed, as well as multiple linear regression analysis for the design of factorial and mixing experiments.

Regression analysis is used for several adsorption models such as: pseudo-1st-order kinetics model, pseudo-2nd-order kinetics model, Langmuir adsorption isotherm model, Freundlich, Redlich-Peterson, Temkin, Toth, and rupture curve models (Mccoy & Liapis, 1991; Menoud et al., 1998; Sousa et al., 2010; Tsibranska & Hristova, 2010; Moura et al., 2011; Neto et al., 2011; Vidal et al., 2011, 2012, 2016; Lima et al., 2012). Through regression analysis, it is possible to estimate the parameters of the models, thus being able to infer about the types of mechanisms involved in the adsorption processes, such as reaction rates, adsorption capacity, diffusion coefficients, etc.

LINEAR REGRESSION ANALYSIS

A simple linear regression model is a model in which we have the relationship between a dependent variable y, called the response variable, and an independent variable x, called the regressor or predictor variable. Normally, regression equations are valid only for the region of the regressor variable contained in the observed data. The simple linear regression model can be represented by the Equation 8.1:

$$y = \beta_0 + \beta_1 x + \varepsilon \tag{8.1}$$

Where, β_0 e β_1 are the regression coefficients. The coefficient β_1 represents the change in the distribution of the mean of y produced by change in the unit of x. If the data range includes $x = 0$, then the intercept β_0 is the mean of the distribution of the answer y when $x = 0$. If the track does not include zero, $\beta_{0,}$ there is no practical interpretation. The coefficient ε represents the fit error.

ESTIMATION OF SIMPLE LINEAR REGRESSION PARAMETERS OF A LANGMUIR ADSORPTION ISOTHERM MODEL

Consider a batch adsorption process for experimental data involving nickel studies in cashew meal, where the data are shown in Table 8.1, relate the values of q_e, response variable, and C_e, predictor variable:

In Langmuir's equation (8.2), the parameters to be estimated are q_{max}, which is the maximum adsorption capacity, and K_L, which is Langmuir's constant.

$$q_e = \frac{q_{máx} K_L C_e}{(1 + K_L C_e)} \tag{8.2}$$

TABLE 8.1
Data from C_e and q_e

Sample	C_e	q_e
1	0.242	0.350
2	0.336	0.486
3	0.557	0.795
4	2.633	2.751
5	9.100	6.334
6	44.949	9.835
7	99.284	10.310
8	130.661	11.215
9	213.384	11.454
10	254.524	12.407

Source: Data from authors.

It is important to know that Equation 8.2 can be linearized in several ways, Equations 8.3–8.6 (Bolster & Hornberger, 2006):

$$\frac{C_e}{q_e} = \frac{1}{q_{máx}K_L} + \frac{C_e}{q_{máx}} \text{(Hannes – Woolf)} \tag{8.3}$$

$$\frac{1}{q_e} = \frac{1}{q_{máx}K_L C_e} + \frac{1}{q_{máx}} \text{(Lineweaver – Burker)} \tag{8.4}$$

$$q_e = q_{máx} - \frac{q_{máx}}{K_L C_e} \text{(Eadie – Hofstee)} \tag{8.5}$$

$$\frac{q_e}{C_e} = q_{máx}K_L - q_e K_L \text{(Scatchard)} \tag{8.6}$$

From these models, it is possible to estimate, by the method of least squares, the parameters of the Langmuir model, q_{max} and K_L. To show how linear regression analysis works, we will choose the linearized form given in Equation 8.4. For the other linearized equations, only the final result will be shown.

- The first step is to create the table in the Microsoft Excel, Table 8.2:

The second step is to select data/data analysis/regression from the menu bar and fill in the values, Figure 8.1:

The results will be generated in Tables 8.3, 8.5–8.7. Table 8.3 presents some measures of adjustment of the regression model, the best known being the coefficient of determination or R-squared, adjusted R-squared, which takes into account the number of degrees of freedom, standard error, and the number of observations in this same table.

TABLE 8.2
Excel Table with Linearized Data

Amostra	C_e	q_e	$1/q_e$	$1/C_e$
1	0.242	0.350	4.132	2.857
2	0.336	0.486	2.976	2.058
3	0.557	0.795	1.795	1.258
4	2.633	2.751	0.38	0.364
5	9.100	6.334	0.11	0.158
6	44.949	9.835	0.022	0.102
7	99.284	10.31	0.01	0.097
8	130.384	11.215	0.008	0.089
9	213.384	11.454	0.005	0.087
10	254.524	12.407	0.004	0.081

Source: Data from the authors.

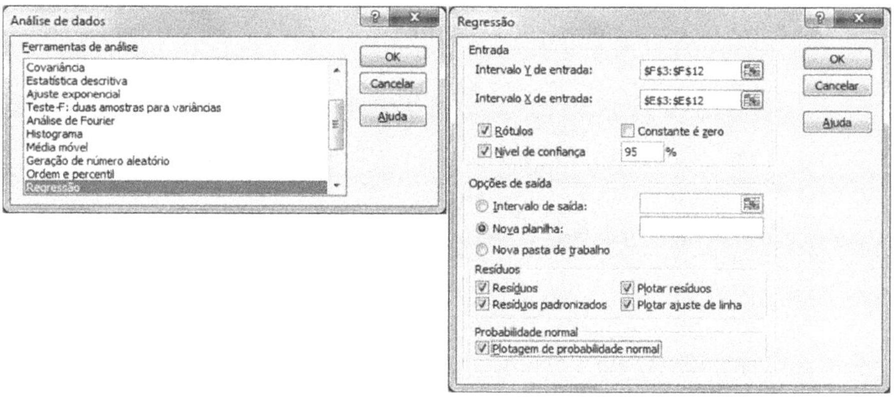

FIGURE 8.1 Tool analysis box. (The authors.)

TABLE 8.3
Regression Model Fit Measures

Regression Statistics	
Multiple R	0.9999
R-Square	0.9998
R-squared set	0.9997
Standard Error	0.0160
Observations	10

Source: The authors.

TABLE 8.4
Description of the ANOVA Table

	SS	GL	F_0
Regression	SS_{Reg}	1	MS_{Reg}/MS_{RES}
Residues	SS_{Res}	n−2	
Total	SS_T	n−1	

Source: The authors.

TABLE 8.5
ANOVA of the Regression Analysis

	GL	SQ	MQ	F_0	F de significação
Regression	1	9.076	9.076	35,351,903	7.16495E-16
Residue	8	0.002	0.000		
Total	9	9.079			

Source: The authors.

TABLE 8.6
Data from the Regression Analysis

	Coefficients	Standard Error	Stat t	value-P	95% IC	95% IC	IP 95,0%	IP 95,0%
$1/Q_{máx}$	0.085	0.006	14.028	6.47E-07	7.12E-02	9.92E-02	7.12E-02	9.92E-02
$1/Q_{máx}\ K_L$	0.667	0.004	188.021	7.16E-16	6.59E-01	6.76E-01	6.59E-01	6.76E-01

Source: The authors.

R-squared and standard error are widely used to evaluate the fit of the model. The coefficient of determination (R-squared) or regression coefficient is the proportion of variation explained by the regressor variable and can be expressed by Equation 8.7:

$$R^2 = \frac{SS_{Reg}}{SS_T} = 1 - \frac{SS_{REs}}{SS_T} \tag{8.7}$$

This coefficient should not be confused with the correlation coefficient, which is a measure of the degree of linear relationship between two variables. $R^2 \times R$

In many articles, (Hameed et al., 2008; Cazetta et al., 2011; Ji et al., 2012; Long et al., 2013; Moradi et al., 2013; Păcurariu et al., 2013) it is common to confuse correlation coefficient (R) with coefficient determination (R^2). The interpretation given to the two is completely different; the correlation coefficient expresses the linear

TABLE 8.7
Table of Residual Analysis

Observation	Estimated (a) $1/q_e$	Residues	Standard Residue
1	2.839	0.021	1.360
2	2.074	−0.016	−1.055
3	1.283	−0.026	−1.722
4	0.339	0.025	1.645
5	0.159	−0.001	−0.045
6	0.100	0.002	0.107
7	0.092	0.005	0.335
8	0.090	−0.001	−0.076
9	0.088	−0.001	−0.069
10	0.088	−0.007	−0.479

Source: The authors.

relationship between the regressor variable and the response variable, in which it can vary from −1 to +1 while the coefficient of determination is between $0 \leq R^2 \leq 1$. The most commonly used correlation coefficient is Pearson's, which expresses covariance of two variables by the product of their standard deviations.

ANALYSIS OF VARIANCE (ANOVA)

Let's test the hypothesis of absence of regression, i.e., coefficient $\beta_1 = 0$. For this, it is necessary to define some quantities. Equations 8.8–8.10 show some sum of squares, where in Equation 8.8, we have the sum of squares of the residuals, which is the sum of the experimental values minus the calculated values. In Equation 8.9, we have the sum of the squares of the regression; in Equation 8.10, we have the sum of the total squares, which is the sum due to the total variation.

$$SS_{RES} = \sum_{i=1}^{n} (y_i - \hat{y}_i)^2 \tag{8.8}$$

$$SS_{Reg} = \sum_{i=1}^{n} (\hat{y}_i - \bar{y})^2 \tag{8.9}$$

$$SS_T = SS_{RES} + SS_{Reg} \tag{8.10}$$

The SST has n − 1 degrees of freedom, SS_{Reg} has 1 degree of freedom, referring to (β̂_1), and SS_{RES} has n − 2 degrees of freedom, referring to both parameters. The test for hypothesis for regression will be an F-test:

$$H_0 : \beta_1 = 0$$

$$H_0 : \beta_1 \neq 0$$

1. SS_{RES} follow a distribution χ^2_{n-2};
2. Se a hipótese nula $H_0 : \beta_1 = 0$ é verdadeira, então $SSReg/\int^2$ follow a distribution χ^2;
3. SS_{RES} e SS_{Reg} are independent.

The value of F_0 will be the sum of the mean squares of the regression divided by the sum of the mean squares of the residuals, Equation 8.11.

$$F_0 = \frac{SS_{Reg}/GL}{SS_{Res}/L} = \frac{SS_{Reg}}{SS_{RES}/(n-2)} = \frac{MS_{Reg}}{MS_{RES}} \tag{8.11}$$

The test follows a distribution of Fischer Snedecor with $F_{1,\,n-2}$.

The expected value of the middle squares is:

$$E(MS_{RES}) = \sigma^2, \quad E(MS_{Reg}) = \sigma^2 + \beta_1^2 S_{xx}$$

The results are presented in Table 8.4.

If F_0 is high, the $\beta_1 \neq 0$.

Then compare the calculated value of F with the tabulated value. Therefore, if $F_{0>}F_{\langle,1,n-2}$, then the regression is significant for a given level of significance. Returning to our example, Table 8.5 presents the analysis of variance (ANOVA). From this table, it is possible to see if the portion of the total variance explained by the regression is statistically significant at a given confidence level. If F_0 is high, then $\beta_1 \neq 0$, that is, the regression is significant at the chosen confidence level, $F_{0>}F_{\langle,1,\,n-2}$. For our example, we have $35{,}351{,}903 > 7.16E^{-16}$.

Table 8.6 shows the regression coefficients with the standard error, the significance test for the estimated parameters, in which they are significant, and the confidence interval. Table 8.7 presents the data of the calculated residuals, which are fundamental in the detection of anomalies in the regression and should be analyzed graphically, Figure 8.2a, and the data should be randomly dispersed in the graph, and the regression line Figure 8.2b.

NONLINEAR REGRESSION ANALYSIS OF THE LANGMUIR ADSORPTION ISOTHERM MODEL

The use of linearization in adsorption models may imply an error in the estimation of parameters (Mazuchelli & Achcar, 2002), and the ideal use of models in their nonlinear forms is ideal. In addition, for some models of adsorption isotherms, linearization is difficult, such as the model with three parameters of Redlich-Peterson. There are several reported articles explaining how to perform nonlinear regression analysis using Excel, including the step-by-step (Brown, 2001). For our example in Excel, we will use the Langmuir model, Equation 8.2. The method can be extended to other models. The parameters to be estimated are q_{max}, which is the maximum adsorption capacity, and K_L, which is Langmuir's constant.

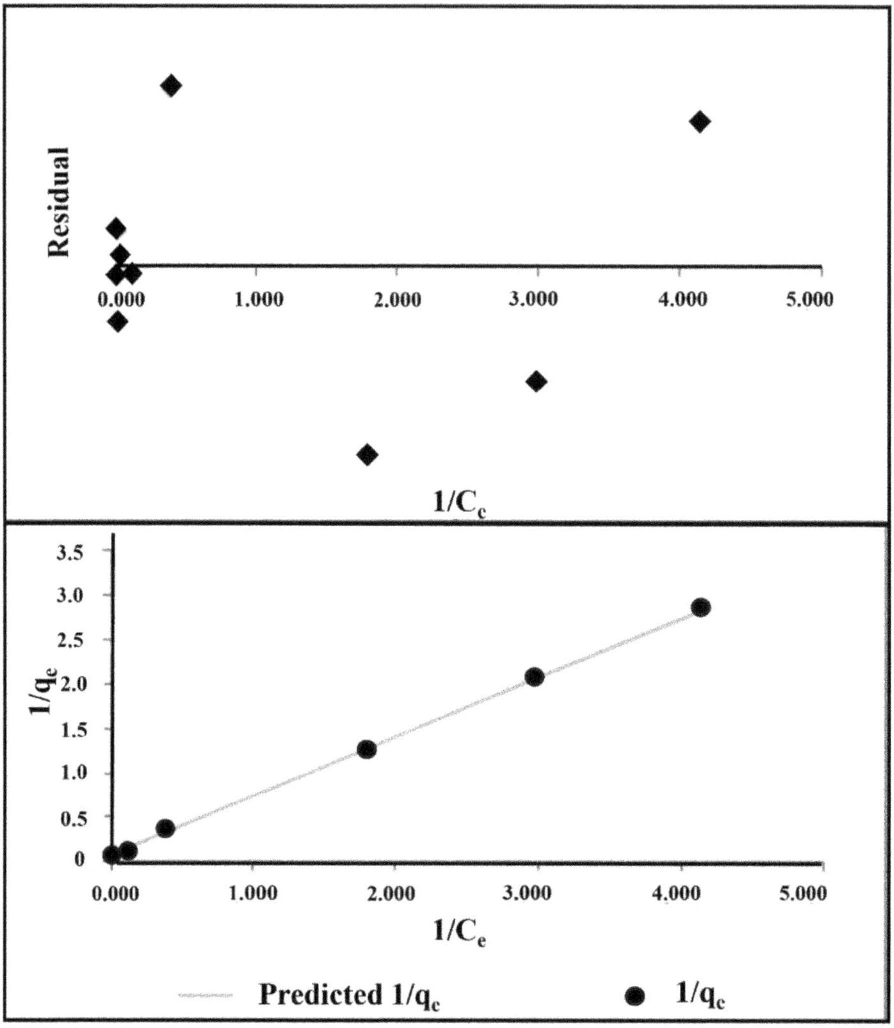

FIGURE 8.2 (a) Residual graph and (b) regression line. (The authors.)

- **Step 1**: Fill in the worksheet as shown in Table 8.8:

  The values for q_{max} and K_L initially used were those obtained from linear regression analysis, or a random value can be chosen, Table 8.9. A bad initial value can lead to a delay in convergence, it can take time or not occur at all. It is important to realize that the error function is already at a small value, Table 8.9.
- **Step 2**: Define the error function for applying the least squares method and the value of R^2, Table 8.10. For this example, the ERRSQ function will be used, where this function represents the sum of the square of the differences between the experimental and calculated values. Other functions can be used, such as sum of squares of errors (ERRSQ), Hybrid fractional error function (HYBRID), Marquardt percentage standard deviation (MPSD),

TABLE 8.8
Filling in the Data of the Langmuir Model in Excel

Sample	C_e	q_e	Q_{eCal}		
1	0.242	0.350			
2	0.336	0.486	2.976	Q_{max}	11.682
3	0.557	0.795	1.795		
4	2.633	2.751	0.380	K_L	0.128
5	9.100	6.334	0.110		
6	44.949	9.835	0.022	R^2	
7	99.284	10.310	0.010		
8	130.384	11.215	0.008	ERRSQ	
9	213.384	11.454	0.005		

TABLE 8.9
Regression Data Initialized by the Values Obtained from the Linear Regression

Sample	C_e	q_e	q_{eCal}		
1	0.242	0.350	0.351		
2	0.336	0.486	0.482	q_{max}	11.682
3	0.557	0.795	0.777		
4	2.633	2.751	2.945	K_L	0.128
5	9.100	6.334	6.286		
6	44.949	9.835	9.952	R^2	0.993
7	99.284	10.310	10.830		
8	130.384	11.215	11.022	ERRSQ	1.547
9	213.384	11.454	11.269		

Source: The authors.

TABLE 8.10
Setting the Error Function in Excel

Sample	C_e	Q_e	q_{eCal}		
1	0.242	0.35	0.352	Q_{max}	11.6822
2	0.336	0.486	2.976	K_L	0.1284
3	0.557	0.795	1.795		{=1-(SOMA((C3:C12-MÉDIA
4	2.633	2.751	0.38	R^2	(C3:C12))^2)/
5	9.1	6.334	0.11		(SOMA(C3:C12-D3:D12)))}
6	44.949	9.835	0.022	ERRSQ	{=(SOMA((C3:C12-D3:D12)^2))}
7	99.284	10.31	0.01		
8	130.384	11.215	0.008		
9	213.384	11.454	0.005		
10	254.524	12.407	0.004		

Source: The authors.

Mean relative error (ARE), sum of absolute error (EABS) (Ho et al., 2001; Allen et al., 2003). When you finish typing the formula by clicking ctrl + shift + enter, this is the Excel command for matrix operations.

- **Step 3**: Create a graph with the experimental values and the calculated values. In Figure 8.3, the experimental and calculated values are far apart. After the inclusion of the parameters obtained from the linear regression analysis, the graphic points are already visually close, Figure 8.4.

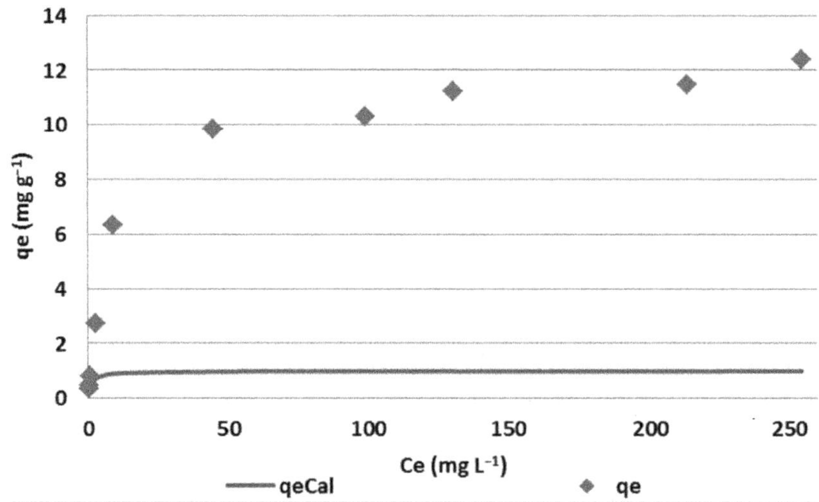

FIGURE 8.3 Graph with initial value $q_{max} = 1$, $K_L = 1$. (The authors.)

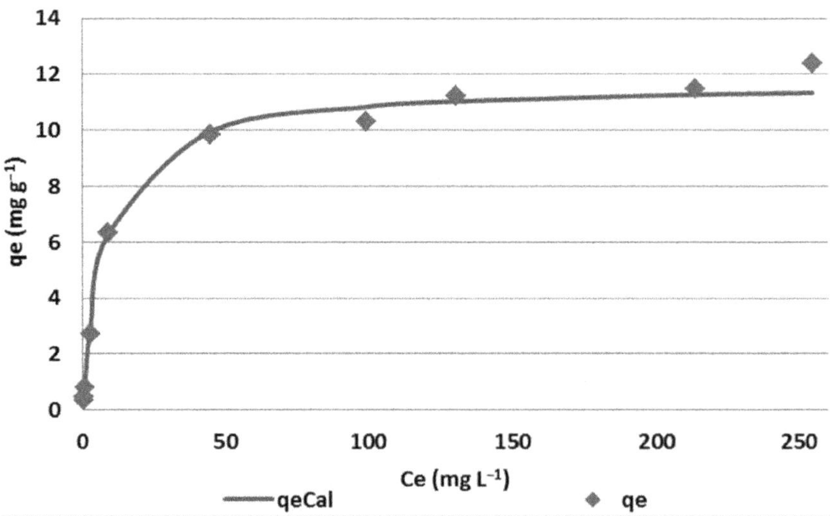

FIGURE 8.4 Graph with initial values for q_{max} and K_L obtained from the linear regression analysis. (The authors.)

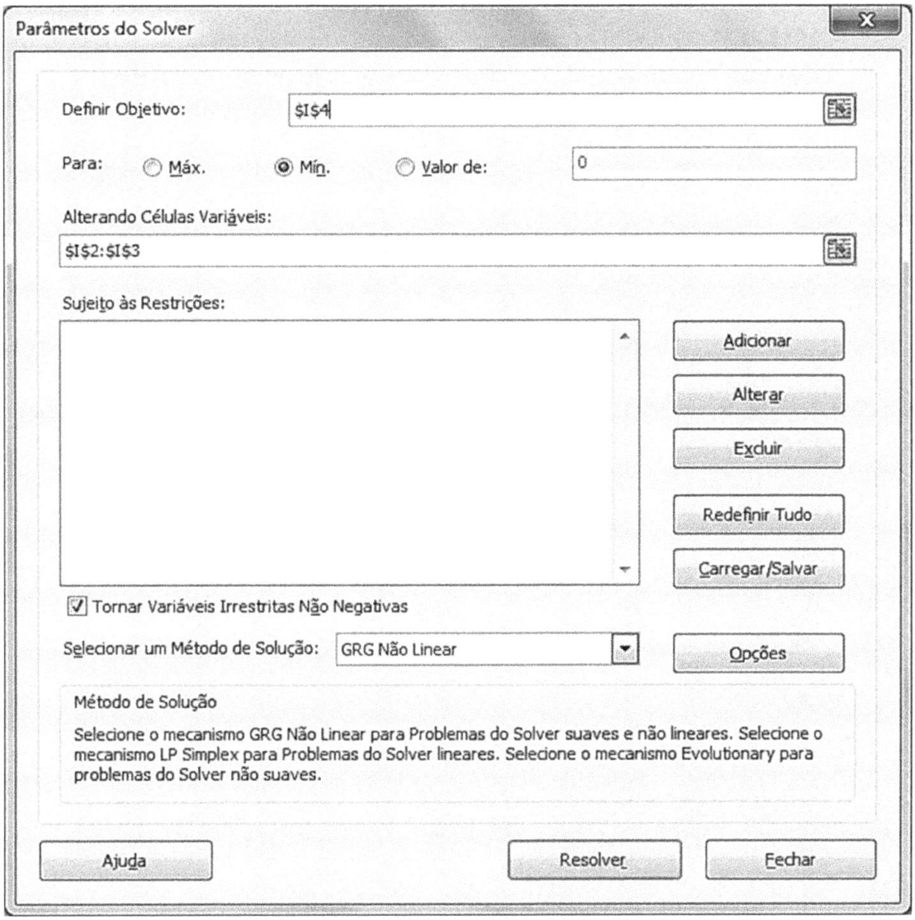

FIGURE 8.5 Text box with the "SOLVER" parameters. (The authors.)

- **Step 4**: After choosing the q_{max} and K_L values, let's use Excel's solver method. We must enter the data<solver menu and fill in the window as shown in Figure 8.5.

Click Resolve. The error value decreases after nonlinear regression analysis (Table 8.11). The result after regression is shown in Figure 8.6.

It is important to verify that the sum of the squared residuals obtained by the nonlinear regression analysis is smaller than that obtained by linear regression analysis, using any type of linearization model, Table 8.12. The model of Eadi-Hofstee did not present a straight line.

Other models of adsorption isotherms, adsorption kinetics, and rupture curves can be studied in a similar manner.

TABLE 8.11
Table and Graph with Data after Nonlinear Regression Analysis

Sample	C_e	q_e	q_{eCal}		
1	0.242	0.350	0.322		
2	0.336	0.486	0.442	Q_{max}	11.990
3	0.557	0.795	0.716		
4	2.633	2.751	2.768	K_L	0.114
5	9.100	6.334	6.105		
6	44.949	9.835	10.032	R^2	0.994
7	99.284	10.310	10.17		
8	130.384	11.215	11.234	ERRSQ	1.271
9	213.384	11.454	11.517		

Source: The authors.

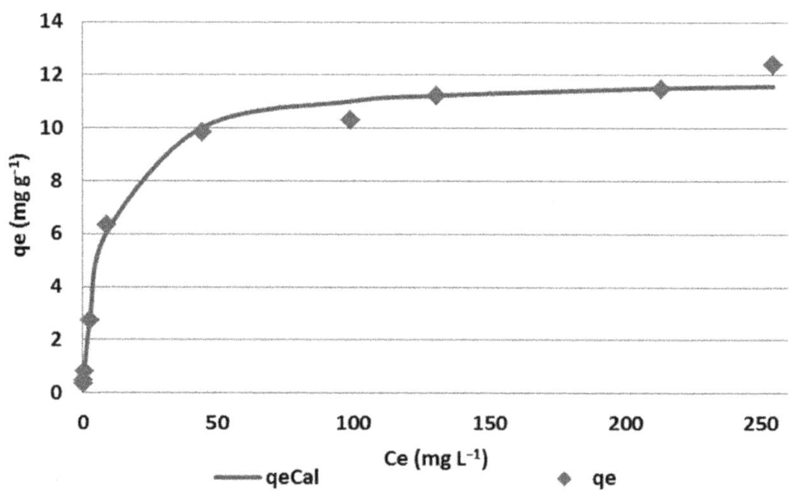

FIGURE 8.6 Graphical result after nonlinear regression analysis with experimental and calculated values. (The authors.)

TABLE 8.12
Comparison between the Nonlinear and Linearized form of the Langmuir Model

	Hannes	Lineweaver-Burker	Eadi-Hofstee	Scatchard	N-Linear
$Q_{máx}$	12.376	11.682	—	11.886	11.99
KL	0.096	0.128	—	0.124	0.114
ERRSQ	1.709	1.548	—	1.346	1.271

Source: The authors.

ADJUSTMENT OF NONLINEAR ISOTHERMS TO EXPERIMENTAL ADSORPTION DATA

Despite the frequent use of Microsoft Office Excel software to obtain data and graphs (Febrianto et al., 2009), a usual alternative tool is Origin, a graphical software used for data analysis and statistics. Having the latter in hand, it is suggested that it be used for the adjustment of nonlinear isothermal models to the data experimental of adsorption, aiming to obtain more reliable data from the application of original nonlinear models (Tran et al., 2017). It will be necessary to configure the nonlinear adjustment function offered by the program. Initially, it is necessary to add a new category to the nonlinear adjustment function of the Origin software, followed by the adjustment of the desired function to the empirical data obtained.

- **Step 1:** Starting the Origin software, the toolbar displays the "Analysis" function, which must be selected. Then the "Nonlinear curve fit" option should be chosen.
- **Step 2:** A new window will open. In the "Settings" tab, the "Function Selection" option is presented with the "Category" option that must be selected. By clicking on the "New" function, it will be possible to add a new category referring to nonlinear isothermal models. The nomenclature "Adsorption" or "Adsorption" is suggested.
- **Step 3:** After saving the new category, the "Function" option should be chosen. In the "Function Name" option, the nomenclature "Langmuir" is suggested. In the "Independent Variables" option, the acronym "Ce" should be added, in the "Dependet Variables" option, "Q_e." In "Parameter Names," the constants found in the Langmuir nonlinear equation (Q_{max}, K_L) will be added. The "Function Form" option should be changed to "Equations." Then, in the "Function" option, Langmuir's nonlinear equation should be written as follows: $Q_e = (Q_{max} * K_L * C_e)/(1 + K_L * C_e)$.
- **Step 4:** Under "Parameters Settings," there is a small "box" on the right that should be selected. A new window will open. In the "Bounds" tab, the values of "LB Value" and "LB Control" should be changed to 0 and <, respectively, for all parameters. In the "Parameters" tab, the value of "Value" should be changed to 1 in all parameters. The "Enable Auto Initialization" option should be checked. The new added function should be saved in "Save" and "OK."

In this way, any other nonlinear equations can be added and saved in the suggested software, and can then be adjusted to the experimental data, helping to choose the model that best represents the adsorption process. As an example, the fitting of Langmuir's nonlinear model using the software previously discussed will be addressed. The objective is to obtain the parameters embedded in the equation, which are maximum adsorption capacity (q_{max}) and Langmuir's constant (K_L).

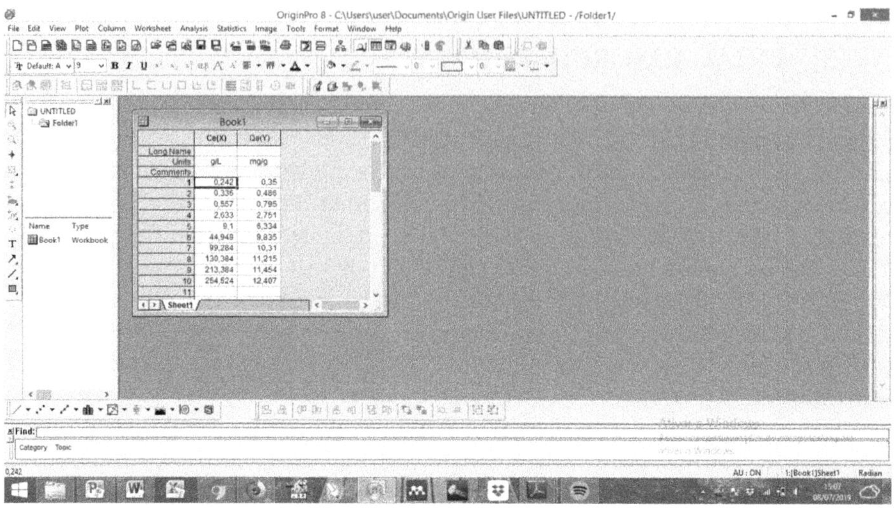

FIGURE 8.7 Representative spreadsheet containing fictitious values. (The authors.)

Initially, the data presented in Table 8.1 were used, arranged in a spreadsheet of the Origin program, as shown in Figure 8.7, for the adjustment of the nonlinear model. Then, the two columns must be selected and in the "Plot" tab, the "Scatter" option must be chosen. The dispersion will be obtained in the form of points. Subsequently, for the adjustment of the nonlinear model to the experimental data presented, in the "Analysis" tab, the "Nonlinear Curve Fit" option should be chosen. A window will open, where the option selected under "Category" will be "Adsorption" or "Adsorption" and the option under "Function" will be "Langmuir." Right below, the last option on the right named "Fit till converged" should be selected making the model fit the experimental data. Finally, the "Fit" option must be selected. Figure 8.8 represents the graph obtained after fitting the nonlinear model to the experimental adsorption data.

In addition to this result, the software also has a "dialog box" (Figure 8.9) containing additional information consisting of the values obtained for maximum adsorption capacity (q_{max}) and Langmuir constant (K_L) and their respective deviations. In addition, a table containing statistical functions is available presenting values of adjusted coefficient of determination (R^2), sum of the square of residual errors (SSE), and reduced Chi-square (x^2) that contribute to the choice of the most appropriate model in the representation of the adsorption phenomenon.

An ANOVA table is also available with additional data. In this table, it is observed that $F_0 > F_{\alpha,1,n-2}$ ($2{,}342.078 > 3.44E^{-12}$), showing that regression is significant at the chosen confidence level.

The Origin program emerges as an alternative tool in the adjustment of nonlinear isothermal models to experimental adsorption data. Thus, the use of linearized equations should be avoided, preventing the risk of misinterpretation of false results. In addition, this program presents itself as an easy-to-use resource for obtaining data and graphs in the various adsorption studies.

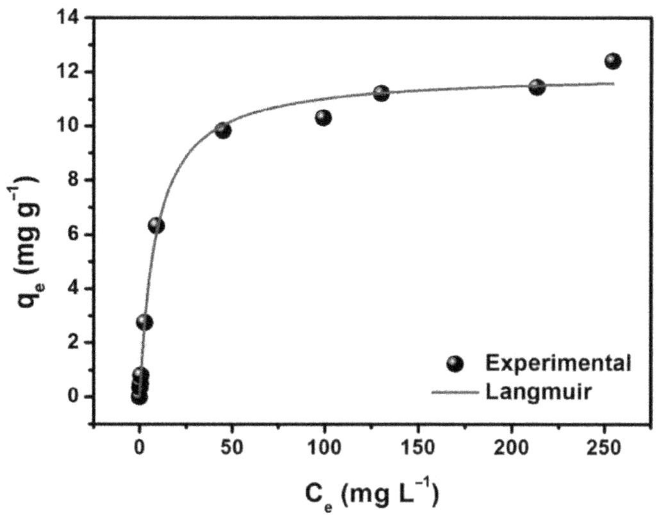

FIGURE 8.8 Adjustment of Langmuir's nonlinear model using the Origin software. (The authors.)

B1

1 Notes

2 Input Data

3 Parameters

		Value	Standard Error
Qe	Qmax	11,99027	0,22343
	Kl	0,11431	0,01274

Iterations Performed = 6
Total Iterations in Session = 6
Fit converged – tolerance criterion satisfied.

4 Statistics

	Qe
Number of Points	11
Degrees of Freedom	9
Reduced Chi-Sqr	0,14119
Residual Sum of Squares	1,27067
Adj. R-Square	0,99472
Fit Status	Succeeded(100)

Fit Status Code:
100 : Fit converged

5 Summary

	Qmax		Kl		Statistics	
	Value	Error	Value	Error	Reduced Chi-Sqr	Adj. R-Square
Qe	11,99027	0,22343	0,11431	0,01274	0,14119	0,99472

6 ANOVA

		DF	Sum of Squares	Mean Square	F Value	Prob>F
Qe	Regression	2	661,33492	330,66746	2342,07831	3,44236E-12
	Residual	9	1,27067	0,14119		
	Uncorrected Total	11	662,60559			
	Corrected Total	10	267,36123			

FIGURE 8.9 Table of data regarding the adjustment of the Langmuir nonlinear model. (The authors.)

REFERENCES

Allen, S. J. et al. (2003). Comparison of optimised isotherm models for basic dye adsorption by kudzu. *Bioresource Technology*, 88, 143–152.

Bates, D. M., & Watts, D. G. (1988). *Nonlinear Regression Analysis and Its Applications.* 1ª. ed. John Wiley & Sons, New York.

Bolster, C. H., & Hornberger, G. M. (2006). On the use of linearized Langmuir equations. *Soil Science Society of America Journal*, 71, 1796–1806.

Brown, A. M. (2001). A step-by-step guide to non-linear regression analysis of experimental data using a Microsoft Excel spreadsheet. *Computer Methods and Programs in Biomedicine*, 65, 191–200.

Cazetta, A. L. et al. (2011). NaOH-activated carbon of high surface area produced from coconut shell: Kinetics and equilibrium studies from the methylene blue adsorption. *Chemical Engineering Journal*, 174, 117–125.

Cornell, J. A. (2011). A Primer on Experiments with Mixture. John Wiley & Sons, Hoboken, NJ.

Draper, N. R., & Smith, H. (1998). *Applied Regression Analysis.* 3ª. ed. John Wiley & Sons, New York.

Febrianto, J. et al. (2009). Equilibrium and kinetic studies in adsorption of heavy metals using biosorbent: A summary of recent studies. *Journal of Hazardous Materials*, 162(2–3), 616–645.

Hameed, B. H., Tan, I. A. W., & Ahmada, L. (2008). Adsorption isotherm, kinetic modeling and mechanism of 2,4,6-trichlorophenol on coconut husk-based activated carbon. *Chemical Engineering Journal*, 144, 235–244.

Ho, Y. S., Porter, J. F., & Mckay, G. (2001). Equilibrium isotherm studies for the sorption of divalent metal ions onto peat: Copper, nickel and lead single component systems. *Water, Air, & Soil Pollution*, 141(1–33), 1–33.

Ji, F., Li, C., Tang, B., Xu, J., Lu, G., & Liu, P. (2012). Preparation of cellulose acetate/zeolite composite fiber and its adsorption behavior for heavy metal ions in aqueous solution. *Chemical Engineering Journal*, 209, 325–333.

Lima, A. C. A. et al. (2012). Modified coconut shell fibers: A green and economical sorbent for the removal of anions from aqueous solutions. *Chemical Engineering Journal*, 185–186, 274–284.

Long, C., Yu, W., & Li, A. (2013). Adsorption of n-hexane vapor by macroporous and hypercrosslinked polymeric resins: Equilibrium and breakthrough analysis. *Chemical Engineering Journal*, 221, 105–110.

Mazuchelli, J., & Achcar, J. A. (2002). Algumas considerações em regressão não linear. *Maringá*, 24, 1761–1770.

McCoy, M. A., & Liapis, A. I. (1991). Evaluation of kinetic models for biospecific adsorption and its implications for finite bath and column performance. *Journal of Chromatography*, 548, 25–60.

Menoud, P., Calvin, L., & Renken, A. (1998). Modelling of heavy metals adsorption to a chelating resin in a fluidized bed reactor. *Chemical Engineering and Processing: Process Intensification*, 37, 89–101.

Montgomery, D. C., Peck, E. A., & Vining, G. G. (2006). *Introduction to Linear Regression Analysis.* 4ª. ed. John Wiley & Sons, New Jersey.

Moradi, O., Fakhri, A., Adami, S., & Adami, S. (2013). Isotherm, thermodynamic, kinetics, and adsorption mechanism studies of ethidium bromide by single-walled carbon nanotube and carboxylate group functionalized single-walled carbon nanotube. *Journal of Colloid and Interface Science*, 395, 224–229.

Moura, C. P. et al. (2011). Adsorption of BTX (benzene, toluene, o-xylene, and p-xylene) from aqueous solutions by modified periodic mesoporous organosilica. *Journal of Colloid and Interface Science*, 363, 626–634.

Neto, B. B., Scarminio, I. S., & Bruns, R. E. (2003). *Como fazer experimentos: pesquisa e desenvolvimento na ciência e na indústria.* 3ª. ed. Unicamp, São Paulo.

Neto, V. O. S. et al. (2011). Use of coconut bagasse as alternative adsorbent for separation of copper (II) ions from aqueous solutions: Isotherms, kinetics, and thermodynamic studie. *Bioresources and Technology*, 6, 3376–3395.

Păcurariu, C. et al. (2013). Adsorption of phenol and p-chlorophenol from aqueous solutions on poly (styrene-co-divinylbenzene) functionalized materials. *Chemical Engineering Journal*, 222, 218–227.

Rodrigues, M. I., & Iemma, A. F. (2009). *Planejamento de Experimentos e Otimização de Processos.* 2ª. ed. Casa do pão, Campinas.

Seber, G. A. F., & Wild, C. J. (2003). *Nonlinear Regression.* Wiley, New York.

Sousa, F. W. et al. (2010). Green coconut shells applied as adsorbent for removal of toxic metal ions using fixed-bed column technology. *Journal of Environmental Management*, 91, 1634–1640.

Tran, H. N. et al. (2017). Mistakes and inconsistencies regarding adsorption of contaminants from aqueous solutions: A critical review. *Water Research*, 120, 88–116.

Tsibranska, I., & Hristova, E. (2010). Modelling of heavy metal adsorption into activated carbon from apricot stones in fluidized bed. *Chemical Engineering and Processing: Process Intensification*, 49, 1122–1127.

Vidal, C. B. et al. (2011). Adsorption of polycyclic aromatic hydrocarbons from aqueous solutions by modified periodic mesoporous organosilica. *Journal of Colloid and Interface Science*, 357, 466–473.

Vidal, C. B. et al. (2012). BTEX removal from aqueous solutions by HDTMA-modified Y zeolite. *Journal of Environmental Management*, 112, 178–185.

Vidal, C. B., Raulino, G. S. C., Melo, D. Q., & Nascimento, R. F. (2016). Biosorption of metal ions using a low cost modified adsorbent: Experimental design and mathematical modeling. *Environmental Technology*, 37, 1–42.

9 Characterization of Adsorbent Materials

Vicente de Oliveira Sousa Neto, Ari Clecius Alves de Lima, Breno Aragão dos Santos, Antonia Mayza de Morais França, Carla Bastos Vidal, Diego de Quadros Melo, Giselle Santiago Cabral Raulino, Janaina Sobreira Rocha, José Marcos Sasaki, and Ronaldo Ferreira do Nascimento

INTRODUCTION

CHARACTERIZATION OF ADSORBENT MATERIALS: KEY STAGES

In the development of adsorbent materials, characterization typically occurs at two distinct stages:

- During the preparation phase of the adsorbent material;
- After the adsorption test, to evaluate how the adsorbate interacts with the adsorbent material.

During the preparation phase, characterization aims to confirm the physicochemical properties of the material, such as surface area and pore distribution, through gas adsorption analysis. Additionally, it involves assessing changes in surface charges via zeta potential or zero charge potential measurements, as well as evaluating the thermal stability of the adsorbent materials.

Thermometric analyses; incorporation of a specific functional group as an adsorption site via spectroscopic techniques; crystallinity determination; phase identification through diffraction techniques; and topological assessment of the adsorbent material using electron microscopy.

Following the adsorption stage, it is essential to evaluate how the adsorbate interacts with the adsorbent. Various techniques can be employed for this purpose, with spectroscopic methods being the most widely used (Neto et al., 2012).

DOI: 10.1201/9781003654391-9

GAS ADSORPTION

Gas adsorption analysis enables the determination of pore distribution, pore size, and surface area. The most commonly used gases for this process include nitrogen, carbon dioxide, helium, argon, and oxygen.

The Brunauer–Emmett–Teller (BET) theory (1938) extends the fundamental principles of Langmuir's theory, allowing for higher adsorption and the formation of multiple layers on the surface. The theory considers that the forces involved in physical adsorption are similar to those governing liquefaction processes, such as Van der Waals forces. Unlike monolayer formation, physical adsorption can proceed further, leading to the development of multilayers on the adsorbent's surface.

The BET equation can be expressed as Equation 9.1.

$$q_e = \frac{BC_e q_{BET}}{(C_s - C_e)\left[1 + (B-1)\left(\frac{C_e}{C_s}\right)\right]} \quad (9.1)$$

BET isotherm and adsorbate properties

where C_5 ($mol.L^{-1}$) represents the saturation concentration of an adsorbate.

B is a constant related to the energy of interaction with the surface.

q_{BET} ($mol.g^{-1}$) denotes the maximum adsorption capacity of the adsorbate in monolayer formation.

Based on the shape of the BET isotherm, it is possible to obtain information regarding the specific surface area and porous structure of the material. These isotherms are derived from nonspecific adsorption using nitrogen. The objective is to express the amount of gas adsorbed through its volume (V_a) under specific conditions.

The standard temperature and pressure (STP) conditions are defined as 0°C and 760 Torr, while pressure is expressed as the relative pressure (P/P_0)—the ratio between the working pressure and the gas vapor pressure at the applied temperature.

Regarding the type of isotherm obtained through gas adsorption, the IUPAC classifies isotherms into six types (Figure 9.1). The ideal Type Ib isotherm is applicable to microporous materials, whereas Type IVa and Type V isotherms are suitable for mesoporous materials.

CLASSIFICATION OF ISOTHERMS

a. Type I isotherms are characteristic of microporous solids. The use of microporous materials is widely applied for gas storage, including CO_2, CH_4, NH_4, and H_2, due to their high selectivity. In this context, the adsorbent material functions as a true molecular sieve.
b. Type II and Type IV isotherms are typical of non-porous solids and solids with relatively large pores, respectively. Type IV isotherms are specifically associated with mesoporous materials. The example presented in Figure 9.2 demonstrates a mesoporous material, where the adsorption curve exhibits hysteresis.

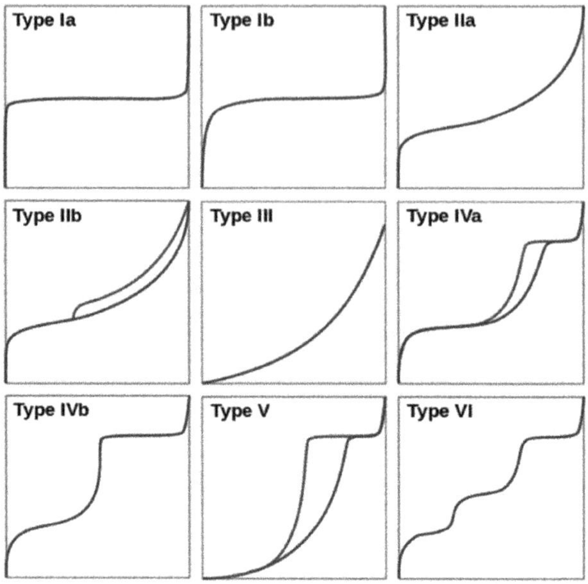

FIGURE 9.1 Classification of Isotherms according to IUPAC. (This Figure by Bref68 is licensed under the Creative Commons Attribution-Share Alike 4.0 International license. https://upload.wikimedia.org/wikipedia/commons/e/e0/Isothermes_iupac.png)

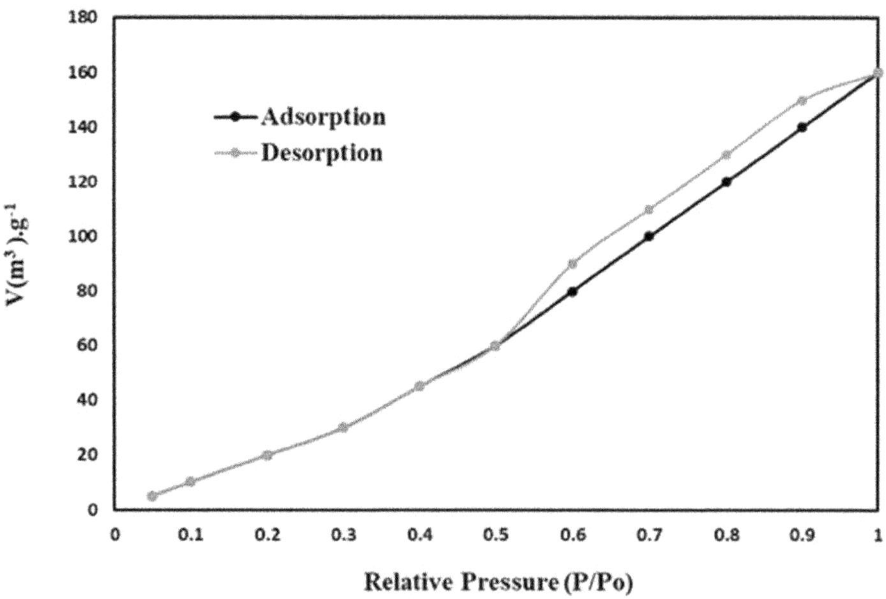

FIGURE 9.2 BET Isotherm. (This figure by Vicente Neto is licensed under the Creative Commons Attribution 4.0 International License. https://commons.wikimedia.org/wiki/File:BET_Isotherm.jpg

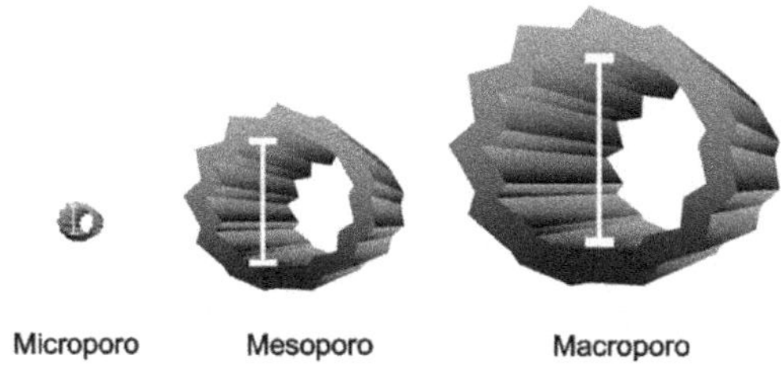

FIGURE 9.3 Comparative pore size scale for micropores, mesopores, and macropores. (This figure by Vicente Neto is licensed under the Creative Commons Attribution 4.0 International License. https://commons.wikimedia.org/wiki/File:Pore_size_scale.jpg)

c. Type III and Type V isotherms are characteristic of systems in which adsorbate molecules exhibit stronger interactions with each other than with the solid surface. These two isotherms are not relevant for porous structure analysis.
d. Type VI isotherm is obtained through gas adsorption on a non-porous solid with an almost uniform surface.

According to the IUPAC, pores can be classified as macropores (>500 Å), mesopores (200–500 Å), and micropores (<200 Å). A comprehensive understanding of the relationship between pore size and the kinetic diameter of the analyte is crucial in selecting the appropriate adsorbent material. In Figure 9.3, we can observe the relative scale and magnitude of micropores, mesopores, and macropores.

Microporous materials such as covalent organic frameworks can have pore sizes ranging from 7 to 47 Å and a surface area of up to 4,000 $m^2 g^{-1}$ (Ding & Wang, 2012). As previously mentioned, they are suitable for gas storage. In liquid media, microporous materials such as metal-organic frameworks can be used for the removal of nitrogen-containing compounds from fuels or in ion-exchange systems (Ding & Wang, 2012).

Most materials referenced in the literature for applications in liquid media are macro- or mesoporous, where selectivity is more related to interactions with surface functional groups rather than pore size. Among these materials, biomass is widely used in adsorption processes due to its high availability and low cost. Silica-based materials are also extensively utilized for adsorption in liquid media. Table 9.1 provides further details.

A material with a large surface area does not necessarily imply that the entire area is available for adsorption. For example, factors such as steric hindrance and nonspecific adsorption sites can limit the material's usability. Additionally, surface modifications may obstruct the pores, thereby reducing the effective surface area.

Determination of Surface Area Using the BET Method

A straightforward regression analysis procedure can be employed to determine the surface area using the BET method:

TABLE 9.1
Comparison of Surface Area and Pore Size of Adsorbents

Biomass	A (m^2g^{-1})	Tp (Å)	Adsorbate	References
Raw coconut bagasse	218	43	NO_3^-, SO_4^{2-}, PO_4^{3-}	Lima et al. (2012)
Coconut bagasse modified with QUAT 188	221	43	NO_3^-, SO_4^{2-}, PO_4^3	Lima et al. (2012)
- SiO_2	382.7	92.6	Pb^{2+}, Cd^{2+}, SO_4^{2-}, $H_2\,SO_4^-$	Gomes et al. (2013)
Si-APTS	251.2	82.8	Pb^{2+}, Cd^{2+}, SO_4^{2-}, $H_2\,SO_4^-$	Gomes et al. (2013)
Si-APTS-EDTA	177.6	73.6	Pb^{2+}, Cd^{2+}, SO_4^{2-}, $H_2\,SO_4^-$	Gomes et al. (2013)
-Zeolite Y	650.3	7.4	BTEX	Vidal et al. (2012)
Mesoporous organosilica	484.03	74.89	HPA's	Vidal et al. (2011)

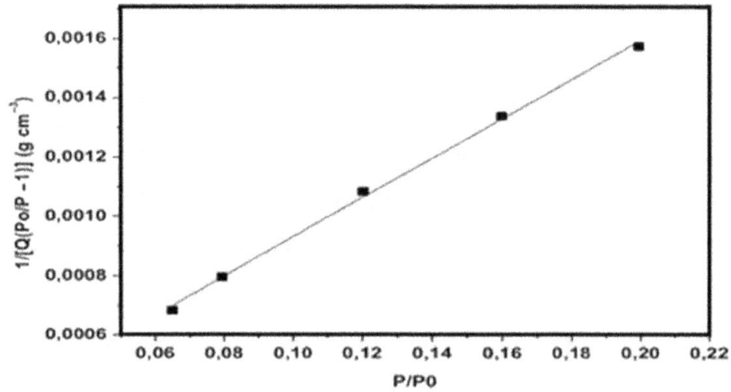

FIGURE 9.4 Linear BET Isotherm. (This figure by Vicente Neto is licensed under the Creative Commons Attribution 4.0 International License. https://commons.wikimedia.org/wiki/File:Linear_BET_Isotherm_Illustration.jpg.)

- **Step 1:** Construct the graph based on the following equation:

$$\frac{\frac{P}{P^0}}{Q_m\left(1-\frac{P}{P^0}\right)}=\frac{1}{Q_mC}+\frac{C-1}{Q_mC}\left(\frac{P}{P^0}\right) \tag{9.2}$$

$$\frac{\frac{P}{P^0}}{Q_m\left(1-\frac{P}{P^0}\right)}X\left(\frac{P}{P^0}\right) \tag{9.3}$$

The graph should be a straight line with slope $s=\frac{C-1}{Q_mC}$ and intercept $i=\frac{1}{Q_mC}$ (Figure 9.4).

We determine the maximum adsorption capacity using the equation $Q_m = 1/(s+i)$ e $c = s/i + 1$, (Equation 9.4). In their original work, BET identified a linear relationship within the (P/P^0) range of 0.05–0.35 (Rouquerol et al., 1999).

$$Q_m = \frac{1}{0.01712 + 0.00179} = 52.88\ cm^3\ g^{-1} \tag{9.4}$$

BET Area, A(BET)

$$A(BET) = Q_m\, L\, \sigma \tag{9.5}$$

Where L is the Avogadro's constant and σ represents the average area occupied by each molecule in a complete monolayer. The cross-sectional area can be calculated using Equation 9.6.

$$\sigma = f\left(M/\rho L\right)^{2/3} \tag{9.6}$$

Where f is the packing factor, which is 1.091 for a close-packed hexagonal system, ρ is the absolute density of the adsorbed liquid at the operating temperature, and M is the molar mass. For nitrogen at 77 K, this value is approximately 162 nm^2. When volume data is used, the surface area can be calculated using Equation 9.7:

$$A(BET) = Q_M * 4.35 \tag{9.7}$$

The constant 4.35 is the Neimark constant for nitrogen in $m^2 cm^{-3}$, obtained by substituting the surface area into Equation 9.6, yielding the surface area calculation in Equation 9.8:

$$A(BET) = 52.88 * 4.35 = 230.03\ m^2\ g^{-1} \tag{9.8}$$

Using the surface area value, the particle size can be calculated using Equation 9.9 (Spurr & Myers, 1957; Klug & Alexander, 1974):

$$d_{BET} = \frac{6000}{\rho * A(BET)} \tag{9.9}$$

Where ρ is the theoretical density.

Surface Charge Determination (PZC and Zeta Potential)

The point of zero charge (PZC) refers to the potential at which the surface charge of the adsorbent is neutral (Fiol & Vilaescusa, 2009). For pH values below the PZC, the material exhibits positively charged surface behavior, whereas for pH values above the PZC, the material possesses negative surface charges. Various methods are available for determining the PZC (Vakros et al., 2002; Fiol & Vilaescusa, 2009),

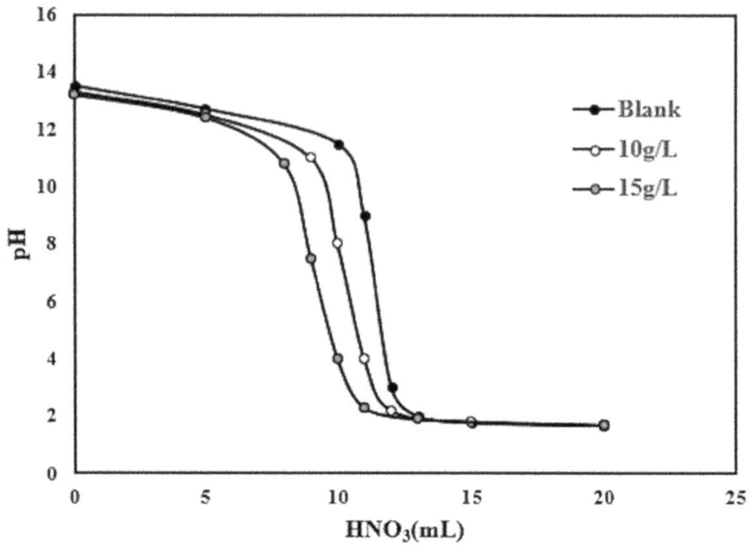

FIGURE 9.5 Zero charge potential graph, titrimetric method. (Figure prepared by authors https://commons.wikimedia.org/wiki/File:Zero_Charge_Potential_Illustration.jpg)

including potentiometric mass titration, mass titration, and the immersion technique (Fiol & Vilaescusa, 2009; Tabak et al., 2010).

The potentiometric mass titration method is illustrated in Figure 9.5. In this example, the PZC is 3.5, and the graph indicates that above this value, the material's surface charge is negative. For anion adsorption, the ideal pH is below this threshold, where the material's surface charge is positive.

Zeta potential is a physical property exhibited by a particle in suspension and plays a crucial role in adsorption studies, as it allows for the determination of the adsorbent's surface charge. Gomes et al. (2013) conducted research using silica modified with EDTA and observed a zeta potential at pH 5.0. Based on this data, anions such as SO_4^{2-} and $H_2 SO_4^-$ were successfully adsorbed at pH 3.0, while cations such as Pb^{2+} and Cd^{2+} were adsorbed at pH 5.5.

Similarly, Lima et al. (2012) utilized biomass for the removal of nitrate, sulfate, and phosphate, obtaining a zeta potential of pH 4.76. This result indicated that the optimal pH for adsorption was above this value, where the material's surface charge was positive, thereby facilitating electrostatic interactions with anions (Figure 9.6).

Thermal Analyses

Thermal analysis encompasses a set of techniques aimed at assessing the material degradation profile, tracking both time and temperature (Vahedi et al., 2022). Important thermodynamic parameters can also be derived from this analysis, such as phase transitions and even the monitoring of chemical reactions. Exposure to high temperatures can sometimes alter the chemical structure and, consequently, the physical properties of materials (Sharma et al., 2021).

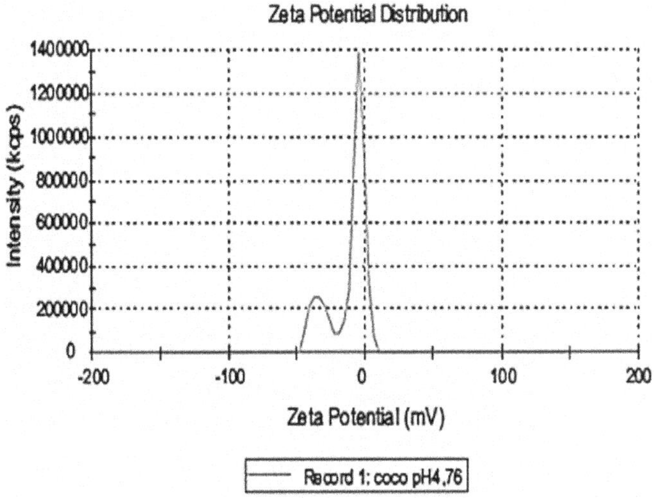

FIGURE 9.6 Zero charge potential of green coconut husk, pH 4.76. (Reproduced and adapted with permission from Lima et al. (2012).)

A crucial application of thermogravimetric techniques is the monitoring of synthesis processes involving calcination, often necessary for material structure formation or the elimination of synthesis intermediates. Examples include the physical activation of activated carbon through calcination (Yang et al., 2010) and the synthesis of zeolites, where the removal of the synthesis surfactant is essential to unblock the pores.

Additionally, thermogravimetric analysis provides insights into the structural stability of materials for specific applications. For instance, adsorbents used in liquid media, which operate at lower temperatures, do not require significant thermal resistance, whereas adsorbents intended for gas separation or catalysis must exhibit thermal stability.

Moura et al. (2011) monitored TG (Thermogravimetry) and DTG (Derivative Thermogravimetry) curves during the synthesis of mesoporous silica. At temperatures of 35°C–100°C, the synthesis solvent was removed; between 180°C and 300°C, the surfactant was eliminated; from 300°C to 400°C, residual surfactant was further removed; between 500°C and 670°C, organic groups were eliminated; and at 670°C–880°C, vicinal silanol groups condensed, leading to the formation of silicon oxides or silica (Moura et al., 2011). Similarly, Lima et al. (2012) studied lignocellulosic material modified with quaternary ammonium salt in an oxidative atmosphere. TGA (Thermogravimetric Analysis) and DSC (Differential Scanning Calorimetry) curves were obtained (Figure 9.7), showing mass loss up to 100°C due to water evaporation. A second event occurred between 200°C and 300°C, where lignocellulosic material and surfactant were lost, stabilizing at 400°C. The first event was an endothermic evaporation process, whereas the second was an exothermic process, releasing energy (Lima et al., 2012).

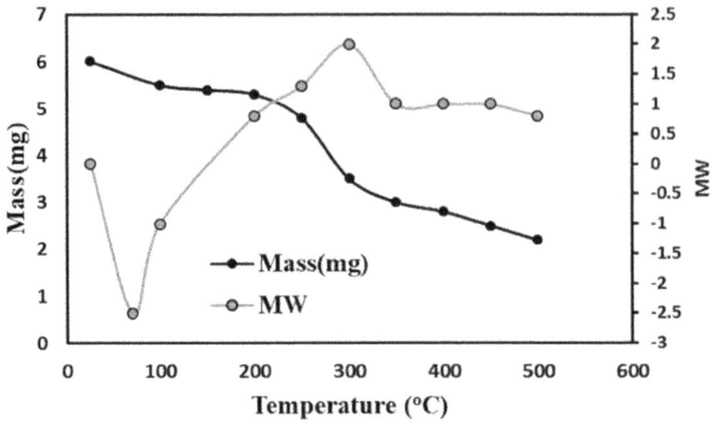

FIGURE 9.7 TGA/DSC of lignocellulosic material. (Reproduced and adapted with permission from Lima et al. (2012).)

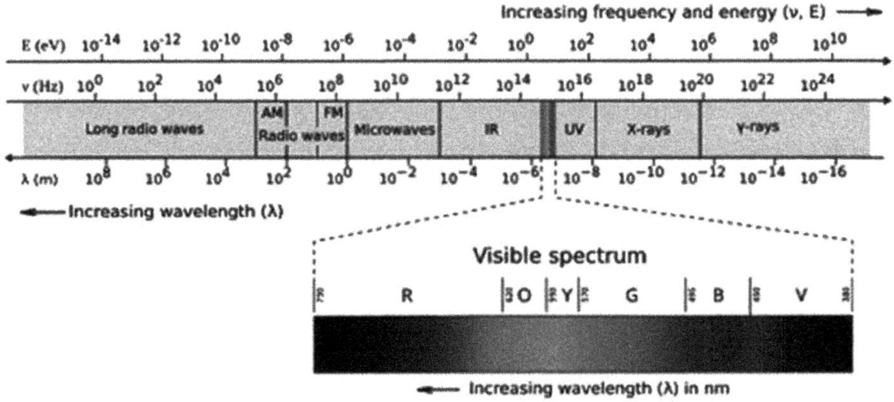

FIGURE 9.8 The electromagnetic spectrum. This figure by Philip Ronan is licensed under the Creative Commons Attribution-Share Alike 2.5. (This figure was originally in color. It has been adapted to grayscale under the conditions permitted by the license. https://upload.wikimedia.org/wikipedia/commons/6/6d/Electromagnetic_radiation_spectrum_-_en.svg)

Fourier-Transform Infrared Spectroscopy Analysis

The presence of functional groups in adsorbent materials plays a crucial role in understanding adsorption mechanisms. Infrared spectroscopy enables the identification of compounds containing covalent bonds, whether organic or inorganic (Silverstein et al., 2005). Infrared radiation corresponds to the segment of the electromagnetic spectrum located between the visible and microwave regions (Figure 9.8).

The most commonly used infrared region for adsorbent characterization is the mid-infrared, which lies within the wavenumber range of 4,000–400 cm^{-1}

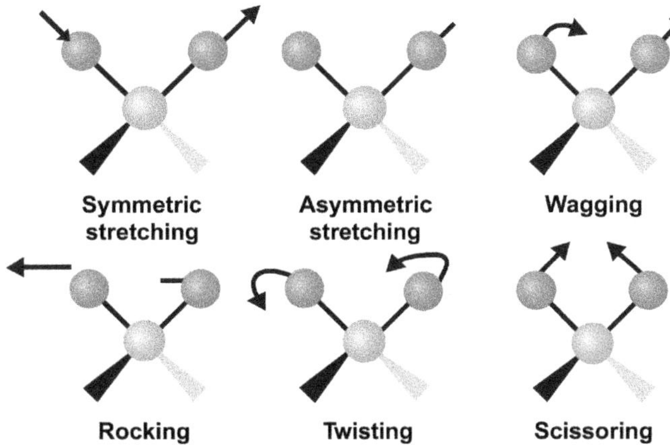

FIGURE 9.9 Vibrational modes by excitation of the molecule with IR. (This figure by Vicente Neto is licensed under the Creative Commons Attribution-Share Alike 2.5. https://commons.wikimedia.org/wiki/File:Vibrational_modes.png)

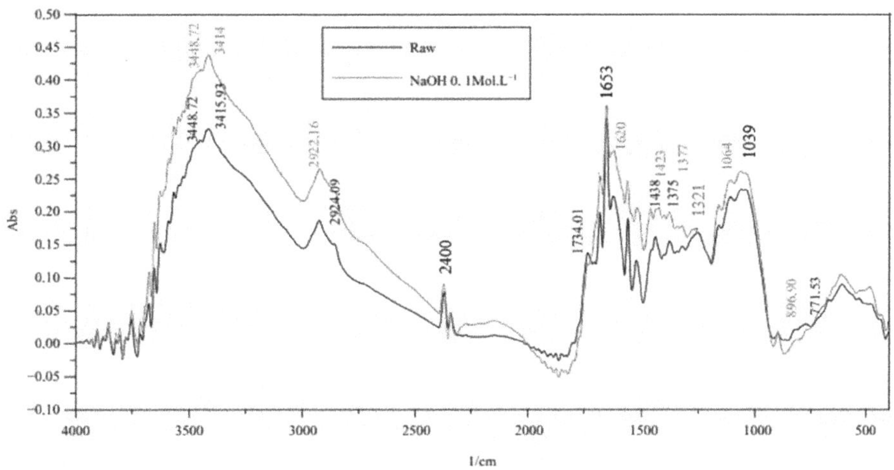

FIGURE 9.10 Infrared spectrum of raw and NaOH-modified green coconut husk powders. (Reproduced and adapted with permission from Sousa et al. (2011). This figure was originally in color. It has been adapted to grayscale.)

(2.5–25 μm in terms of wavelength). Within this range, changes in the vibrational and rotational energies of molecules occur, generating a spectrum (Silverstein et al., 2005) (Figure 9.9).

Based on infrared analysis, it is possible to identify organic functional groups that can act as adsorption sites (Sousa et al., 2010, 2011). By working with treated and untreated green coconut husk powder (Figure 9.10), the main organic groups present in the material were determined (Table 9.2).

TABLE 9.2
Principal Bands of the Infrared Spectrum Observed in NaOH-Treated and Untreated Green Coconut Husk Powder

Wavenumber, cm^{-1}	Functional Groups
3448 e 3415	Axial OH stretch of alcohol or phenol
2922 e 2924	Groups –CH
1734	Aldehyde and ketone groups
1681 e 1653	Carbonyls, carboxylic acids, and esters
1560 e 1520	Amines and amides
1423 e 1377	Axial -CN deformation of amino and amide groups
1064 e 1039	Groups C-N and C-O
897	Polysaccharides

Source: Sousa et al. (2011).

The infrared spectrum of both treated and untreated material revealed the presence of weak and strong peaks within the 3,448–607 cm^{-1} range, indicating, among others, the presence of carbonyls, carboxyls, phenols, esters, and polysaccharides.

The bands at 3,448 and 3,415 cm^{-1} are characteristic of the axial stretching of O-H from alcohols or phenols; the bands at 2,922 and 2,924 cm^{-1} correspond to –CH groups; the band at 1,734 cm^{-1} is characteristic of aldehydes and ketones; the bands at 1,681 and 1,653 cm^{-1} are associated with carbonyl groups, carboxylic acids, and esters; the bands at 1,560 and 1,520 cm^{-1} correspond to amines and amides; the bands at 1,423 and 1,377 cm^{-1} indicate axial deformation of –CN in amino and amide groups; the bands at 1,064 and 1,039 cm^{-1} are related to C–N and C–O groups; and the band at 897 cm^{-1} corresponds to polysaccharide structures.

The functional groups observed in both treated and untreated material are the primary agents responsible for the adsorption of metal ions in solution. The study on the effect of chemical treatment applied to green coconut husk powder demonstrated that the basic treatment was generally more effective than other treatments (Abushammala & Mao, 2020; Reyes et al., 2022).

Similar results were obtained by Moreira et al. (2009), aiming to identify the functional groups present in the cashew peduncle bagasse before and after treatment with 0.1 mol L^{-1} NaOH. The biosorbent is a natural polymer composed mainly of cellulose, hemicellulose, and lignin. Due to the mixture of compounds, the infrared absorption spectra are quite complex (Figure 9.11).

The alkaline treatment of lignocellulosic material is an irreversible process that induces changes in its supramolecular structure and morphology, facilitating its "solubilization." Two different cellulose chain structures (types I and II) may form (Kroon-Batenburg & Kroon, 1997). Alkaline treatment also removes components such as lignin and hemicellulose. The aqueous NH_4OH/thiourea cellulose solution enhances cellulose solubility (Jin et al., 2007; Abushammala & Mao, 2020; Yu et al., 2024) due to its ability to maintain intramolecular bonding. It is well

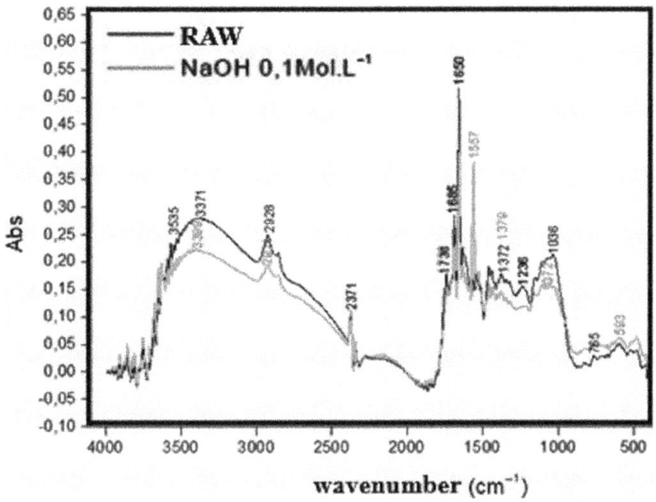

FIGURE 9.11 Infrared spectra of treated and untreated cashew peduncle bagasse. (This figure was originally in color. It has been adapted to grayscale under the conditions permitted by Moreira et al. (2009). Química Nova (QN) is a fully open access journal. All articles published by the QN are made freely and permanently accessible online immediately upon publication, without any subscription charges or registration barriers.)

established that the conversion of type I cellulose to type II is more efficient in the presence of the NaOH/thiourea ionic solution (Neto et al., 2012). Consequently, the increased adsorption capacity observed in coconut bagasse subjected to alkaline treatment via IL (ionic liquid) can be attributed to the formation of type II cellulose, which possesses hydroxyl groups that are more available for metal interaction.

Neto et al. (2012) investigated the effect of cadmium adsorption capacity on the surface of coconut when subjected to alkaline treatment using ionic liquids (thiourea/ammonia). They obtained Fourier-Transform Infrared spectroscopy (FT-IR) spectra of raw coconut bagasse (CB) and coconut bagasse treated with thiourea (CBT), as shown in Figure 9.12. The transmittance values observed at 3,390, 2,922, 1,514, 1,435, 1,377, 1,247, 1,049, and 898 cm^{-1} in the raw material align with previous findings for sugarcane bagasse (Sun et al., 2004; Reyes et *al.*, 2022). The prominent band at 3,390 cm^{-1} corresponds to hydroxyl group stretching (alcohol/phenol), while the absorption at 2,934 cm^{-1} is linked to C–H stretching. The minor absorbance peaks at 1,607, 1,511, 1,427, and 1,323 cm^{-1} represent aromatic skeletal vibrations and ring breathing, associated with C–O stretching in lignin (Sun et al., 2005; Yu et al., 2024).

The bands at 1,377 and 1,247 cm^{-1} are related to C–H and C–O stretching in acetyl groups within hemicellulose. Additionally, the pronounced band at 1,049 cm^{-1} is attributed to C–O stretching present in cellulose, hemicellulose, and lignin, or alternatively, C–O–C stretching within cellulose and hemicellulose (Sun et al., 2005; Yu et al., 2024). The distinct small peak at 898 cm^{-1} arises from β-glucosidic linkages between sugar units in hemicellulose and cellulose (Sun et al., 2005; Yu et al., 2024).

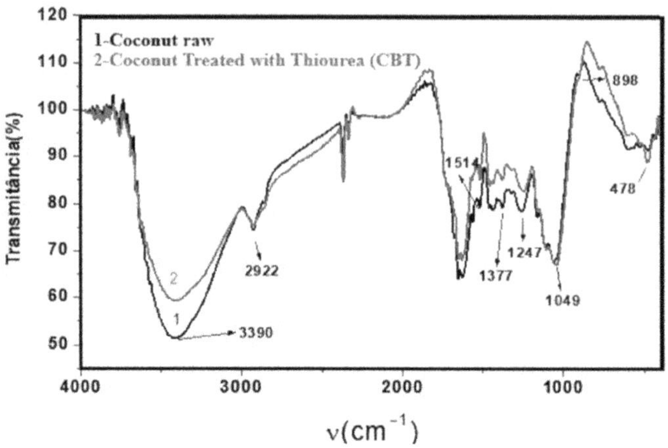

FIGURE 9.12 FT-IR spectrum of raw coconut bagasse (1) and coconut bagasse treated with an ionic solution (2). Samples prepared in KBr pellets at 1% (m/m). (Figure adapted with permission from Neto et al. (2012).)

Treatment with thiourea solution did not induce chemical modifications in CB. However, a noticeable reduction in band intensities following thiourea-ammonia treatment suggests the partial removal of constituents such as lignin and hemicellulose (Zhang et al., 2002).

X-Ray Diffraction Analysis

Adsorbent materials can exist in either an amorphous or crystalline form, where "crystalline" refers to materials exhibiting regular repetition in three-dimensional space, composed of molecules or molecular groups arranged in a structured manner.

X-ray diffraction (XRD) is a material characterization technique that utilizes diffraction phenomena in crystalline materials. It provides insight into the structure of sufficiently crystalline materials, enables qualitative and quantitative analysis, and determines particle size. X-rays interact with the electrons of the chemical elements present in a sample, and under specific crystalline and experimental conditions, it is possible to locate light atoms. Neutron diffraction, in contrast, involves the interaction between neutrons and atomic nuclei and is applied in cases where XRD does not yield optimal results. Electron diffraction is a technique particularly suited for surface investigations, known as LEED—Low-Energy Electron Diffraction (Atkins & de Paula, 2014). The fundamental principle of XRD relies on the coherent scattering of electromagnetic radiation, caused by the interaction between an incident X-ray beam and the electrons of atoms within a material. This analysis allows for phase identification and semi-quantitative phase assessment. Figure 9.13 presents an X-ray diffractogram of a clay sample before and after modification with tin dioxide.

The identification of crystalline phases can be performed using specialized software, such as X'Pert High Score Plus™ with the specific ICDD-PDF2 database.

The average particle size can be calculated using the Scherrer equation (Williamson & Hall, 1953):

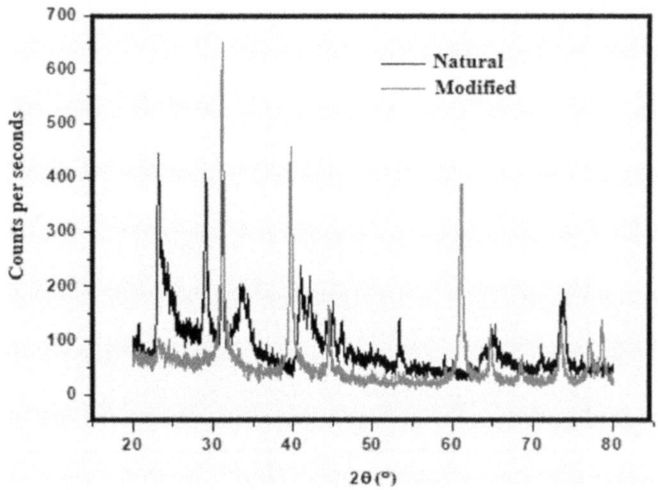

FIGURE 9.13 Natural Montmorillonite Clay and Tin Oxide-Modified Clay. (This figure by Vicente Neto is licensed under the Creative Commons Attribution-Share Alike 2.5. https://commons.wikimedia.org/wiki/File:Natural_montmorillonite_clay_and_tin_oxide-modified_clay.jpg)

$$D = \frac{k\lambda}{\beta\cos(\theta)} \tag{9.10}$$

where λ is the X-ray wavelength, θ is the Bragg angle, and β is the full width at half maximum; the value of k depends on the crystallite shape, symmetry, and type of reflection (James, 1962).

The Williamson-Hall plot (1953), for the titanium dioxide phase shown in Figure 9.14, was obtained by plotting the calculated values of $\beta\cos(\theta)/\lambda$ versus $\sin(\theta)$. The particle size (D) is determined by the linear coefficient, while the microstrain (ε) is obtained from the angular coefficient when the sample exhibits homogeneity, meaning the values conform to the linear equation (9.12) (Williamson & Hall, 1953; Ribeiro *et al.*, 2012).

$$\frac{\beta\cos(\theta)}{\lambda} = \frac{K}{D} + \frac{4\varepsilon}{\lambda}\text{sen}(\theta) \tag{9.11}$$

For the Williamson-Hall plot, a linear behavior indicates a homogeneous material, with a positive angular coefficient signifying lattice expansion and a negative one indicating contraction. A more detailed explanation of the graph's use can be found in the references (Williamson & Hall, 1953; Ribeiro et al., 2012).

Le et al. (2012), working with titanium dioxide and activated coconut husk charcoal in methylene blue degradation, found a particle size of 22.4 nm for sample P25, a value close to that obtained via the specific surface area method (20.74 nm).

The Williamson-Hall plot (Figure 9.15) obtained for the titanium dioxide phase is typical of a non-homogeneous particle size and microstrain distribution. The full

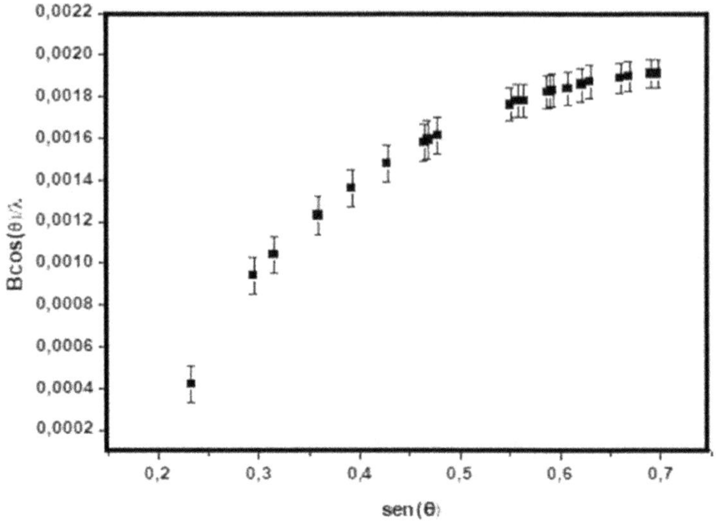

FIGURE 9.14 Williamson-Hall graph for titanium dioxide. (This figure by Vicente Neto is licensed under the Creative Commons Attribution-Share Alike 2.5. Source: https://commons.wikimedia.org/wiki/File:Williamson-Hall_Illustration_for_titanium_dioxide.jpg)

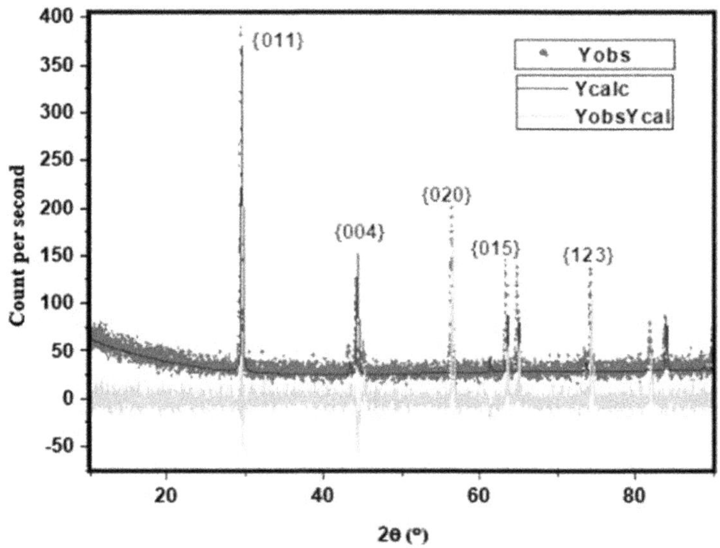

FIGURE 9.15 X-Ray diffractograms: calculated (blue), observed (dotted), and the difference between observed and calculated (green) for a titanium dioxide sample. (This figure by Vicente Neto is licensed under the Creative Commons Attribution-Share Alike 2.5. https://commons.wikimedia.org/wiki/File:Diffraction_pattern.jpg)

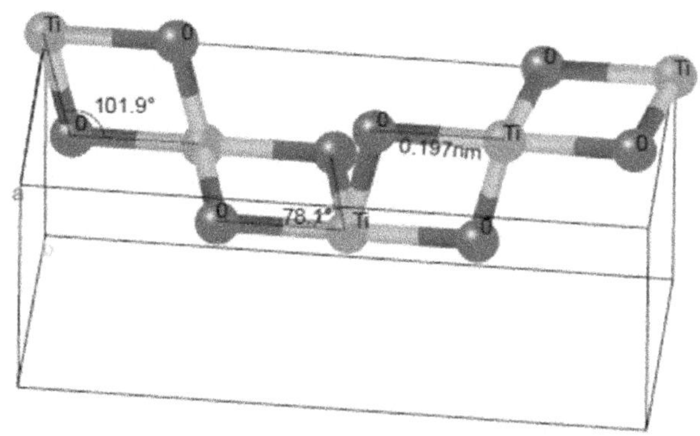

FIGURE 9.16 Representation of the titanium dioxide unit cell (Ti in gray and oxygen in red) showing the bond lengths and bond angles. (This figure by Vicente Neto is licensed under the Creative Commons Attribution-Share Alike 2.5. https://commons.wikimedia.org/wiki/File:Representation_of_the_titanium_dioxide_unit_cell.jpg)

TABLE 9.3
Refinement Data of Titanium Dioxide

Parameters	Unit	Microfiche (ICSD 24276)	Calculated
A	Å	3.776(2)	3.7754(3)
B	Å	3.776(2)	3.7754(3)
C	Å	9.486(6)	9.5035(3)
A	—	90	90
B	—	90	90
Γ	—	90	90
Density	$g.cm^{-3}$	3.92	3.925
Cell volume	10^6 pm^3	135.25	135.25(2)
System	—	Tetragonal	—
Space Group	—	I41/amd	—
R-P	14.23%		
R-WP	20.27%		
R-Expected	16.63%		
S	1.22		

Source: Prepared by Vicente Neto.

widths at half maximum (β) of the diffraction peaks for the families of crystallographic planes {hkl} were obtained after refining the titanium dioxide structure (diffractogram shown in Figure 9.16) using the DBWS program and its graphical interface, DBWS tools (Bleicher et al., 2000; Ramkumar et al., 2018; Baskar & Patania, 2020).

The refinement-calculated values, such as lattice parameters, density, and unit cell volume, are presented in Table 9.3 alongside values from the literature. Additionally,

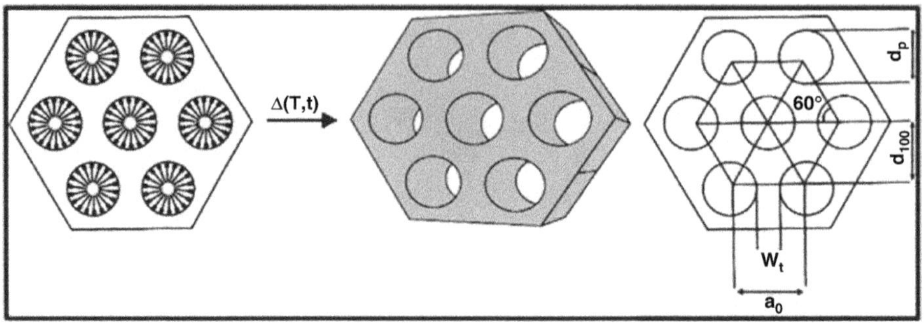

FIGURE 9.17 Illustration of the pore wall thickness. (This figure by Vicente Neto is licensed under the Creative Commons Attribution-Share Alike 2.5. https://commons.wikimedia.org/wiki/File:Illustration_of_the_pore_wall_thickness.jpg)

the crystallographic system, space group of the titanium dioxide phase, and refinement convergence parameters (R-P, R-WP, R-Expected, and S) are listed. Figure 9.15 presents the graph of the observed and calculated values for the refinement.

The three-dimensional representation of the unit cell, bond angles, and bond lengths between the atoms of titanium dioxide (extracted from the ICSDTM database) can be obtained using the software Jmol: an open-source Java viewer for chemical structures in 3D (Herraez, 2006; Madsen & Good, 2023; see Figure 9.16).

In addition to the microstructure of a crystal, the XRD technique allows for the calculation of pore wall thickness (Vidal et al., 2011; Moura et al., 2011; Castro et al., 2007; Pang et al., 2025). To achieve this, the lattice parameter a_0 must be determined using Equations 9.12 and 9.13. Then, with an estimate of the pore diameter, the pore wall thickness is calculated using Equation 9.14. Equation 9.13 is valid for materials with a hexagonal structure (see Figure 9.17).

$$d_{100} = \frac{\lambda}{\text{sen}(\theta)} \tag{9.12}$$

$$a_0 = \frac{2\,d_{100}}{\sqrt{3}} \tag{9.13}$$

$$a_0 = W_t + d_p \Rightarrow W_t\, a_0 - d_p \tag{9.14}$$

For materials with low crystalline order, it is possible to identify the phase; however, the diffractogram exhibits broad peaks. For instance, Figure 9.18 presents the XRD pattern of a lignocellulosic material, lignin, and microcrystalline cellulose. The crystalline phase of cellulose is determined by the presence of the most intense characteristic peak at $2\theta = 26°$, identified by the ICDD database (00-0050-2241). Some authors determine the degree of crystallinity of the material using Equation 9.15, where φ represents the volume fraction, and ρc and ρ correspond to the mass concentration of the sample and the crystalline phase, respectively. However, according to the IUPAC

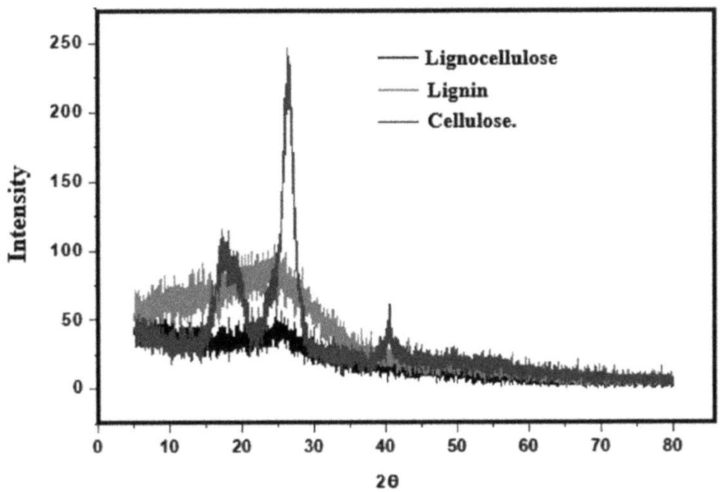

FIGURE 9.18 X-Ray diffractogram of lignocellulosic material. (This figure by Vicente Neto is licensed under the Creative Commons Attribution-Share Alike 2.5. https://commons.wikimedia.org/wiki/File:X-ray_diffractogram_of_lignocellulosic_material.jpg)

(Meille et al., 2011; Napolitano et al., 2024), this approach should only be applied when the phases have the same chemical composition and are not considered interfaces. Therefore, it is not suitable for lignocellulosic materials. Other techniques, such as calorimetry and infrared spectroscopy, can also be employed to determine the degree of crystallinity of the material.

$$w_c = \varphi_c \frac{\rho_c}{\rho} \tag{9.15}$$

Scanning Electron Microscopy

Scanning electron microscopy (SEM) is a characterization technique that utilizes an electron beam instead of the photons used in a conventional optical microscope. Modern devices allow magnifications of up to 300,000 times or more for most solid materials. This technique can provide information such as composition, topography, texture, and surface characteristics of a solid sample, forming a three-dimensional image (Dedavid et al., 2007; Vieira et al., 2021).

For image formation, interaction between the electron beam and the sample is required (see Figure 9.19).

Therefore, the sample must be conductive. For non-conductive samples, they can be made conductive by depositing conductive material. The deposited layers enhance electron emission, as they emit more electrons than the sample material, aiding image construction. The most common method is coating by deposition of metallic ions of gold (Au), gold-palladium alloy (Au-Pd), platinum (Pt), or graphite. Graphite coating is used to cover surface regions where gold atoms do not provide effective coverage, as they are larger atoms, or when the incorporation of atoms into the surface is not

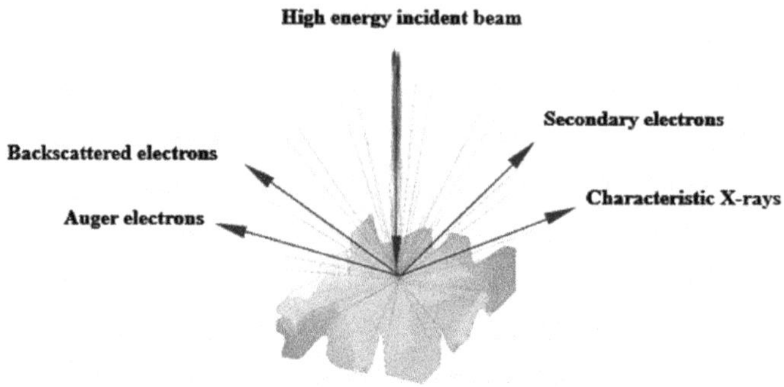

FIGURE 9.19 Interaction of the electron beam with the sample. (This figure by Vicente Neto is licensed under the Creative Commons Attribution-Share Alike 2.5. https://commons.wikimedia.org/wiki/File:Interaction_of_the_electron_beam_with_the_sample.jpg)

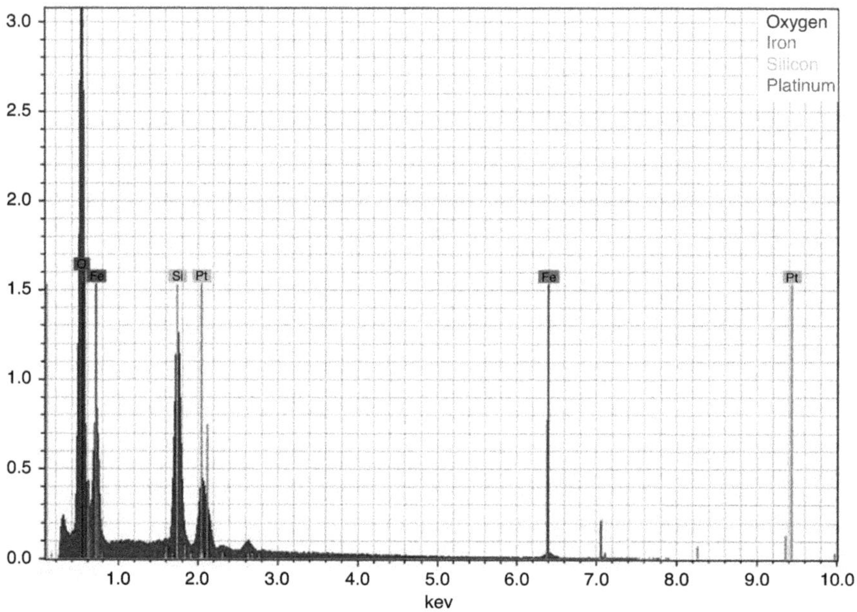

FIGURE 9.20 ESD spectrum: the peaks related to platinum originate from the coating. (https://commons.wikimedia.org/wiki/File:ESD_spectrum.jpg)

desired, such as in X-ray analysis for determining gold or elements close to it in the periodic table (Lii-Rosales et al., 2020).

Coupled with SEM, elemental analysis can be performed through energy-dispersive spectroscopy (EDS). A detector installed in the microscope's vacuum chamber measures the energy associated with this electron. Since the electrons of a given atom have distinct energies, it is possible, at the point of beam incidence, to determine which chemical elements are present in the selected area (Vieira et al., 2021). In Figure 9.20, the peaks related to platinum originate from the coating.

Through the material's topography, it is possible to assess its homogeneity and the formation of pores. Figure 9.21 presents the powder from raw green coconut husk, both untreated and treated with NaOH (Sousa, 2007), where the material is highly porous and exhibits a significantly irregular surface with tubular holes distributed throughout the entire adsorbent surface.

Figure 9.22 presents a microscopy image of a cashew peduncle treated with NaOH 0.1 mol L^{-1} for 3 h (Moreira et al., 2009). In general, it was observed that the material

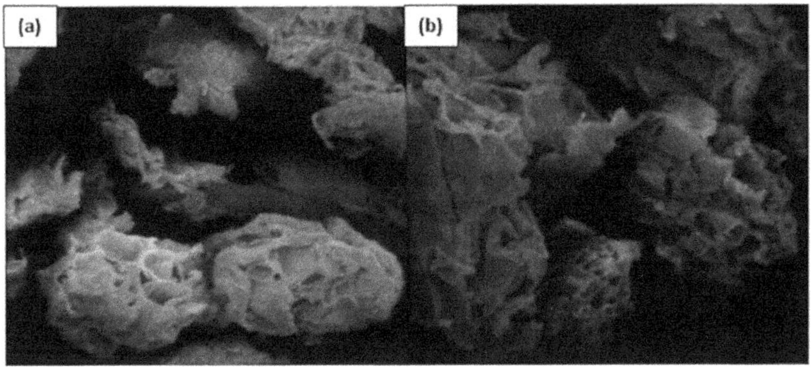

FIGURE 9.21 (a) Surface of 60–99 mesh raw material, (b) surface of 60–99 mesh NaOH-treated material 0.1 mol L^{-1}. (Adapted from Sousa (2007). Master's dissertation in Civil Engineering, Environmental Sanitation, Portuguese version. Technology Center, Federal University of Ceará, Fortaleza.)

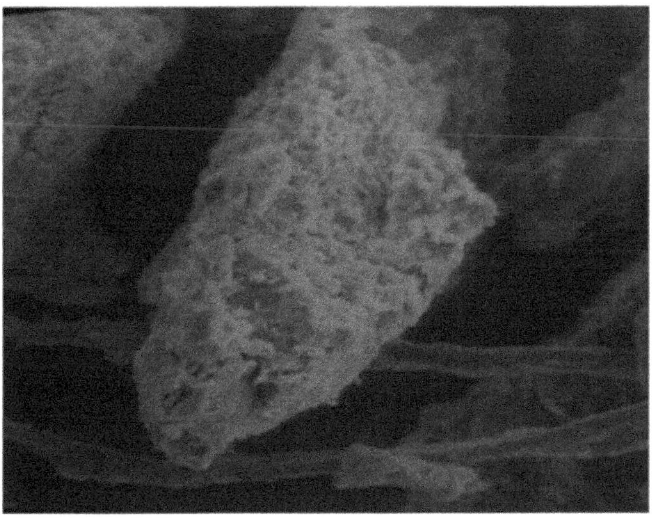

FIGURE 9.22 Micrograph of cashew peduncle treated with 0.1 mol NaOH L^{-1}/3 h with particle size of 20–59 mesh (500×). (Reproduced, adapted with permission from Moreira (2008), Master's dissertation in Civil Engineering, Environmental Sanitation, Portuguese version. Technology Center, Federal University of Ceará, Fortaleza.)

exhibited different morphology along its surface, appearing porous with varying pore sizes, characterizing a heterogeneous surface.

Lima et al. (2012), when treating green CB with 1% sodium chloride/sodium carbonate and 50% sodium hydroxide, observed that, in the treated material, the fibers appeared as aggregates, indicating that the chemical treatment—most likely due to sodium hydroxide—degraded the lignin, releasing the fibrils (Figure 9.23).

Through microscopy it is possible to observe if, during the synthesis of zeolites, the conversion was partial with the presence of amorphous material; in Figure 9.24, the micrograph of a zeolite A shows a crystalline material.

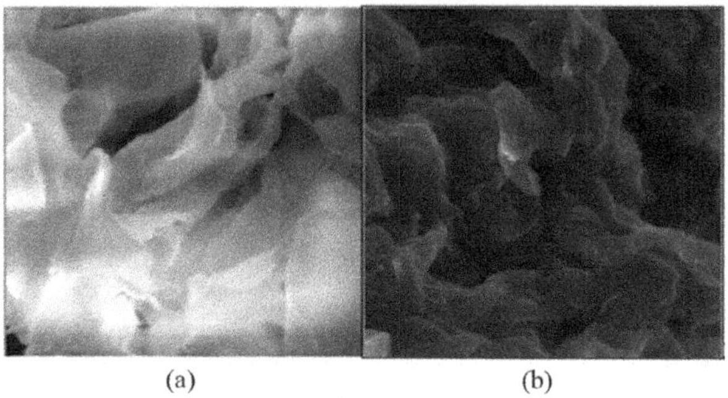

FIGURE 9.23 Raw coconut husk (a) and modified coconut husk (b). (Figure adapted with permission from Lima et al. (2012)).

FIGURE 9.24 Micrograph of a zeolite A with a crystalline defect. (https://commons.wikimedia.org/wiki/File:Micrograph_of_a_Zeolite_A_with_a_crystalline_defect.jpg)

Raman Spectroscopy Analysis

Raman spectroscopy is a technique that encompasses various types of materials (Zhang et al., 2016) and is fundamental for two-dimensional materials, stimulating the exploration of different types of layered materials (Deshmukh & Hussain, 2021). The extensive potential of this characterization technique allows for the distinction of the number of layers, as well as the structural coupling and stacking of layers forming different materials (Cunha et al., 2018).

Currently, with the advancement of Raman spectroscopy equipment, it is possible to overcome the inaccuracies and instrumental limitations inherent in early devices, enabling detailed studies of spectra for different materials under experimental conditions (Bumbrah & Sharma, 2016).

Raman spectroscopy is a versatile method for analyzing a wide range of samples, capable of resolving most of the constraints imposed by other spectroscopic techniques. It can be used for both qualitative and quantitative purposes. Qualitative analysis is performed by measuring the frequency of scattered radiation to establish the sample's identity, while quantitative analysis is conducted by measuring the intensity of scattered radiation to estimate the concentration of the analyte present in the sample (Sikirzhytski et al., 2011; Zhang et al., 2019).

Raman spectroscopy is based on the inelastic scattering of incident radiation through its interaction with vibrating molecules. In this spectroscopy, the sample is illuminated by a monochromatic laser beam, which interacts with the molecules of the sample, causing the dispersion of light at a different frequency than the incident light (inelastic scattering) (Lombardi & Birke, 2008). Figure 9.26a and b illustrate the main Raman microscopy apparatus as well as the inelastic scattering occurring in the analyzed molecules.

When a portion of the scattered radiation has a frequency equal to that of the incident radiation, it is called Rayleigh scattering. However, only a small fraction of the scattered radiation has a frequency different from the incident radiation, which constitutes Raman scattering.

When the frequency of the incident radiation is higher than that of the scattered radiation, Stokes lines appear in the Raman spectrum (Figure 9.25c). Conversely, when the frequency of the incident radiation is lower than that of the scattered radiation, anti-Stokes lines appear in the Raman spectrum.

The shifts observed in the analysis of Raman bands of certain compounds involve transitions from lower to higher vibrational energy levels. Therefore, Stokes bands are more intense than anti-Stokes bands and are measured in conventional Raman spectroscopy, whereas anti-Stokes bands are measured with fluorescent samples. The extent of the shifts observed in Raman spectra does not depend on the wavelength of the incident radiation (Altun et al., 2017).

Several spectroscopic techniques are employed as a general screening method for the analysis of various compounds, while others provide the identity and specific nature of the material. Raman spectroscopy is used to investigate the chemical structure of the material, making it possible to obtain information regarding the phases and polymorphisms of compounds, as well as detect contaminations, impurities,

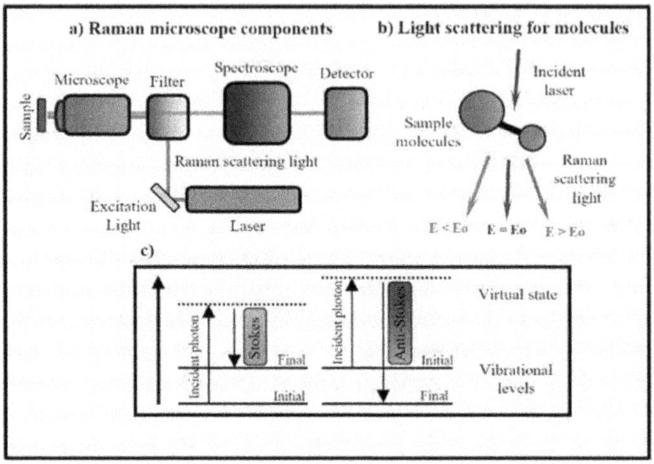

FIGURE 9.25 Representations of the functioning of Raman spectroscopy. (a) Components corresponding to the Raman microscope; (b) demonstration of the interaction between the incident laser energy and the molecules composing the samples; and (c) energy diagram for Raman Stokes and anti-Stokes scattering. (https://commons.wikimedia.org/wiki/File:Representations-Raman_Spectroscopy.jpg.)

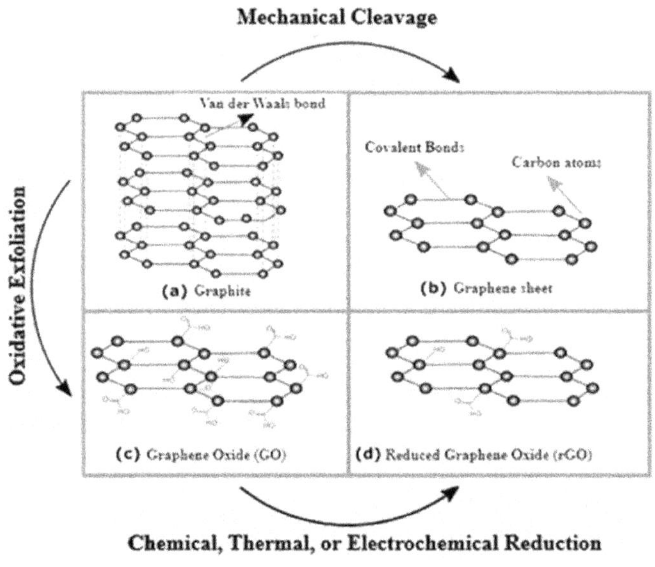

FIGURE 9.26 Atomic and schematic structure of Graphite (a), Graphene (b), Graphene Oxide (GO) (c), and Reduced Graphene Oxide (rGO) (d). (https://commons.wikimedia.org/wiki/File:Atomic_and_schematic_structures.jpg.)

intrinsic stresses, and bond tensions within the samples. Additionally, it allows for predictions of interaction models among different compounds.

Although Raman spectroscopy is a powerful and highly significant technique, in many cases, it can be integrated with other methodologies to achieve a more comprehensive study of the system. Both infrared spectroscopy and XRD analysis serve as complementary techniques to Raman spectroscopy and are often discussed together to enable a complete interpretation of the mechanisms of interaction among the compounds.

Adsorption Study via Raman Spectroscopy

As previously reported, the textural properties of materials, such as surface area, pore volume, and average pore diameter in nanostructures, are critical factors in determining adsorption capacity (Bergmann & Machado, 2015). Nanometer-scale compounds are highly reactive structures, potentially possessing active adsorption surface sites or dangling bonds on their outer layers, playing a significant role in the adsorption process (Ly et al., 2019).

Carbon allotrope nanocomposites are promising systems for various applications, particularly for molecular adsorption, due to their intrinsic dimensionality, distinct textural and electronic properties, and differentiation in the coordination number of carbon atoms or the stacking sequence of crystalline lattice layers. These variations, along with different degrees of sp^2 hybridization, result in carbonaceous nanomaterials with unique atomic arrangements. These topological modifications lead to diverse interactions between adsorbents and the adsorbate molecule (Ly et al., 2019; Sun et al., 2019; Ihsanullah, 2019).

Properties such as surface area, pore volume, and average pore diameter, as well as active surface sites and dangling bonds within the structure, play a crucial role in determining the adsorption capacity of the nanosystem.

In this context, graphene has emerged as the most relevant nanomaterial today, both in industrial applications and advanced research, with its large-scale production potentially generating a significant impact due to its remarkable properties.

No other material possesses the extensive range of attributes that graphene exhibits, making it ideal for numerous applications. It is incredibly lightweight and flexible, electrically and thermally conductive, transparent, stronger than steel, and thinner than the diameter of a single strand of human hair (Chatterjee et al., 2017; Ly et al., 2019; Sun et al., 2019).

The following discussion will explore relevant studies that aim to provide insights into the adsorption properties of graphene and its derivatives, highlighting Raman spectroscopy as a fundamental tool for understanding the system in each case.

Adsorption in Graphene and Its Derivatives

The incorporation of nanoparticles and functional groups on the surface of graphene can enhance its selectivity, sensitivity, and detection limits, making it a promising candidate for nanoadsorbents aimed at the prevention, control, and reduction of water pollution (Chatterjee et al., 2017). In addition to its adjustable textural and chemical properties, the high number of delocalized π-electrons makes graphene highly attractive for environmental decontamination applications (Lee et al., 2019).

Both pure graphene and its derivatives, graphene oxide (GO) and reduced graphene oxide (rGO), have garnered significant attention due to their remarkable physicochemical properties. These materials exhibit distinct morphological and chemical characteristics, closely tied to the synthesis methods employed, which influence various properties and applications (Lee et al., 2019).

The oxidation and expansion process of graphite (Figure 9.26a–d) can be achieved through the use of strong oxidizing agents to produce GO, followed by chemical reduction to obtain rGO or thermal exfoliation to yield graphene.

The pores in these nanomaterials are slit-shaped and present between graphene's parallel layers. Their porosity distribution indicates the presence of micropores and mesopores (Lee et al., 2019). Their surface areas are considerably smaller compared to that of a single graphene layer ($2{,}630\,m^2g^{-1}$) due to defects resulting from incomplete exfoliation of GO, layer agglomeration during the reduction process, sheet stacking caused by van der Waals interactions, as well as wrinkling and rolling of graphene sheets.

Since the isolation of defect-free graphene sheets for large-scale applications remains a challenge, its oxidized forms, such as GO and rGO, have proven to be highly promising.

GO consists of stacked graphene sheets with a high degree of oxidation, featuring irregular edges and a wrinkled surface. On its surface, functional oxygen-containing groups (–COOH and –C=O) are located at the sheet's edges, while epoxy and hydroxyl groups are present in its basal plane.

GO exhibits high solubility in water due to oxygen-containing groups responsible for forming stable aqueous colloids. It has a high negative charge density, providing reactive sites for the adsorption of various adsorbates, particularly those with a positive charge.

Meanwhile, rGO can establish π-π stacking interactions between sp^2 regions and the aromatic structure of dyes. It also enhances electrostatic interactions between the charged dye and the oxygen-containing groups present on the surface of rGO, offering easily regenerable adsorption properties. This makes rGO a cost-effective and efficient alternative for water treatment, aiding in the adsorption of various dyes.

Raman spectroscopy plays a crucial role in the characterization of graphene and its derivatives, such as GO, particularly in adsorption applications (Figure 9.27). This non-destructive technique provides detailed insights into the structural and chemical properties of carbon-based materials. In the study by Melchor-Durán et al. (2024), Raman spectroscopy was used to identify key vibrational modes—specifically the D and G bands—which reflect the degree of disorder and the presence of oxygenated functional groups introduced during the oxidation of graphite into GO. The characteristic D and G peaks in the Raman spectra of graphite and GO samples were analyzed to assess the degree of disorder and functionalization. The G peak ($\sim 1{,}583\,cm^{-1}$) corresponds to the vibration of sp^2-hybridized carbon atoms, while the D peak ($\sim 1{,}333\,cm^{-1}$) is associated with structural defects arising from sp^3-hybridized carbon atoms (Rasuli et al., 2020; George et al.; 2021, Li et al., 2021).

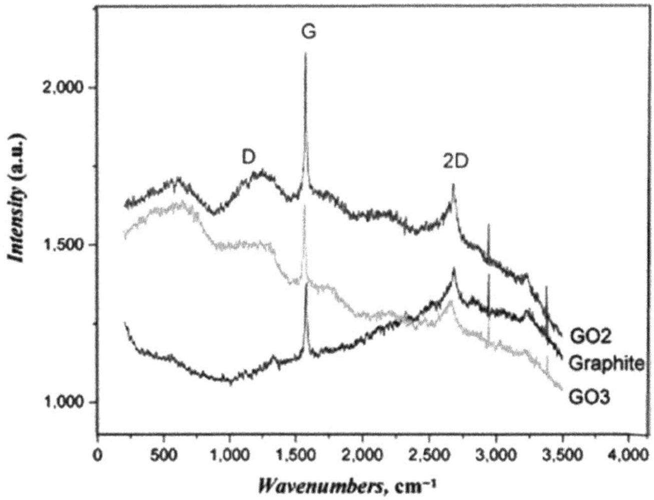

FIGURE 9.27 Results of Raman spectroscopy for graphite and GO samples. (This figure is from Melchor-Durán et al. (2024). This article is an open access article distributed under the terms and conditions of the Creative Commons Attribution (CC BY) license (https://creativecommons.org/licenses/by/4.0/). https://www.mdpi.com/separations/separations-11-00294/article_deploy/html/images/separations-11-00294-g001.png.)

The intensity ratio of these bands (ID/IG) served as a reliable indicator of the level of structural defects and surface functionalization, both of which directly influence the adsorption capacity of GO for heavy metals like Pb^{2+} and Cu^{2+}. The ID/IG ratio was used to quantify the extent of oxidation, showing a progressive increase from graphite (0.417) to GO2 (0.8215) and GO3 (0.9203), confirming that GO3 exhibited the highest degree of oxidation. A higher ID/IG ratio indicated a more oxidized and defect-rich surface, which enhances the material's ability to interact with and capture metal ions (Melchor-Durán et al., 2024). Thus, Raman spectroscopy was essential for correlating the physicochemical properties of GO with its performance in magnetic-assisted adsorption systems.

SUMMARY

Characterization of adsorbent materials is essential at two key stages: during preparation and after the adsorption process. In the preparation phase, physicochemical properties are assessed. These properties include surface area and pore distribution via gas adsorption, surface charge through zeta potential or PZC analyses, and thermal stability using thermogravimetric methods. Additional techniques such as spectroscopy, electron microscopy, and XRD help identify functional groups, assess crystallinity, and evaluate morphology. Gas adsorption, particularly based on BET theory, enables surface area and pore size determination using gases such as nitrogen or carbon dioxide. The shape of BET isotherms provides insights into porosity and adsorptive potential. PZC and zeta potential analyses determine surface charge behavior, which is crucial for understanding the pH-dependent affinity for anions or cations.

After adsorption, the interaction between adsorbate and adsorbent is typically evaluated via spectroscopic techniques. Infrared spectroscopy reveals functional groups involved in adsorption mechanisms. XRD identifies crystalline phases and structural changes post-modification. SEM uses an electron beam to generate high-resolution, three-dimensional images of solid materials, revealing surface morphology. EDS, coupled with SEM, enables elemental analysis by detecting characteristic X-rays. Raman spectroscopy complements SEM by identifying molecular structures through inelastic light scattering. It is highly effective for both qualitative and quantitative analysis, particularly in layered materials, offering insight into structural coupling, composition, and vibrational properties. These characterization methods collectively support the optimization, reproducibility, and application-specific suitability of adsorbent materials.

EXERCISES

Multiple-Choice Questions

1. **Which of the following is a crucial application of thermogravimetric techniques?**
   a. Monitoring synthesis processes involving calcination
   b. Determining the viscosity of liquids
   c. Analyzing electrical conductivity in metals
   d. Measuring pH variations in solutions
2. **What is the primary reason for removing surfactants during the synthesis of zeolites?**
   a. To enhance thermal conductivity
   b. To prevent excessive moisture absorption
   c. To unblock the pores of the material
   d. To improve the mechanical strength of zeolites
3. **According to Lima *et al.* (2012), what type of process occurs between 200°C and 300°C during the thermal analysis of lignocellulosic material?**
   a. Endothermic evaporation
   b. Exothermic surfactant and lignocellulosic material loss
   c. Physical adsorption of gases
   d. Crystallization of silicon oxides
4. **Which infrared region is most commonly used for the characterization of adsorbents?**
   a. Near-infrared (NIR)
   b. Mid-infrared (MIR)
   c. Far-infrared (FIR)
   d. Ultraviolet (UV)
5. **Which functional groups correspond to the infrared bands at 3,448 and 3,415 cm^{-1} in the spectrum of green coconut husk powder?**
   a. Aldehydes and ketones
   b. Amines and amides

   c. Axial O-H stretching of alcohols or phenols
   d. C-N and C-O groups
6. **What is the primary function of infrared spectroscopy in adsorption studies?**
   a. To measure adsorption equilibrium constants
   b. To identify functional groups acting as adsorption sites
   c. To determine the surface area of adsorbents
   d. To quantify the concentration of metal ions in solution
7. How does XRD contribute to the characterization of crystalline materials, and what advantages does it offer compared to neutron and electron diffraction techniques?
8. Explain the significance of the Scherrer equation in determining average particle size. What parameters influence its accuracy?
9. Describe the role of the Williamson-Hall plot in evaluating microstrain and particle size. How does its linear behavior indicate material homogeneity?
10. How does the refinement process using the DBWS program enhance the accuracy of crystallographic parameter determination?
11. Discuss the importance of XRD in assessing the crystallinity of materials with low crystalline order. What challenges arise in phase identification?
12. How does Raman spectroscopy contribute to the study of two-dimensional materials, and what structural characteristics can it reveal?
13. What is the distinction between Rayleigh scattering and Raman scattering, and how do Stokes and anti-Stokes lines differ in a Raman spectrum?
14. Why Raman spectroscopy is considered a versatile method for qualitative and quantitative analysis, and what advantages does it have over other spectroscopic techniques?

REFERENCES

Abushammala, H., & Mao, J. (2020). A review on the partial and complete dissolution and fractionation of wood and lignocelluloses using imidazolium ionic liquids. *Polymers*, 12(1), 195. https://doi.org/10.3390/polym12010195

Altun, A. O., Bond, T., Pronk, W., & Park, H. G. (2017). Sensitive detection of competitive molecular adsorption by surface-enhanced Raman spectroscopy. *Langmuir: The ACS Journal of Surfaces and Colloids*, 33(28), 6999–7006. https://doi.org/10.1021/acs.langmuir.7b01186

Atkins, P. W., & de Paula, J. (2014). *Physical Chemistry: Thermodynamics, Structure, and Change*. 10ª. ed. W. H. Freeman, New York. ISBN-13: 978-1429290197.

Baskar, M., & Patania, A. (2020). X-Ray powder diffraction data—identification and structure elucidation of unknown stone—rietveld refinement approach. *Journal of Biophysical Chemistry*, 11, 51–61. https://doi.org/10.4236/jbpc.2020.114005

Bergmann, C. P., & Machado, F. M. (2015). Carbon nanoadsorbents. In C. P. Bergmann, & F. M. Machado (Eds.), *Carbon Nanomaterials as Adsorbents for Environmental and Biological Applications* (pp. 11–32). Springer, Cham.

Bleicher, L., Sasaki, J. M., & Santos, C. O. P. (2000). Development of a graphical interface for the Rietveld refinement program DBWS. *Journal of Applied Crystallography*, 33, 1189–1190. https://doi.org/10.1107/S0021889800005410

Bumbrah, G. S., & Sharma, R. M. (2016). Raman spectroscopy—Basic principle, instrumentation, and selected applications for the characterization of drugs of abuse. *Egyptian Journal of Forensic Sciences*, 6(3), 209–215.

Castro, F. L. de et al. (2007). Solid state fluorescence of a 3,4,9,10-perylenetetracarboxylic diimide derivative encapsulated in the pores of mesoporous silica MCM-41. *Microporous and Mesoporous Materials*, 102(1–3), 258–264. https://doi.org/10.1016/j.micromeso.2006.12.042

Chatterjee, A., Gale, D. J. G., Grebennikov, D., Whelan, L. D., & Merschrod, E. F. S. (2017). Surface potential and morphology mapping to investigate analyte adsorption effects on surface-enhanced Raman scattering (SERS). *Chemical Communications*, 53(88), 12024–12027. https://doi.org/10.1039/C7CC05473D

Cunha, R., et al. (2018). Raman spectroscopy revealing noble gas adsorption on single-walled carbon nanotube bundles. *Carbon*, 127, 312–319.

Dedavid, B. A., Gomes, C. I., & Machado, G. (2007). *Scanning Electron Microscopy: Applications and Sample Preparation—Polymeric, Metallic, and Semiconductor Materials [Electronic Resource]* (Portuguese version). EDIPUCRS, Porto Alegre.

Deshmukh, K., & Hussain, C. M. (Eds.). (2023). *Advanced Fluoropolymer Nanocomposites: Fabrication, Processing, Characterization, and Applications*. 1st ed. Woodhead Publishing. https://doi.org/10.1016/C2021-0-02570-0

Ding, S.-Y., & Wang, W. (2012). Covalent organic frameworks (COFs): From design to applications. *Chemical Society Reviews* 42, 548–568. https://doi.org/10.1039/C2CS35072F

Fiol, N., & Vilaescusa, I. (2009). Determination of sorbent point zero charge: Usefulness in sorption studies. *Environmental Chemistry Letters*, 7, 79–84. https://doi.org/10.1007/s10311-008-0139-0

George, J., Upadhyay, S., & Balachandran, M. (2021). Exploration of graphene layers in various carbon materials by Raman spectroscopic techniques. *Mapana Journal of Sciences*, 20, 65–81.

Gomes, E. C. C., et al. (2013). Synthesis of bifunctional mesoporous silica spheres as potential adsorbent for ions in solution. *Chemical Engineering Journal*, 214, 27–33. https://doi.org/10.1016/j.cej.2012.10.053

Herraez, A. (2006). Biomolecules in the computer: Jmol to the eescue. *Biochemistry and Molecular Biology Education*, 34(4). https://doi.org/10.1002/bmb.2006.494034042644

Ihsanullah, I. (2019). Carbon nanotube membranes for water purification: Developments, challenges, and prospects for the future. *Separation and Purification Technology*, 209, 307–337.

James, R. W. (1962). *The Optical Principles of the Diffraction of X-Rays*. G. Bell and Sons Ltd., London.

Jin, H., Zha, C., & Gu, L. (2007). Direct dissolution of cellulose in NaOH/thiourea/urea aqueous solution. *Carbohydrate Research*, 342(6), 851–858. https://doi.org/10.1016/j.carres.2006.12.023

Klug, H. P., & Alexander, L. E. (1974). *X-Ray Diffraction Procedure for Polycrystalline and Amorphous Materials*. 2nd ed. Reinhold Publishing Corporation, New York.

Kroon-Batenburg, L. M., & Kroon, J. (1997). The crystal and molecular structures of cellulose I and II. *Glycoconjugate Journal*, 14(5), 677–690. https://doi.org/10.1023/a:1018509231331

Le, H. A., Linh, L. T., Chin, S., & Jurng, J. (2012). Photocatalytic degradation of methylene blue by a combination of TiO_2-anatase and coconut shell activated carbon. *Powder Technology*, 225, 167–175. https://doi.org/10.1016/j.powtec.2012.04.004

Lee, X. J., et al. (2019). Review on graphene and its derivatives: Synthesis methods and potential industrial implementation. *Journal of the Taiwan Institute of Chemical Engineers*, 98, 163–180. https://doi.org/10.1016/j.jtice.2018.10.028

Li, Z., et al. (2021). One-pot exfoliation and synthesis of hydroxyapatite-functionalized graphene as multifunctional nanomaterials based on electrochemical approach. *Composites Part A: Applied Science and Manufacturing*, 149, 106583.

Lii-Rosales, A., Han, Y., Jing, D., Tringides, M. C., & Thiel, P. A. (2020). Search for encapsulation of platinum, silver, and gold at the surface of graphite. *Physical Review Research*, 2(3), 033175. https://doi.org/10.1103/PhysRevResearch.2.033175

Lima, A. C. A., Nascimento, R. F., de Sousa, F. F., Filho, J. M., & Oliveira, A. C. (2012). Modified coconut shell fibers: A green and economical sorbent for the removal of anions from aqueous solutions. *Chemical Engineering Journal*, 185–186, 274–284. https://doi.org/10.1016/j.cej.2012.01.037

Lombardi, J. R., & Birke, R. L. (2008). A unified approach to surface-enhanced Raman spectroscopy. *The Journal of Physical Chemistry C*, 112(14), 5605–5617. https://doi.org/10.1021/jp800167v

Ly, N. H., Nguyen, T. H., Nghi, N. Đ., Kim, Y. H., & Joo, S. W. (2019). Surface-enhanced Raman scattering detection of fipronil pesticide adsorbed on silver nanoparticles. Sensors, 19(6), 1355. https://doi.org/10.3390/s19061355

Madsen, A. T., & Good, D. J. (2023). In silico examination of single nucleotide missense mutations in $NHLH_2$, a gene linked to infertility and obesity. *International Journal of Molecular Sciences*, 24(4), 3193. https://doi.org/10.3390/ijms24043193

Meille, S., et al. (2011). Definitions of terms relating to crystalline polymers (IUPAC Recommendations 2011). *Pure and Applied Chemistry*, 83(10), 1831–1871. https://doi.org/10.1351/PAC-REC-10-11-13

Melchor-Durán, A. P., Moreno-Virgen, M. R., Bonilla-Petriciolet, A., Reynel-Ávila, H. E., Lucio Ortiz, E., & González-Vázquez, O. F. (2024). Heavy metal removal from water using graphene oxide in magnetic-assisted adsorption systems: Characterization, adsorption properties, and modelling. *Separations*, 11(10), 294. https://doi.org/10.3390/separations11100294

Moreira, S. A. (2008). Adsorption of metallic ions from aqueous effluent using cashew peduncle bagasse: Batch and fixed-bed column study (Master's dissertation in Civil Engineering, Environmental Sanitation, Portuguese version). Technology Center, Federal University of Ceará, Fortaleza.

Moreira, S. A., Sousa, F. W., & Oliveira, A. G. (2009). Removal of metals from aqueous solution using cashew bagasse [Article in Portuguese]. *Química Nova*, 32(7), 1717–1722. https://doi.org/10.1590/S0100-40422009000700007

Moura, C. P., et al. (2011). Adsorption of BTX (benzene, toluene, o-xylene, and p-xylene) from aqueous solutions by modified periodic mesoporous organosilica. *Journal of Colloid and Interface Science*, 363(2), 626–634. https://doi.org/10.1016/j.jcis.2011.07.054

Napolitano, C., et al. (2024). A new semicrystalline polymer from renewable Source: Poly-1, 4-(3-methylene-cyclopentene) as an example of three-dimensional topotactic transition triggered by temperature. *Macromolecules*, 57(15), 7409–7417. https://doi.org/10.1021/acs.macromol.4c01563

Neto, V. d. O. S., et al. (2012). Coconut bagasse treated by thiourea/ammonia solution for cadmium removal: Kinetics and adsorption equilibrium. *BioResources*, 7(2), 1504–1524.

Pang, J., Huang, L., Liu, H., & Yi, X. (2025). Melt pool dynamics and pore formation in selective laser melting: Mechanisms and microstructural insights. *Materials & Design*, 254, 114043. https://doi.org/10.1016/j.matdes.2025.114043

Ramkumar, T., Selvakumar, M., Vasanthsankar, R., Sathishkumar, A. S., Narayanasamy, P., & Girija, G. (2018). Rietveld refinement of powder X-ray diffraction, microstructural and mechanical studies of magnesium matrix composites processed by high energy ball milling. *Journal of Magnesium and Alloys*, 6(4), 390–398. https://doi.org/10.1016/j.jma.2018.08.002

Rasuli, H., Rasuli, R., Alizadeh, M., & BoonTong, G. (2020). Microwave-assisted exfoliation and tearing of graphene oxide in the presence of TiO_2 nanoparticles. *Results in Physics*, 18, 103200.

Reyes, G., King, A. W. T., Koso, T. V., Penttilä, P. A., Kosonen, H., & Rojas, O. J. (2022). Cellulose dissolution and gelation in NaOH(aq) under controlled CO_2 atmosphere: Supramolecular structure and flow properties. *Green Chemistry: An International Journal and Green Chemistry ReSource: GC*, 24(20), 8029–8035. https://doi.org/10.1039/d2gc02916b

Ribeiro, T. S., Sasaki, J. M., & Vasconcelos, I. F. (2012). Structural disorder of ball-milled, nanosized, Fe-doped SnO_2: X-ray diffraction and Mössbauer spectroscopy characterization. *Journal of Materials Science*, 47(7), 2630–2636. https://doi.org/10.1007/s10853-011-6088-1

Rouquerol, F., Rouquerol, J., & Sing, K. S. W. (1999). *Adsorption by Powders and Porous Solids*. Academic Press, San Diego, CA. https://doi.org/10.1016/C2010-0-66232-8

Sharma, A., Sharma, V. K., & Mausam, K. (2021). Analysis on thermal distinctiveness of compound polymers by micro filling erbium oxide. *Materials Today: Proceedings*, 45(2), 2751–2754. https://doi.org/10.1016/j.matpr.2020.11.605

Sikirzhytski, V., Sikirzhytskaya, A., & Lednev, I. K. (2011). Multidimensional Raman spectroscopic signatures as a tool for forensic identification of body fluid traces: A review. *Applied Spectroscopy*, 65(11), 1223–1232. https://doi.org/10.1366/11-06455

Silverstein, R. M., Webster, F. X., Kiemle, D., & Brice, D. L (2005). *Spectrometric Identification of Organic Compounds*. 8th Revised ed. John Wiley & Sons, Hoboken, NJ.

Sousa, F. W. (2007). Adsorption of toxic metals in aqueous effluent using treated green coconut husk powder (Master's dissertation in Civil Engineering, Environmental Sanitation, Portuguese version). Technology Center, Federal University of Ceará, Fortaleza.

Sousa, F. W., Oliveira, A. G., Ribeiro, J. P., De Keukeleire, D., Sousa, A. F., & Nascimento, R. F. (2011). Single and multielementary isotherms of toxic metals in aqueous solution using treated coconut shell powder. *Desalination and Water Treatment*, 36(1), 289–296. https://doi.org/10.5004/dwt.2011.2597

Sousa, F. W., Oliveira, A. G., Ribeiro, J. P., Rosa, M. F., Keukeleire, D., & Nascimento, R. F. (2010). Green coconut shells applied as adsorbent for removal of toxic metal ions using fixed-bed column technology. *Journal of Environmental Management*, 91(8), 1634–1640. https://doi.org/10.1016/j.jenvman.2010.02.011

Spurr, R. A., & Myers, H. (1957). Quantitative analysis of anatase-rutile mixtures with an X-ray diffractometer. *Analytical Chemistry*, 29(5), 760–762. https://doi.org/10.1021/ac60125a006

Sun, J. J., et al. (2019). Role of adsorption orientation in surface plasmon-driven coupling reactions studied by tip-enhanced Raman spectroscopy. *The Journal of Physical Chemistry Letters*, 10(10), 2306–2312. https://doi.org/10.1021/acs.jpclett.9b00203

Sun, J. X., Sun, X. F., Zhao, H., & Sun, R.-C. (2004). Isolation and characterization of cellulose from sugarcane bagasse. *Polymer Degradation and Stability*, 84(3), 331–339. https://doi.org/10.1016/j.polymdegradstab.2004.02.008

Sun, J. X., Xu, F., Sun, X. F., Xiao, B., & Sun, R.-C. (2005). Physico-chemical and thermal characterization of cellulose from barley straw. *Polymer Degradation and Stability*, 88(3), 521–531. https://doi.org/10.1016/j.polymdegradstab.2004.12.013

Tabak, A., Baltas, N., Afsin, B., Emirik, M., Caglar, B., & Eren, E. (2010). Adsorption of reactive red 120 from aqueous solutions by cetylpyridinium-bentonite. *Journal of Chemical Technology & Biotechnology*, 85, 1199–1207. https://doi.org/10.1002/jctb.2416

Vahedi, B., Golab, E., Nasiri Sadr, A., & Vafai, K. (2022). Thermal, thermodynamic, and exergoeconomic investigation of a parabolic trough collector utilizing nanofluids. *Applied Thermal Engineering*, 206, 118117. https://doi.org/10.1016/j.applthermaleng.2022.118117

Vakros, J., Kordulis, C., & Lycourghiotis, A. (2002). Potentiometric mass titrations: A quick scan for determining the point of zero charge. *The Royal Society of Chemistry*, 17, 1980–1981. https://doi.org/10.1039/B205398E

Vidal, C. B., et al. (2011). Adsorption of polycyclic aromatic hydrocarbons from aqueous solutions by modified periodic mesoporous organosilica. *Journal of Colloid and Interface Science*, 357(2), 466–473. https://doi.org/10.1016/j.jcis.2011.02.013

Vidal, C. B., et al. (2012). BTEX removal from aqueous solutions by HDTMA-modified Y zeolite. *Journal of Environmental Management*, 112, 178–185. https://doi.org/10.1016/j.jenvman.2012.07.026

Vieira, M. E. M., Silva, M. L. S., Oliveira, L. F. C., Perrone, Í. T., & Stephani, R. (2021). X-Ray energy dispersive spectroscopy (EDS) coupled with scanning electron microscope (SEM): Fundamentals and applications in dairy products. *Research, Society and Development*, 10(10), e262101018622. https://doi.org/10.33448/rsd-v10i10.18622

Williamson, G. K., & Hall, W. H. (1953). X-Ray line broadening from filed aluminium and wolfram. *Acta Metallurgica*, 1(1), 22–31. https://doi.org/10.1016/0001-6160(53)90006-6

Yang, K., Peng, J., Srinivasakannan, C., Zhang, L., Xia, H., & Duan, X. (2010). Preparation of high surface area activated carbon from coconut shells using microwave heating. *Bioresource Technology*, 101(15), 6163–6169. https://doi.org/10.1016/j.biortech.2010.03.001

Yu, Y., Wang, W., & Liu, Y. (2024). Enhancing the mechanical properties of regenerated cellulose through high-temperature pre-gelation. Materials, 17(19), 4886. https://doi.org/10.3390/ma17194886

Zhang, L., Ruan, D., & Gao, S. (2002). Dissolution and regeneration of cellulose in NaOH/thiourea aqueous solution. *Journal of Polymer Science, Part B: Polymer Physics*, 40, 1521–1529. https://doi.org/10.1002/polb.10215

Zhang, X., Tan, Q. H., Wu, J. B., Shi, W., & Tan, P. H. (2016). Review on the Raman spectroscopy of different types of layered materials. *Nanoscale*, 8(12), 6435–6450. https://doi.org/10.1039/c5nr07205k

Zhang, Y., Zhang, R., Jiang, S., Zhang, Y., & Dong, Z. C. (2019). Probing adsorption configurations of small molecules on surfaces by single-molecule tip-enhanced Raman spectroscopy. *Chemphyschem: A European Journal of Chemical Physics and Physical Chemistry*, 20(1), 37–41. https://doi.org/10.1002/cphc.201800861

Index

N

O

P

Q

R

S

T

V

Y

Z